Strata Title Property Rights

Multi-owned properties make up an ever-increasing proportion of commercial, tourist and residential development, in both urban and rural landscapes around the world. This book critically analyses the legal, social and economic complexities of strata or community title schemes. At a time when countries such as Australia and the United States turn ever-larger areas into strata title/condominiums and community title/homeowner associations, this book shows how governments, the judiciary and citizens need to better understand the ramifications of these private communities.

While most strata title analysis has been technical, focusing on specific sections of legislation, this book provides higher-level analysis, discussing the wider economic, social and political implications of Australia's strata and community title law. In particular, the book argues that private by-laws, however desirable to initial parties, are often economically inefficient and socially regressive when enforced against an ever-changing group of owners. The book will be of particular interest to scholars and legal practitioners of property law in Australia, but as the Australian strata title model has formed the basis for legislation in many countries, the book draws out lessons and analysis that will be of use to those studying privately owned communities across the world.

Cathy Sherry is a member of the Faculty of Law, University of New South Wales, and a leading Australian expert on strata and community title law. She is a General Editor of the international property journal, *Property Law Review*.

Routledge Complex Real Property Rights Series

Series editor: Professor Spike Boydell
University of Technology, Sydney, Australia

Real Property Rights are central to the global economy and provide a legal framework for how society (be it developed or customary) relates to land and buildings. We need to better understand property rights to ensure sustainable societies, careful use of limited resources and sound ecological stewardship of our land and water. Contemporary property rights theory is dynamic and needs to engage thinkers who are prepared to think outside their disciplinary limitations.

The Routledge Complex Real Property Rights Series strives to take a transdisciplinary approach to understanding property rights and specifically encourages heterodox thinking. Through rich international case studies our goal is to build models to connect theory to observed reality, allowing us to inform potential policy outcomes. This series is both an ideal forum and reference for students and scholars of property rights and land issues.

Land, Indigenous Peoples and Conflict
Edited by Alan Tidwell and Barry Zellen

Beyond Communal and Individual Ownership
Indigenous Land Reform in Australia
Leon Terrill

Strata Title Property Rights
Private governance of multi-owned properties
Cathy Sherry

Strata Title Property Rights

Private governance of multi-owned properties

Cathy Sherry

Routledge
Taylor & Francis Group
LONDON AND NEW YORK

First published 2017
by Routledge
2 Park Square, Milton Park, Abingdon, Oxon OX14 4RN

and by Routledge
605 Third Avenue, New York, NY 10017

First issued in paperback 2021

Routledge is an imprint of the Taylor & Francis Group, an informa business

British Library Cataloguing in Publication Data
A catalogue record for this book is available from the British Library

Library of Congress Cataloging in Publication Data
Names: Sherry, Cathy, author.
Title: Strata title property rights : private governance of multi-owned properties / Cathy Sherry.Description: New York : Routledge, 2016. | Includes bibliographical references and index.
Identifiers: LCCN 2016022857| ISBN 9781138914926 (hardback : alk. paper) | ISBN 9781315690575 (ebook : alk. paper)
Subjects: LCSH: Common interest ownership communities–Law and legislation–Australia. | Common interest ownership communities–Law and legislation–United States. | Right of property–Australia. | Right of property–United States. | Real property–Australia. | Real property–United States.
Classification: LCC K738 .S525 2016 | DDC 346.7304/32–dc23LC record available at https://lccn.loc.gov/2016022857

ISBN 13: 978-0-367-35568-5 (pbk)
ISBN 13: 978-1-138-91492-6 (hbk)

Typeset in Times New Roman
by Wearset Ltd, Boldon, Tyne and Wear

Publisher's Note
The publisher has gone to great lengths to ensure the quality of this reprint but points out that some imperfections in the original copies may be apparent.

For Roddy, Emily, Ruby and Sarah

Contents

Figures

Foreword

Real Property Rights are central to the economy and provide a legal framework for how society (be it developed or customary) relates to land and buildings. Property rights are both institutional arrangements and social relations. We need to better understand property rights to ensure sustainable societies, careful use of limited resources and sound ecological stewardship of our land and water.

Land conflict is all around us – from corporate and political corruption over land dealings in the developed world, through to land grab in developing countries, compromised indigenous property rights and resource exploitation. At a time where global food security, water security and shelter are paramount, an understanding of property rights is key to sustainability.

Contemporary property rights theory is dynamic and this series strives to engage thinkers who are prepared to step beyond their disciplinary limitations. 'Property Rights' is a broad term that is fundamentally about social relations. Real property rights, obligations and restrictions can be found in, and change across, the full range of human societies, both in time and space. Property rights research has emerged from a broad range of disciplines including (but not limited to) archaeology, anthropology, ethics, sociology, psychology, law, geography, history, philosophy, economics, planning and business studies. What makes this series special is that it facilitates a transdisciplinary approach to understanding property rights and specifically promotes heterodox thinking.

One of the challenges of researching real property rights is that many of us bring our own preconceptions to the table. We ground these on the cultural, political, environmental, economic and legal frameworks in which we grow and work. In this vein, I brought my own preconceptions when I relocated from England to Queensland, Australia in the early 1990s. Hitherto all my financial models were based on institutional leases that ran for 21 or 25 years and often demonstrated significant value for tenants, whereas in Australia the leases were much shorter. Likewise, English law precluded the creation of 'flying freeholds' or commonhold property, whereas in Australia strata and community title with their complex by-law structures have become a dominant form of contemporary urban tenure.

In *Strata Title Property Rights: Private governance of multi-owned properties*, Cathy Sherry highlights the dynamism of property rights to adapt and evolve in their attempt to best serve society at any given time. Indeed, she identifies that it was the inability of positive obligations to endure with conventional freehold title that prevented maintenance obligations being imposed in a collectively owned building, and which consequently created a need for an alternative tenure model. Drawing on examples of urban growth responses in post-war Europe and America, it was New South Wales that took the lead with legislation in Australia in the early 1960s to enable the vertical subdivision of land into registered freehold titles. Quickly followed in other states, this reconceptualisation of a freehold interest into strata title satisfied banks in terms of loan collateral, while imposing maintenance levies and a raft of (often onerous) obligations in the form of by-laws on a development, overcoming the limitations of common law.

There is a sense in this book that we have to be careful what we wish for, as Cathy Sherry analyses the contemporary ramifications of the by-laws in strata title and community title (in the case of horizontal subdivision), as forms of legislation that empower bodies corporate to craft by-laws over both collectively and privately owned property within a development scheme. *Strata Title Property Rights: Private governance of multi-owned properties* provides a timely analysis of this by-law-making power over private property, drawing on notions of democracy, morality, entitlement and self-justification as theoretical lenses of legal interpretation to explore the physical landscape of prevailing political and social systems.

Strata Title Property Rights: Private governance of multi-owned properties fills an important knowledge gap in an area of law that is prevalent in the property law of contemporary urban society, but which has hitherto evaded adequate scrutiny by the judiciary or academe. Strata title has typically been analysed as though it were a law unto itself, rather than the reality that it is part of the general property system. Using high-level analysis, Cathy Sherry highlights the need to understand and engage with the judicial deliberations from other jurisdictions, notably the lessons from the United States on positive and restrictive covenants and association rules, rather than states within countries like Australia attempting to reinvent the wheel in what appear to be flawed notions of open-ended bundles of property rights. Such negative liberties are evinced in the blurring of the public/private divide created by strata and community title schemes that challenge notions of parenting and children's safety, as well as racism, autonomy and free speech, in the face of privately created rules. In response to this, Sherry argues the case for Australia and other countries to use the United States case law as our guide, to ensure that powers to craft by-laws are limited by judicial or legislative requirements of reasonableness that prioritise community values and well-being over simplistic satisfaction of individual preference.

Spike Boydell, General Editor
Sydney, July 2016

Preface

This book began life as a PhD thesis written in the Faculty of Law at the University of New South Wales. It was supervised by Professor Brendan Edgeworth, UNSW and Dr Scott Grattan, University of Sydney, and I am indebted to both for their consistent support and guidance. I also owe particular thanks to Professor Edgeworth for his encouragement, collegiality and generous sharing of knowledge over the course of my academic career.

The strata title industry in Australia owes much of its existence to the energy and efforts of private lawyers, many of whom have been extraordinarily generous with their time and expertise during my research. Particular thanks to Michael Allen, Gary Bugden, Bob Browne, Neville Allen, Col Meyers, Colin Grace, Michael Teys, Michael Kleinschmidt and Tim Graham. My thanks also to the Australasian College of Community Association Lawyers (ACCAL), and the work of Nina Psaltis. Special thanks to Carolyn Chudleigh for years of friendship, in addition to community title expertise.

My academic colleagues have also made valuable contributions to this work. My thanks to Dr Hazel Easthope and Professor Bill Randolph, City Futures Research Centre, UNSW; Nicole Johnston, Faculty of Business and Law, Deakin University; Professor Chris Guilding, Griffith Business School, Griffith University; Professor Sarah Blandy, University of Sheffield; and Professor Susan Bright, New College, Oxford. My thanks also to Emeritus Professor Peter Butt, for the 8.15 am, 1989 Land Law tutorial which stimulated my interest in property law, and taught me that some of the best people are land lawyers.

My colleagues in the Faculty of Law, UNSW have provided me with years of friendship and academic support, particularly Dr Chris Rossiter and Denis Harley; Associate Professor Lisa Toohey; Professor Ross Buckley, who specifically encouraged me to write this book; and Associate Professor Lyria Bennett-Moses, whose intellectual energy, enthusiasm and generosity know no bounds.

My thanks to talented photographer, Lola Hubner, for taking photographs for the book.

And finally, thanks to my family and friends, in particular Emma McDonald, my parents Dr Beverley and Sean Sherry, and above all, my husband Roddy Baird and daughters, Emily, Ruby and Sarah.

Introduction

Strata and community title are the fastest growing forms of property title in Australia. With successive state government policies in favour of urban consolidation, Australian cities have experienced a sharp rise in high and medium density development, in particular in Sydney, Melbourne, and the Gold Coast–Brisbane–Sunshine Coast corridor.[1] In 2011, over a quarter of Sydney's population, just over one million people, lived in strata title apartments.[2] The New South Wales government predicts that within 20 years, half of the state's population will live or work in a strata or community scheme.[3] In Queensland, there are almost 50 000 community schemes, and just under 400 000 individual lots.[4] Nationally, it is estimated that there are now close to two million strata lots in Australia.[5]

Strata title raises extremely complex questions in property law. It creates individual and collective titles to land/air space, as well as governing the ongoing rights and obligations between people in the resulting communities. It is not uncommon to hear real estate agents and even lawyers use the term 'Torrens title' or even 'freehold title' to mean '*not* strata title'. While technically incorrect (as strata is both Torrens title and freehold),[6] this misconception no doubt stems from an appreciation of the limits that strata title places on the freedoms normally associated with property ownership in liberal democracies. In particular, it evinces an understanding of the intrusive nature of strata by-laws.

All strata schemes have by-laws lodged for registration with the strata plan of subdivision. These may be the 'model' by-laws written by the legislature, or by-laws written by the developer's lawyer specifically for the scheme. Whether model or bespoke, all by-laws can subsequently be altered by the owners with appropriate majority. While by-laws often relate to noise, garbage, parking and other matters relevant to high-density living, they are not limited to these. In most states, the power to write by-laws is extraordinarily expansive, with the only general requirement being that by-laws relate to the control, management, administration, use or enjoyment of lots or common property. By-laws bind all owners, mortgagees, lessees and occupiers, as well as the body corporate, and breach of by-laws can lead to the imposition of fines. In short, the strata legislation empowers private citizens (either the developer or subsequent owners) to write rules for their fellow citizens that regulate activity not only on collectively owned property, but also inside private lots.

When by-laws predominantly governed six-lot strata schemes, this private legislative power was possibly not a great cause for concern. However, many strata schemes are now much larger, with hundreds of lots and even more residents. From the 1970s, the strata legal form was extended to horizontal subdivisions, variously called 'cluster', 'group' or 'community' titles. Like strata title, these titles subdivide land into individually owned lots and common property, regulated by private by-laws, but the common property is likely to be roads, parks, sewers, recreational facilities and bushland. In size and/or population, many horizontal schemes are the equivalent of a suburb, and they result in large swathes of what would otherwise be publicly owned and managed land and infrastructure, vesting in the hands and regulatory control of private citizens.

Despite the significant increase in the scale and use of strata and community title, and the corresponding increase in the proportion of our city and the number of citizens both subject to, and wielding, the private by-law-making power, the by-law-making power has received little academic or judicial scrutiny. This book aims to fill some of that research gap.

The book is divided into two parts, each with three chapters. Part I establishes the historic, legal, theoretical and global context for Australian strata and community title law. Part II then considers the current Australian law, both legislative and case law.

Part I, Chapter 1 begins with a discussion of the history of low- and high-rise housing in Australia and the property doctrines that underpinned that development. It highlights the impediment that Anglo-Australian property law presented to the creation of freehold title to apartments: the common law and equitable prohibition on positive burdens running with freehold land prevented the imposition of maintenance obligations, essential in collectively owned buildings. The chapter notes that changes in the United States' common law in the mid-nineteenth century allowed positive obligations to run with freehold land, but that it was not until the enactment of strata legislation in the mid-twentieth century that the limitation was overcome in Australian law. The chapter flags the possibility that insufficient attention was paid to the rationale for the common law prohibition on positive obligations and the essential function the prohibition may still serve.

Chapter 2 establishes the theoretical framework for the book by examining the work of contemporary property theorists. One of the problems that the book highlights is a tendency on the part of government, industry, and even courts and lawyers, not to consider strata and community title law as property law. This is despite the fact that as law that allocates registered Torrens titles to land and regulates their use, strata and community title are irrefutably a subset of property. As one of the oldest and most analysed areas of law, it is a mistake to disregard the insights of either property theory or doctrine when considering strata and community title. Thus, one of the first aims of the book is to establish an appropriate *property* theory with which to analyse strata and community title law. The theory chosen suggests that by their consistent application through time and space, property entitlements help to define our social, economic and political order. In order to avoid undermining hard-won gains in modern democracies, we must ensure that property entitlements remain consistent with the plural values of our liberal democracy.

Chapter 3 moves the book from Australia to the United States. As noted, mid-nineteenth-century developments in American property law meant that positive obligations could run with freehold land, which in turn led to the widespread creation of private communities with extensive shared facilities and private governing associations.

To a large degree, Australian strata title developments, and in particular community title developments, are modelled on these American communities. For this reason alone, the American literature and jurisprudence merits attention. However, the American material is accorded considerable space in the book because of the importance of the legal, political, economic and social issues that it addresses. Chapter 3 sets out United States doctrinal debates about enduring obligations on freehold land; tensions between freedom of contract and the more limited boundaries of property law; debates about the limits of small group autonomy in liberal democracies; empirical studies on the lived reality of private communities; and the multiple forces that drove the rise of these communities in the United States. All of these discussions are centrally relevant to the by-law-making power in strata and community title, and the documentation of these debates is a vital contribution to Australian research. It will help to establish that strata and community title law is not simply concerned with legislative technicality, but raises complex questions about the legitimacy of widespread private regulation of land.

Chapter 4 begins Part II of the book and the analysis of Australian state law in light of both the property theory and United States debates, set out in the preceding chapters. It discusses the leading higher court decisions on the by-law-making power, arguing that there appears to be a limited appreciation of the implications of by-laws in the legislation, and consequently, case law. Chapter 4 has a particular focus on the economic implications of by-laws when they are permitted to burden freehold land with monetary obligations.

Consistent with the theoretical framework for the book, which argues that property law must be consistent with the *multiple* values of liberal democracy, Chapter 5 expands the analysis of by-laws beyond economics, to their impact on personal autonomy. In particular, it examines the power to regulate private homes with collectively made by-laws, and the social and political implications of that power being multiplied thousands of times, across our physical landscape and through time.

Finally, Chapter 6 broadens the analysis again to consider the effect of by-laws on the more amorphous concept of 'well-being'. The chapter includes a case study on children in strata and community schemes, examining the impact of privately written by-laws on children's safety and play.

The final section of the book draws together the insights gained from the preceding chapters, arguing that the current by-law-making power is far too broad, and fails to take into account the serious consequences of granting private citizens the power to write rules for their neighbours. It argues that the by-law-making power needs to be limited by both legislatively and judicially devised 'boundary rules', which prohibit obligations and restrictions that compromise land's social and economic utility, and which are at odds with the basic values of a pluralist democracy. The book concludes that when devising and assessing property law, which includes the strata and community title by-law-making power, we are forced to confront the question, 'in which world would we rather live?'[7]

Before moving to the first chapter, the following clarifications should be noted. Strata title law is state legislation in Australia, and while there are frequent calls for common legislation, that is unlikely to ever eventuate. In the constitutional division of power in Australia, land falls in the state not federal sphere, and without constitutional amendment or the referral of state powers to the Commonwealth, state governments will remain free to review their own legislation without reference to other states. Sometimes there are good reasons for legislative variations. For example, Queensland is Australia's premier tourist destination, and the physical, social and economic differences between the Gold Coast and the suburbs of Adelaide in South Australia are marked. It makes sense for these differences to be reflected in each state's property legislation.

That said, the basic framework and concepts behind strata title legislation are common to all states, and rather than trying to deal exhaustively with the detail of every state's legislation, this book will use New South Wales law as the starting point for discussion. New South Wales has been chosen as it was essentially the first state to enact strata legislation,[8] which was then used as the model for others states, as well as overseas jurisdictions. New South Wales is Australia's most populous state and there are roughly as many people living in apartments in New South Wales as all of the other states combined. However, the most significant reason for using the New South Wales legislation as a template for discussion is that it represents a middle ground. Queensland's legislation is the most detailed and most frequently amended, but it is strikingly accommodating of developers' interests, while Victoria's legislation is arguably the most minimalist and restrictive. Legislation in Western Australia and the ACT resembles that of New South Wales, more closely than they do Queensland or Victoria, while the legislation of Tasmania and South Australia resembles earlier incarnations of the New South Wales Acts. The New South Wales Acts will be used to discuss core concepts, but other states' Acts will be used wherever relevant to highlight desirable and less desirable alternatives.

Ultimately, irrespective of variations, all states' Acts perform the same function – they mediate the complex interactions of multiple parties with respect to a single piece of land.

The New South Wales legislation suffers from an excessive variation in terminology but this has one advantage for this book. New South Wales makes a distinction between vertical and horizontal subdivision, generally using the term 'strata title' for apartment buildings and 'community title' for master planned estates. There are separate 'strata title' and 'community title' Acts. Legally, this is unnecessary as all schemes are simply subdivisions of land (which includes airspace), into individually owned lots and collectively owned common property. Other states, with the exception of South Australia and the ACT, do not make a distinction, using a single piece of legislation to regulate vertical and horizontal subdivisions. However, while these developments may be *legally* identical, owing to the scale of many community title developments there are social, economic and political differences. A six-lot, walk-up apartment building is not the same as a 500-lot master planned estate and the New South Wales terminology 'strata' and 'community' title helps to identify and address these differences.

The downside to this distinction is the aforementioned excess in terminology in New South Wales, in relation to key elements of the legislation. First, the registration of all schemes automatically creates a separate legal entity made up of all owners, not tenants. New South Wales initially called this legal entity a 'body corporate' in the original *Conveyancing (Strata Titles) 1961 (NSW)* ('*CSTA*'). The legislation then split the development and management aspects of strata title into the *Strata Schemes (Freehold Development) Act 1973* ('*SSFDA*') and the *Strata Schemes Management Act 1996* ('*SSMA 1996*'), with the *SSMA* introducing a new term 'owners corporation', while the *SSFDA* confusingly retained the old term 'body corporate'. Thankfully, this anomaly has been removed in the *Strata Schemes Development Act 2015* ('*SSDA*'), which uses the term 'owners corporation'. However, the community titles legislation adopted the terms 'community association', as well as 'precinct association' and 'neighbourhood association', echoing terminology from the United States. All of these are simply bodies corporate, that is, they are separate legal entities, created by legislation, and made up of all lot owners, not tenants. While Queensland and Tasmania use the term 'body corporate', Victoria uses 'owners corporation', South Australia uses 'strata corporation', Western Australia uses 'strata company' and the Northern Territory uses 'corporation'. For ease of reference, this book uses the generic term 'body corporate' for all, however, it occasionally uses 'owners corporation' or 'community association' when discussing judgments that used those specific terms.

There is inconsistency within New South Wales and across Australia in relation to the label given to the private regulation of a scheme, the primary subject of this book. The New South Wales strata Acts use the term 'by-laws', while the community title Acts use the terms 'community', 'precinct' and 'neighbourhood' 'management statement'. Queensland also uses the term 'management statement', while South Australia and the Northern Territory use 'articles', Victoria and the ACT use 'rules' and Tasmania uses 'by-laws'. While there are variations between the states in relation to what these documents must and must not contain, conceptually, they are identical. They are the regulations written by private citizens (developers or the ultimate lot owners) that attach to land, binding anyone with an interest in the land. This book will use the generic term 'by-laws' for all.

Finally, the book is not intended to be limited to particular kinds of strata and community title schemes, but rather takes its lead from the prominence of particular schemes in our landscape and in our courts. The majority of schemes in all states are residential, and thus particular attention will be paid to homes. However, the schemes that have been the greatest source of litigation in higher courts are those in which significant sums of money are at stake. These tend to be mixed commercial and residential schemes, residential schemes with professional management, and tourist or serviced apartment schemes.

Notes

1 New South Wales Government, 'Draft Metropolitan Strategy for Sydney to 2031' (2013); New South Wales Government, 'A New Planning System for New South Wales: White Paper' (2013); Victorian Government (Department of Infrastructure), 'Melbourne 2030: Planning for Sustainable Growth' (2002); Queensland Government, 'South East Queensland Regional Plan 2009–2031' (2009).

2 Hazel Easthope, Bill Randolph and Sarah Judd, 'Governing the Compact City: the Role and Effectiveness of Strata Management' (City Futures Research Centre, 2012), 11.

3 NSW Fair Trading, 'Strata & Community Title Law Reform Position Paper' (November 2013) 2.

4 Anne Wallace, Michael Weir and Les McCrimmon, *Real Property Law in Queensland* (Lawbook Co, 2015), 504.

5 Easthope, Randolph and Judd, above n 2, 11.

6 With the exception of the limited number of developments under the *Strata Schemes (Leasehold Development) Act* 1986 (NSW).

7 Joseph William Singer, *Entitlement: The Paradoxes of Property* (Yale University Press, 2000) 138.

8 Technically, Victoria was the first state to enact strata legislation in the *Transfer of Land (Stratum Estates) Act 1960* but this legislation was not successful, and was replaced with the *Strata Titles Act 1967* (Vic), modelled on the New South Wales *Conveyancing (Strata Titles) Act 1961*.

Part I

1 The genesis of the strata and community title Acts

Introduction

On 1 July 1961 the *Conveyancing (Strata Titles) Act 1961* (NSW) ('*CSTA*') commenced in New South Wales in order 'to facilitate the subdivision of land into strata and the disposition of titles thereto'.[1] The Act allowed for the subdivision of land and airspace into individually owned freehold apartments and collectively owned common property. Others states quickly followed suit with their own special legislation for apartments; Tasmania enacting legislation in 1962, Queensland in 1965, Western Australia and Victoria in 1967, the Australian Capital Territory in 1970 and the Northern Territory in 1974.

All states' legislation essentially followed the New South Wales template. A body corporate, constituted by all owners, was automatically created by the registration of the plan of subdivision, with the power to raise levies for the maintenance, insurance and administration of the building, as well as the power to enforce by-laws. The by-laws were for the 'control, management, administration, use and enjoyment of the lots and the common property',[2] and were provided in the schedules to the Act. However, the First schedule by-laws could be altered by a unanimous resolution of the owners and the Second by special resolution.[3] In other words, the Act gave private citizens the collective power to create rules for their neighbours' use and enjoyment of not only shared property, but also privately owned apartments. This power is the primary subject of this book.

It is often said that strata title is an Australia innovation.[4] While it is true that other jurisdictions have modelled their laws on Australian legislation,[5] the 'Made in Australia' label is inaccurate. Many countries experienced intense housing shortages in the post-War period – most obviously in Europe, in particular Germany – which they ideally wanted to remedy with individual home ownership.[6] A century of consistent urbanisation,[7] coupled with engineering developments which made high-rise building possible,[8] meant that urban apartments, not freestanding houses, were the obvious solution to housing shortages. However, if people were going to be housed in apartments, and governments wanted people to own that housing, the problem of individual apartment ownership within a shared building needed to be solved. A number of countries introduced legislation to facilitate apartment ownership, the earliest being Germany, which enacted the *Wohnungseigentumsgesetz*[9] in 1951. This facilitated the division of a building into individually owned apartments and collectively owned common property, governed by a set of rules.[10]

While Australia may not have been the first country to introduce strata title, we have undoubtedly embraced it. Strata title and its horizontal equivalent, community title, are the fastest growing forms of property title in Australia today. As noted in the Introduction, successive state governments have implemented urban consolidation policies in Australian cities,[11] particularly the New South Wales, Queensland and Victorian governments,[12] which has resulted in a sharp rise in high and medium density development. In New South Wales, the period from 1996 to 2006 saw an average increase of attached dwellings by 10 000 per year,[13] and by 2011, over one-quarter of Sydney's population, just over one million people, lived in strata title apartments.[14] The New South Wales government predicts that within 20 years, half of the state's population will live or work in a strata or community scheme.[15] In South East Queensland, between 2001–2006, apartment stock increased by 76 per cent, and in the period, 2001–2011, 60–70 per cent of new dwellings were contained within a defined urban footprint, exceeding a government target of 50 per cent.[16] Melbourne has experienced significant high-density development close to its Central Business District and current plans aim to make an expanded Melbourne city centre Australia's largest business and residential centre by 2040.[17]

While strata and community title include commercial, retirement, industrial, tourist and mixed-use developments, the overwhelming majority of properties are residential.[18] In order to understand properly the social and legal impetus for strata and community title legislation, it is necessary to consider the history of urban and suburban residential development in Australian cities, as well as the legal doctrines that facilitated that development.

A short history of Australian housing form and law

The sharp increase in people living in high-density housing is in obvious contrast to The Great Australian Dream – the freestanding house on a quarter acre block. It is a truism that residential property and homeownership have long loomed large in the Australian consciousness. Graeme Davison argues that from the moment of white settlement onwards, successive Australian governments and their citizens consistently strove to realise the suburban ideal of freestanding houses set in spacious gardens far from the urban decay of European cities that many colonists, emancipists and refugees had known:

> The suburb was emerging as a solution to the urban ills of the Old World. Like a colony, the suburb was a place of escape or refuge, and it was shaped, therefore, largely by the logic of avoidance. The suburb was, in essence, a mirror image of the slum. While the slum was seen as dense, dirty, unnatural, disorderly and disease-ridden, the suburb was seen as open, clean, natural, orderly and healthy.[19]

From the early twentieth century the Garden Suburb Movement, inspired by Ebenezer Howard's (1850–1928) writing,[20] was hugely influential in the development of Australian cities. Planners, politicians and builders all enthusiastically took up the ideal of well-planned suburbs with wide streets, public parkland and low-rise, freestanding, single family housing.[21] There was intense belief 'in the capacity of a good physical environment to transform, for the better, individual destiny, family life and community spirit'.[22] Although increasingly criticised by intellectuals, feminists and environmentalists,[23] the pursuit of the suburban ideal persisted well into the late twentieth century, with Aiden Davison concluding that:

> The creation of suburbs [became] a defining preoccupation of Australian society in the 20th century. During this period the suburban population increased from less than 1 to more than 11 million in the capital cities alone. Depending upon how they are defined ... somewhere in the order of 70 per cent of Australia's population now live in suburban environments.[24]

One of the most important aspects of Australian suburban living was freehold ownership. While Australians tend to assume *freestanding* suburban housing and *freehold* ownership are synonymous, it is possible to be a suburban citizen and perpetual tenant; British suburbs were often built and remain leasehold.[25] However, in Australia, this was not the case.[26] Many immigrants 'looked back upon the experience of living as tenants in their homeland, and longed to be free of the fear of the landlord and the bailiff'.[27] Governments shared their conviction. John D. Fitzgerald, barrister, Chairman of the New South Wales Housing Board (1912–1917) and Minister for Local Government (1916–1920), argued that: 'If you want to make the working man contented, you must give him better housing conditions than he has hitherto had, and let him own his own house in addition ... human nature is in favour of the freehold'.[28] Reviewing the history of twentieth-century Australian housing legislation, Alex Kondos said, '[l]et me put it this way: irrespective of party, and level of government, Australian governments foster home ownership'.[29]

The dominant *form* of Australian housing (freestanding), and the dominant *tenure* (freehold), have significant implications for strata and community title.

Freehold title is extremely difficult to regulate through private property law. This concept will become a central theme of the book, discussed in more detail later, but it is necessary to lay the basic groundwork here. The fundamental difference between a leasehold interest and the freehold fee simple is that a leasehold interest is limited by time and that a wide range of covenants can be attached to the leasehold estate, creating significant positive obligations (e.g. rent and repair), as well as restrictions on the way in which the leaseholder may use the land. The ability to enforce positive and negative covenants means that for those who own a significant area of land, long-term leases can be used as an effective system of private planning.[30] Prior to the introduction of public planning law in late-nineteenth-century Britain,[31] leases were used for exactly that, with entire suburbs in London and other British cities being developed through long-term leases granted by the freehold-owning aristocracy and private institutions. Initial 'building leases' mandated building materials and styles, set-backs, central garden squares, and hydraulic infrastructure, theoretically ensuring a high standard of initial construction. Subsequent 99-year leases were used to control the social make-up of neighbourhoods, maintain their residential character, and prevent further subdivision.[32]

In contrast to leasehold interests, the freehold fee simple is characterised by freedom of use. Once a grantor passes a fee simple to a grantee, unlike a landlord, the grantor loses control of the land and is unable to restrict its use or compel the grantee to perform any act, such as repair the land or the payment of money. The one narrow exception to this general rule is the freehold covenant. The anomalous case of *Tulk* v. *Moxhay*[33] held that a purchaser of land could be bound by a promise, made by a predecessor in title, never to build on the land (Leicester Square in London), and to maintain it as a pleasure garden. The purchaser was bound because he had notice of the covenant. The decision flew in the face of the seminal cases of *Keppell* v. *Bailey*[34] and *Hill* v. *Tupper*[35] by allowing an incident of a 'novel kind' to be enforced against a subsequent purchaser. Unlike leasehold interests, fees simple theoretically last forever; as a result, if a covenant attaches to a fee simple interest, like the fee simple, it will be perpetual.[36] For reasons that will be discussed in detail in the next chapter, English and Australian courts subsequently did their best to radically curb *Tulk*'s application, insisting that covenants benefit land, not a business or person,[37] and most significantly, that covenants only include restrictions, not positive obligations.[38] As a result, freehold covenants had limited content and consequently could not be used to create detailed systems of private planning in the way of leasehold estates.

Despite their limited planning power, freehold covenants were used widely in Australia for two reasons. The first was that owing to the political preference for freehold ownership described above, leasehold planning was simply not an option.[39] The second was that the restrictive nature of freehold covenants was actually relatively effective for the initial implementation, (if not consistent enforcement[40]), of the Garden Suburb ideal of single-family dwellings set in spacious gardens.[41] Minimum allotment size covenants prevent subdivision, keeping gardens large; single-dwelling-only covenants prevent terraces, duplexes and granny flats; height-limiting covenants prevent multi-storey dwellings; residential-only covenants prevent businesses and industry; and building material covenants (for example, 'brick and tile' covenants), and minimum building cost covenants, prevent building in substandard materials like fibro and corrugated iron. In order to increase their marketability, Australian developers routinely applied restrictive covenants to their suburban developments throughout the twentieth century.

However, restrictive covenants cannot compel the payment of money and Australian courts have consistently enforced this limitation, so much so that the default rule for whether a covenant is positive is whether it requires the

Figure 1.1 Low-rise, freestanding house built and maintained in accordance with a 'brick and tile' restrictive covenant.[42]

expenditure of money. If it does so, it is unenforceable.[43] The consistent judicial restriction of covenants to negative and not positive obligations has a fundamental significance for the freehold development of high-rise buildings and low-rise suburban estates with facilities.

First, of necessity, high-rise buildings have shared property (lifts, stairwells, halls, car park, roof and surrounding land), and of necessity, this must be maintained. This requires the payment of money. If apartment owners have leasehold titles, the landlord, as the sole owner of the freehold, can maintain the building, or, via the leases the landlord and/or the tenants can be compelled to pay for maintenance. However, if the apartments are freehold title, it is not possible to impose positive obligations to repair on those titles, binding all successive owners of apartments. This inability to compel positive obligations to maintain freehold property has been identified as the sole insurmountable impediment to the freehold development of apartments in English and consequently Australian law.[44] The English have avoided this problem with long-term leasehold title, a form of property ownership to which they have long been accustomed,[45] but in Australia, as we have seen, long-term leasehold titles have never been socially or politically acceptable, and thus other solutions, discussed below, had to be found.

Second, for marketing and amenity purposes, high-rise buildings and low-rise suburban estates can have shared facilities and infrastructure like gardens, boardwalks, clubhouses, parks, marinas, pools and tennis courts, as well as more modern innovations like gyms, air-conditioning, telecommunications cables and black and grey water treatment plants. Shared property means that people can enjoy services and facilities that would be uneconomical for individuals alone. But, if it is not possible to compel people to pay for these facilities and infrastructure, there is no point building them in the first place. In Australia, it was not possible to compel payment and so our apartment buildings, and more significantly, our suburban development, were never built with extensive, shared, private facilities. In contrast to English neighbourhoods, which often had a private central square, accessible only by the adjoining leasehold residents who were compelled to pay for its upkeep via their leases,[46] Australian suburban development only had *public* space in the form of playing fields, parks and bush reserves, maintained by the local council through rates, and open to residents and non-residents alike.

In this sense, Australian suburban development also stood in marked contrast to suburban development in the United States of America, a jurisdiction whose law and housing form will become central to this book. The United States, like Australia, is a land of plentiful space and extensive suburban development. As in Australia, the Garden Suburb Movement was extremely influential,[47] in addition to the work of America's own planned suburb visionary, Fredrick Law Olmsted (1822–1903).[48] America also shared Australia's political aversion to long-term leases,[49] and thus suburban development was invariably freehold. However, there was one significant physical and social difference between some American suburbs and all suburbs in Australia – the presence of extensive collectively owned private property. This difference can be traced to a divergence in Australian and American freehold covenant law.

Prior to the decision in *Tulk*, the United States had developed a doctrine of 'real covenants', which allowed *positive* obligations to attach to freehold land.[50] This in turn made it possible to build suburban housing with shared private facilities (for example, parks, boardwalks, clubhouses, marinas, pools), which freehold owners could be compelled to maintain. The result was the gradual development in the late nineteenth and early twentieth century of private communities with facilities, and their exponential development in the post-World War II period.[51] It was inevitably discovered that if communities had private facilities to be maintained by freehold owners, a body was needed to levy and administer the necessary money. If that body could enforce positive covenants to pay levies, it could also usefully enforce any other covenants as well. In *Neponsit Property Owners' Association, Inc* v. *Emigrant Industrial Savings Bank*,[52] the New York Court of Appeals confirmed the legitimacy of a private association both levying money and enforcing covenants, heralding the rise of private associations, commonly now known as homeowner associations (HOAs).

The consequence of Australian law not being able to compel the maintenance of shared facilities meant that not only did we not have such facilities in our suburbs, we also did not have any need for private residential associations. This meant that restrictive covenants continued to depend on orthodox enforcement by individuals, via complicated rules of equity. Although restrictive covenants operate through their mass application to residential subdivisions (ideally, all lots are benefited and all lots are burdened by the restriction[53]), they are not enforced by the collective owners; they are enforced by an individual owner whose land has the benefit of the covenant against an individual owner whose land is burdened by the covenant. This requires a preparedness on the part of owners to incur the expense and stress of mounting a Supreme Court case against their neighbour and not surprisingly, restrictive covenants are often more honoured in their breach than observance. Rosebery, a suburb of Sydney built for working- and middle-class families in 1912, and planned with restrictive covenants,[54] is a good illustration of this phenomenon. A heritage study of the suburb conducted in 2006 classified houses as either 'contributory' (retained original form and detailing), 'neutral' (some alterations but still recognisable as from the period Rosebery was developed), or 'detracting' (so altered that their original construction is not recognisable).[55] It concluded that Rosebery's integrity, (that is, compliance with the restrictive covenant), was only 43 per cent. Neighbours had obviously not been prepared to enforce the covenant on the subdivision.[56] Sometimes restrictive covenants were more successful in controlling development over time, such as those in Haberfield, in Sydney's inner west, but it is the exception rather than the rule. Many Australian suburbs were planned with restrictive covenants; few still look as uniform as Haberfield.[57]

The upshot of this divergence between United States and Australian law is that urban planning through private property law was relatively unintrusive and erratic in Australia.[58] There are many suburbs in Australia that are affected by old 'brick and tile' covenants of which owners are blissfully unaware. Covenants are simply not enforced[59] and even if they are, they regulate very little, only being restrictions and not positive obligations. Further, all states have legislation that empowers courts to declare that covenants have become 'obsolete' and unenforceable,[60] and some states, like New South Wales, have enacted legislation to ensure that public planning instruments override private restrictive covenants. Section 28 of the *Environmental Planning and Assessment Act 1979* (NSW) suspends the operation of covenants if they conflict with public planning instruments, ensuring that decisions about suburb and city-wide planning are made by elected governments, not by private developers and subsequent landowners. In Queensland, restrictive covenants have always been peculiarly ineffective, being unable to be recorded on the Torrens register.[61] Despite their consistent use by developers, they are almost never enforceable beyond the first purchaser, minimising their ability to have any long-term impact on land.

In contrast, in the United States, because covenants could have both negative and *positive* content, they had much greater potential to compel and control landowners. Their positive aspects, which included the payment of money, facilitated the provision of extensive common property. Further, the existence of a private association with the power to enforce covenants allowed impersonal, collective and thus more thorough enforcement of covenants and association rules, which in turn blurred the line between private and public law. An association with the power to create and enforce rules over a geographically defined area looks distinctly like a form of government. As a result, United States' courts have been actively considering the economic, social and political implications of privately created covenants on land for over a century, producing considerable case law. This in turn sparked deep and challenging academic debate.

No similar level of judicial or academic debate has occurred in Australia. This would not matter if Australian law had continued to limit private regulation of land to the 'soft' regulation of restrictive covenants and short-term leases. However we did not. The enactment of the *CSTA* began a series of legislative enactments that now grant private citizens, both developers and subsequent owners, enormous power to regulate land and the people who use it. It is to this legislative change we now turn.

The path to strata and community title legislation

The corollary of consistent preference for freehold ownership of freestanding houses in leafy, green suburbs was a deep antipathy to high-density apartment living on the part of governments and planners. Despite some initial attempts at the turn of the twentieth century to house working-class families in apartments in Sydney,[62] and a brief flirtation with high-rise public housing in Sydney and Melbourne in the 1960s, apartments were never accepted as an option for mass housing of Australian citizens. New South Wales planners and governments consistently characterised apartments as 'the slums of the future',[63] ironically preferring to house working-class families in the slums of the present – the dilapidated terraces of inner Sydney.[64] In Melbourne and Brisbane, planning rules limited site coverage, making the wide-spread construction of apartment blocks uneconomical until the post-war period.[65] In other state capitals, small populations and plentiful land minimised the need for high-density development for much of the twentieth century.

However, despite official resistance to apartment construction and the dogged pursuit of the Great Australian Dream for the masses, the existence of pre-1960s apartment buildings in Australian cities makes it clear that some people, at least, were living outside The Dream. Apartments were built in South Yarra and Toorak in Melbourne, Fortitude Valley, South Brisbane, New Farm and Hamilton in Brisbane, and in and around Perth city. Owing to the city's more lax building laws, apartment construction was most marked in Sydney, with 70 000 apartments built between the wars, predominantly in the inner, middle ring and beachside suburbs.[66]

Residents of apartments fell roughly into two groups and their tenure is significant.

The first group was young, single and often bohemian; writers, artists, and performers, people prepared to explore a life beyond suburbia.[67] Robin Slessor recalled that when his brother, famous Australian poet Kenneth Slessor, found a flat in the Kings Cross area of Sydney in the 1920s, for their mother, 'the idea of taking a flat was only one step away from announcing he was going to shack up with a prostitute'.[68] Of course not all apartment dwellers were bohemian;[69] many were simply single men and women who preferred the freedom of apartment life to the mean regulation of privately run boarding and lodging houses.[70] For example, Riviera Flats, completed in 1938 in Perth, were billed as a chance to escape from boarding house communal lavatories, creaking iron bedsteads and 'hideously' papered and carpeted rooms. The flats met a demand from young singles and married couples to live in 'clean, bright, independent circumstances ... [with] well-planned rooms with built-in modern furniture ... available at moderate rentals'.[71] From the 1930s onwards, apartments in Sydney, from Kings Cross to the eastern beaches, and in St Kilda in Melbourne, increasingly housed European refugees for whom apartment living had been the norm.

This first group of apartment dwellers were tenants, and in the character of Australian urban leases, their tenure was relatively short and insecure. Although technically incorrect, governments and citizens would not characterise these leases as 'ownership'.

The second group of apartment dwellers were a different class altogether. They lived in company title units,[72] and owing to a refusal of lending institutions to finance the purchase of company title units (discussed below), they were 'people who had sufficient independent resources to buy their flats – a well-to-do stratum of society'.[73]

Some of the very first apartment buildings to be constructed in Sydney and Melbourne, such as Amesbury House built on Domain Road in South Yarra in 1921 and The Astor, built on Sydney's Macquarie Street in 1923, were company title. Physically, socially and legally, these apartment blocks were modelled on

Figure 1.2 Apartment block in a beachside suburb of Sydney built in the inter-war period.[74]

overseas developments. Physically, early Sydney apartments resembled those in New York and Chicago (which were in turn modelled on Paris), while Melbourne's were more British: discreet pseudo-mansions, externally identical to large, single homes. Socially, most were marketed and occupied as elite urban living.[75] For the landed gentry, apartments were a glamorous city bolt hole, and for wealthy city residents, they presented a solution to the lack of availability of servants necessary to run huge Victorian houses.[76] The Astor had a café and restaurant in the basement which sent up food to apartments via a dumb waiter. It also had a caretaker, who attended to residents' every need, from arranging funerals to sending a boy down to David Jones, an upmarket city department store. The resident caretaker of The Astor in the 1930s eventually had a nervous breakdown, quite possibly brought on by the demands of his wealthy clientele,[77] (an experience with which modern strata and building managers can no doubt sympathise).

Legally, Australian company title buildings were modelled on New York's 'co-operative' apartments, which were a corporate share-holding model, with which the wealthy business people to whom they were marketed were familiar.[78] A company owned the land, and apartments were purchased by buying shares in the company; the company then granted purchasers a lease of their apartment.[79] A board of directors managed the building through the company's articles of association, which included the power to veto any transfer of shares, thus controlling the class of potential owners. While The Astor's company title form was promoted as a 'fine investment and a permanent home', as noted above, banks did not agree. The power of the board of directors to veto sales interfered with the ease of a mortgagee's power of sale and as a result, banks would not lend money for the purchase of a company title apartment.[80] Consequently, company title could never provide housing for the masses. Vetting or outright banning of tenants, facilitated by the articles of association, intensified this problem.

Leasehold and company title apartments operated relatively independently of each other in the first half of the twentieth century, attracting distinct groups of investors and residents. However, in the post-World War II housing crisis, the two forms of title intersected in an unpredictable way, intensifying the housing shortage and precipitating the enactment of Australia's first strata title legislation.[81]

To prevent landlords exploiting the housing shortage, the New South Wales Labor government passed the *Landlord and Tenant (Amendment) Act 1948* (NSW) ('*LTA Act*'). This capped rents and drastically limited powers of ejectment, protecting tenants. However, the Act rendered apartment buildings less attractive to investors, suppressing private sector incentives to build apartments for rental purposes, worsening the housing shortage.[82]

Developers did not stop building apartments altogether. They discovered a useful loophole in s 62(5)(m) of the *LTA Act*, which allowed eviction if the premises were 'reasonably required by the lessor for reconstruction or demolition'. As a result, if a company were floated to own the old or reconstructed flats, the tenants could be evicted. For developers, often working in concert with real estate agents, this became an irresistible way of freeing land of rent-control provisions. Interestingly, conversion of buildings to company title, or 'co-operatives', had been used in exactly the same way in New York, the city from which Australia copied the company title form. Conversions occurred in New York in both the post-World War I and post-World War II periods, proving disastrous for tenants who did not want to, or could not afford to, buy the new co-operative flat. Resistance to conversions was strong.[83] Conversions in Sydney had exactly the same consequence for tenants and the files of the New South Wales Department of the Attorney General and Minister for Justice contained scores of representations from residents who were being forced from their homes. Kondos described the 1950s as an 'intensive period of housing conflict' in Sydney, with ever-increasing pressure on government to find a workable solution to the problem of the freehold ownership of apartments.[84]

In 1950, the New South Wales Labor government established the Property Law Revision Committee, with L W Taylor, a solicitor as Chairman, J Baalman, the Chief Examiner of Titles as Executive Member, and C D Monahan, a barrister, who later became a judge.[85] The Committee reported to the Minister in 1957, with both Taylor and Monahan satisfied with the existing company title method of owning flats and Baalman strongly disagreeing. Baalman insisted that the only way to overcome the major deficiency of company title shares being inadequate security for a mortgage was to introduce legislation for the creation of freehold title to apartments. Kondos argues that many people involved in wider apartment title discussions, including the Local Government Association, the Law Society, leading solicitors' firms and the Registrar General, remained convinced that the problem of apartment ownership could be solved with legislation standardising procedures and the articles of associations for company title. However, the balance was tipped in favour of strata title legislation when a consortium of construction companies weighed into the debate in 1958. Baalman was the consortium's legal counsel.

Naturally, the consortium's interest in the issue was to increase the market of potential purchasers for their product. Without finance to assist in the purchase of apartments, the pool of purchasers would be forever limited to those of substantial independent means. Further, the *LTA Act* made the option of purchasing (or in the case of developers themselves, retaining) freehold title to rent very unattractive indeed.

On 18 December 1958, the managing director of a large construction company representing the consortium wrote to the Minister for Justice stating that his company was building several home unit projects and had more on the drawing board. He stated that the company had 'specialised knowledge, therefore, of the disadvantages of the present situation regarding home unit titles, more particularly in respect of the provision of long-term finance to the purchase of such units'.[86] He revealed that the company had employed counsel to consider the issue and that this had led to a solution that they would be prepared to share with the Minister. On 3 February 1958, the consortium did exactly that, presenting the Minister with a draft '*Stratified Titles*' Bill.

The Bill was assessed by the Senior Legal Officer of the Registrar General's Department, drawing a 'devastatingly critical' report. This led to the consortium hastily organising a luncheon at the Wentworth Hotel on 11 March 1959, to which they invited a 'who's who of the managers of the urban system'.[87] If it was not extraordinary enough that a private consortium took it upon itself to organise such an event, the consortium proceeded to instigate, supervise *and pay* the drafting committee that spent the next eight months considering the Bill. As Kondos notes, this was 'a remarkable arrangement, to say the least, in the law-making process of any society'.[88] It also began a long pattern of developer involvement in the enactment of state legislation designed to facilitate development.[89]

The *Conveyancing (Strata Titles) Bill 1959* (NSW) was presented to Parliament on 26 November 1959, but only after the Senior Legal Officer met with the senior officers of eight lending institutions, on the explicit instruction of the Minister for Justice. The Minister wanted to ensure that the Bill had the uncritical support of these institutions and Kondos argued that 'the event stands as a symbolic confirmation that the *CSTA* was primarily geared to meet the needs of financial institutions'.[90]

After the first reading in 1959, the passage of the Bill was suspended for over a year so that interested bodies could comment. Rath, Grimes and Moore said that many did so and that 'as a result the Act represents not so much a scheme devised by a drafting committee as one embodying the considered views of a community'.[91] This also began a pattern of community consultation in strata title legislation, with sometimes dubious results. Continual 'community' or 'stakeholder' consultation in the creation of complex legislation can produce a product that looks similar to the proverbial horse designed by a committee.[92]

The *CSTA* finally came into force in July 1961, allowing the creation of the desired freehold title to apartments by the registration of a strata plan of subdivision. A body corporate was automatically created by the legislation, made up of all lot owners, not tenants, and it was responsible for the financial and administrative management of the scheme. The body corporate's duties included insuring and keeping the common property in a state of good and serviceable repair,[93] and its powers included the ability to raise 'levying contributions' to discharge those obligations. In other words, legislation had been used to overcome the common law and equitable limitation that positive covenants to pay money could not be attached to freehold titles. All apartment titles were burdened with statutorily created positive obligations to pay for maintenance.

The body corporate was also given the power to enforce by-laws. As noted at the beginning of the chapter, the by-law-making power was for 'the control, management, administration, use and enjoyment of the lots and common property', and while there were by-laws provided in the First and Second Schedule, both could be repealed or amended by an appropriate majority. The only express limit on the by-law-making power was a prohibition on restrictions on transfer, lease and mortgage of lots, to prevent the re-emergence of the very problems the Act was designed to solve.[94] Other than this, the legislature had given private citizens an almost carte blanche power to write rules for their neighbours by virtue of ownership of property.

It is important to note that in the history that Kondos records about the creation of the Act, there is seemingly a complete absence of concern among any of the people involved in the drafting of the Bill about private citizens being given this power. Kondos does make reference to the '[d]ifficulties associated with the "house management" aspect of home unit living' being 'the chief area of [public] complaint as reflected by the large number of letters that poured into the Attorney General's office in the latter half of the 1950s'.[95] He notes with disapproval that these difficulties did not seem to constitute a problem in the eyes of Baalman, the Executive Member of the Property Law Revision Committee. However, the complaints made by members of the public seem to relate to the *lack* of 'house rules' needed for harmonious living; they do not extend to any concern with the limits that might need to be placed on the power to create such rules. Even Kondos himself, writing thoughtfully and critically 20 years later, using 'power' and 'the pursuance of one's interest in preference to the interests of others'[96] as the explicit basis for his analysis, did not consider the questions that arise when private citizens are given a quasi-law-making power by virtue of property ownership.

One obvious explanation for the lack of attention paid to this new power was that at the time the *CSTA* was drafted, it only applied to buildings and not land, and that those buildings were relatively small. It is necessary to clarify this statement immediately, because technically, the first part is incorrect. The *CSTA* allowed for the subdivision of land with a strata plan. Owing to the maxim 'cuius est solum eius est usque ad coelum et ad inferos', 'land' stretches from the heavens to the centre of the earth below,[97] and it was (and still is), airspace 'land' that a strata plan subdivides. Realistically, the only way to subdivide air is by reference to structures built in that air and that is precisely what the strata plan did; it defined the boundaries of each lot by reference to floors, walls and ceilings of the building in which the lot was contained.[98] Although technically the strata plan was subdividing 'land', in reality what it was doing was subdividing a building with minimal land around it.

That building was also likely to be relatively small – six to 12 apartment 'walk-ups' being the norm. This was because the vast majority of Sydney apartment buildings were constructed on freestanding housing lots, purchased by small builders who redeveloped one housing lot at a time.[99] In Kings Cross and Darling Point, those housing lots were the site of former Georgian and Victorian mansions, with sandstone walls and wrought iron gates often still standing as testament to past architectural glories. From Manly to Bondi to Bankstown, apartment buildings were built on the site of former Great Australian Dreams, Federation or California bungalows surrounded by gardens. Harry Seidler described Sydney's apartment construction thus:

> The total effect of this demolition of individual houses for replacement on the same site by now quite standard three-storey flats is truly horrifying. The

> results are barrack-type buildings, their long dimensions filling the depth of the narrow allotment. What used to be yards at the back and on the sides ... are denuded of vegetation and paved for cars. On floors above, the living rooms of adjacent blocks face each other across the five metre wide canyon.[100]

The physical form of apartment buildings was legally, socially and politically significant. The physical area that the new by-law-making power applied to was small, and the number of people regulated was limited. This is in contrast to the United States, where the doctrine of 'real' (positive) covenants had led to the regulation of large horizontal areas of land, not just single buildings; great swathes of suburbia and entire planned towns were regulated by privately written and enforced covenants. As will become clear in the detailed exploration of United States law and academic debate in Chapter 3, the scale of this regulation (measured by both land area and population), raises questions that the regulation of a six-lot strata scheme does not. Thus, it is not surprising that the by-law-making power did not cause initial concern.

However, today, the landscape of strata title in New South Wales and other Australian states has changed dramatically. First, many schemes are no longer three-storey walk-ups. They are very high-rise buildings with hundreds of strata lots and even more owners and tenants, all of whom are regulated by the by-laws. By way of example, Melbourne's Eureka Tower is 92 storeys high, with 84 floors of apartments, 10 floors of parking, and 556 apartments.

Second, while large schemes are still a minority, there are thousands more smaller schemes today than there were in the 1960s, radically increasing the total number of strata lots. There are now close to two million strata lots in Australia.[101] Chapter 2 will discuss the necessity of assessing the legitimacy of property law not simply by reference to how it operates between individuals but by reference to how those individual entitlements, multiplied thousands of times through time and space, constitute our social and political regime. But for now, suffice to say, there is a qualitative difference between a few thousand people being regulated by by-laws and having the power to regulate others, and millions of people being subjected to and wielding that power.

Third, as explained in the next section, the strata title form has been extended to horizontal subdivisions, allowing developers to copy the private planned communities that have existed in the United States for a century. These include low-rise master planned residential estates with extensive recreation facilities, like Macquarie Links[102] and Harrington Grove[103] in Sydney's south west and Hope Island and Sanctuary Cove on Queensland's Gold Coast. Others are a conglomeration of high-rise strata schemes, like Jacksons Landing in Sydney's inner city, a community scheme comprised of 20 strata schemes with 3000 residents.[104] Then there are 'tiered' schemes, like Breakfast Point, a development on the Parramatta River in Sydney that contains both low-rise and high-rise buildings, with its country club and 'village green' bearing testimony to its United States' antecedents.[105]

Figure 1.3 Typical small-scale, walk-up early strata schemes, Eastlakes, Sydney.[106]

Figure 1.4 Community title estate with low-rise freestanding housing and high-rise strata apartments in a single over-arching community scheme, Sydney.[107]

Finally, the stakeholders in strata schemes are no longer limited to owners and tenants. Physically large and populous schemes and/or schemes that provide tourist accommodation or serviced apartments, invariably employ professional strata managers and caretakers, adding valuable third party businesses to the already complex legal, social and property mix. Australia's largest tourist destination, the Gold Coast in Queensland, is now largely dependent on strata title, and nationally, the strata industry employs an estimated 20 000 people who manage $500 billion of property assets.[108]

In short, a resident of a six-lot, 1960s strata scheme, transported into a modern 500-lot, master planned strata estate, with swimming pools, a hotel and a golf course, might be tempted to say, 'Toto, I've a feeling we're not in Kansas anymore'.[109]

The successive legislation

From the foregoing it is becoming clear that an exploration of strata title might be somewhat more complex that initial impressions suggest. Before moving to the next two chapters on property theory and United States law, which will delve deeply into that complexity, the current state of Australian strata and community title legislation will be explained.

In the decade following the passage of the *CSTA*, all other states passed similar legislation to facilitate the development of apartments and to overcome the prohibition on positive obligation on freehold land. First generation Acts were typically relatively short, dealing with the basic subdivision of land and buildings into individually owned lots and collectively owned common property. From a contemporary perspective, a significant characteristic of the early Acts was that many matters that relate to the running of strata schemes – repair of common property, meetings, quorum, voting – were placed in 'First Schedule' by-laws which could be altered with unanimous resolution. The 'Second Schedule' by-laws provided by the legislation were extremely brief, covering the illegal or noisy use of a lot or common property, but could be altered or added to by special resolution.

By the 1970s, difficulties had emerged with first generation strata legislation, with the 'rules of internal management … and the enforcement of those rules [becoming] the principal concern'.[110] In New South Wales in 1973, the original *CSTA* was replaced with the *Strata Titles Act 1973 (NSW)* ('*1973 Act*'), which contained substantially more detail and regulation, particularly in relation to the body corporate. Significantly, large parts of the alterable First Schedule by-laws, which related to the repair of common property, meeting procedures and to the body corporate 'council' (now known as a 'strata'), were moved into the body of the new *1973 Act*. Presumably this was in recognition of the fact that even with unanimous agreement, it was not appropriate for private citizens to determine these matters, and that procedural provisions and obligations of the body corporate should be set by the legislature and applied uniformly to all schemes. The necessity for the move highlights confusion about the proper role of freedom contract in strata schemes that continues to this day, and which will become a theme of the book. Despite these changes to the sch 1 by-laws, the power to create new by-laws remained the same. Section 58(2) of the *1973 Act* continued to grant the body corporate power to make by-laws for 'the control, management, administration, use or enjoyment of the lots or the lots and common property', with prohibitions on restrictions on transfer, lease and mortgage of lots being the only remaining limit.

At the same time as New South Wales enacted its second generation strata legislation, Queensland became the first state to extend the strata form to low-rise subdivisions or 'group titles'.[111] Queensland had previously enacted the *Building Unit Titles Act 1965* for the subdivision of buildings into apartments, but the *Group Titles Act 1973* (Qld) allowed for the subdivision of land into individual housing lots with collectively owned open space and infrastructure, governed by a body corporate with the power to enforce and create private by-laws. It was the first legislation to facilitate the development of highly regulated, low-rise communities with private facilities hitherto only common in the United States.

The legislation was significant for two reasons. First, like the original New South Wales strata titles Act, it was primarily driven by developers, starting a pattern that would prove consistent in Queensland, and arguably the source of many of their current body corporate disputes. During the Parliamentary debates on the Group Titles Bill 1972,[112] the Opposition revealed that the Minister for Justice's Second Reading speech was taken almost word for word from a letter sent to the Queensland Law Commission by the lawyers of one of the state's largest developers, Hooker Centenary.[113] Hooker was the developer of Brisbane's Centenary Estates, a new residential development made up of freestanding homes on large lots, built around privately owned cul-de-sac 'courts'. Writing about the development, E McD Batt, the manager of Centenary Estates, stated that 'greater economy is achieved by eliminating conventional roads and methods of drainage and reducing maintenance costs to the City';[114] in other words those assets were being privatised and the costs passed on to homeowners in perpetuity. He also stated that the development structure would increase 'community cohesiveness' and that the concept could be adapted to 'medium-density housing schemes, waterway and golf course developments'.[115] In the ensuing 40 years, this is exactly what occurred.

To facilitate the privately owned roads and infrastructure in Centenary Estates, Hooker had been forced to use a complicated company title and easement structure, with all purchasers being allocated a parcel of shares in a company that owned the common property. The company was responsible for maintenance of common property and homeowner/shareholders had to pay the company's costs. The company title structure was used to overcome the limitations in covenant law, most marked in Queensland, where neither positive nor negative covenants were permissible on the Torrens register.[116] Covenants could not be attached to the freehold titles, compelling all subsequent owners to pay for maintenance of the common property. Mr Batt concluded his description of the development structure with the statement, 'no doubt in the future special legislation similar to the strata titles Acts will be drafted to cater for projects of this nature'.[117] This is precisely what occurred when the Minister for Justice stood up to give his Second Reading speech for the Group Titles Bill 1972, a speech that seems to have been largely informed by Hooker Centenary's lawyer.

The second significance of the *Group Titles Act 1973* was that it was enacted with an insufficient, or possibly completely absent, understanding of the function of the limitations in freehold covenant law that strata and group titles legislation was overriding. This is evident in the Second Reading speech for the *Sanctuary Cove Resort Act 1985*, one of a succession of Acts passed in Queensland to facilitate site specific and mixed-use development.[118] The *Sanctuary Cove Resort Act* was passed for a complicated US-style tourist, retail and residential scheme that a developer wanted to build at Sanctuary Cove, on the Gold Coast, despite environmental objections. The Hon R J Hinze, Minister for Local Government, Main Roads and Racing, (popularly known as the Minister for Everything), said that the,

> provisions are based upon existing Canadian and American condominium resort legislation.... The important thing to recognise at an early stage is that legislation of this type is not new in the global context. It is only new to Queensland and Australia. Many members may have visited resorts of this type in Hawaii, Palm Springs, Phoenix, and other places.[119]

Mr Hinze went on to say that,

> The primary mechanism used for controlling such developments in the American scene is by way of imposing constraints, covenants on title, and restraints on the developments. As this House will appreciate, the concept of covenants on title is unacceptable to this Government, and this legislation is designed to allow development to be proceeded without the imposition of such encumbrances.

While this statement is technically correct – neither the *Sanctuary Cove Resort Act 1985*,[120] nor any other strata or community title legislation authorises the creation of *common law* positive covenants of the United States kind – the legislatively created levying and by-laws provisions perform exactly the same function. They allow freehold land to be subdivided with enduring maintenance, architectural and behavioural obligations and restrictions, enforceable against successors in title of the fee. If the legislation did not have the same effect as the common law in the United States, it would not have succeeded in facilitating the kind of developments developers wanted to copy. The crucial difference between the United States common law and the Australian legislation is that the former was developed on a case-by-case basis by judges who were mindful of the dangers of limitations on freehold land, whereas the latter was done in a single legislative sweep, driven by developers whose sole motivation was profit from property sales and ongoing earnings from developer-made, body corporate-performed contracts. These issues will be explored in detail in Chapters 3 and 4.

The Queensland *Group Titles Act 1973* paved the way for similar legislation in all other states, which allowed the horizontal subdivision of land into low or medium density individual housing lots and collectively owned common property, now typically marketed as 'lifestyle' or 'resort developments'. Horizontal subdivisions also include townhouse developments, and in disused industrial areas of inner cities, sites with multiple high-rise buildings.

In 1980, the *Group Titles Act 1973* was amalgamated with the original Queensland strata Act to create the delightfully named *BUGTA*, the *Building Units and Group Titles Act*. New South Wales chose to enact separate community title legislation in the late 1980s, the *Community Land Development Act 1989* and the *Community Land Management Act 1989*, and like South Australia,[121] and the ACT,[122] but unlike other states, has maintained a legislative distinction between vertical and horizontal subdivision. As noted in the Introduction, legally there is no need for this separation, although the terminology helps to distinguish the social, economic and political differences between single high-rise buildings (strata title) and large planned estates (community title).[123]

Current legislation and the by-law-making power

By the turn of the twenty-first century, most states were on to their second or third generation of strata and community title legislation. At the time of writing, Western Australia and Queensland are in the process of creating third and fourth generation legislation respectively, and New South Wales has just passed fourth generation Acts. One generalisation can be confidently made about all of the states' legislation – all Acts have consistently become longer and more complicated, endlessly trying to plug holes and address problems that have relentlessly arisen in the complex legal and social environment of multi-owned properties. By way of example, the original New South Wales *CSTA* was 38 pages long. The recently enacted *SSDA* and *SSMA 2015* are a combined 274 pages. It has transpired that shared ownership of land is more challenging than initially assumed.

The basic content of all states' current Acts is essentially the same, but as noted in the Introduction, the New South Wales provisions will be used as a template for explanation. First, with the exception of a small number of leasehold strata schemes, all strata and community titles are freehold Torrens titles, and like all Torrens titles, they are dependent on surveys to define the boundaries of the lot created by registration. In the strata scheme, once a building is almost complete, a surveyor draws up a strata plan of subdivision, which includes a 'location plan' and a 'floor plan'. The location plan shows the building in relation to the boundaries of the land and the floor plan shows the boundaries of the individual strata lots within the building. Low-rise community schemes do not need buildings to be complete before subdivision plans are drawn.

Space that falls outside a lot is 'common property'; for example, stairs, foyers, halls, lifts, driveways, gyms, pools, roads, parks, infrastructure, bushland

and increasingly complex 'green utilities'. Common property is owned by all lot owners as tenants in common, in proportion to their lot or unit entitlements. Unit entitlements are generally based on the unimproved market value of a lot and have been relatively unproblematic in most states, with the exception of Queensland where they have generated considerable litigation and legislative change.[124]

The body corporate is made up of all lot owners, not tenants, and is automatically created on the registration of the plan of subdivision. Significantly, a body corporate does not have the powers of a natural person, but only those expressly or impliedly given to it by legislation, primarily the strata and community title Acts.[125] A body corporate must manage, control, repair and maintain the common property, as well as manage the finances of the scheme, take out insurance, and keep records and accounts. In order to carry out these tasks, the body corporate must levy owners each year in accordance with their unit entitlements, to extract the necessary money for administrative and sinking funds to maintain and run the scheme.

Large schemes can be 'tiered' and have multiple bodies corporate. For example, a large site might have an overarching body corporate, two subsidiary bodies corporate, which in turn have further subsidiary bodies corporate. This allows the site to be physically subdivided into discrete sections and most importantly, to allocate costs appropriately. In the diagram below, which uses the New South Wales community title Acts' terminology, each neighbourhood association may be a group of townhouses around a communal pool. The pool will be the individual neighbourhood association's common property. The owners in that association have exclusive use of the pool, and it is their responsibility to maintain it. The roads, pavements, parks and infrastructure that are used by all residents of the scheme (and might otherwise be public assets), would form part of the over-arching community association common property. Individual owners are members of their own body corporate (the strata owners corporation or

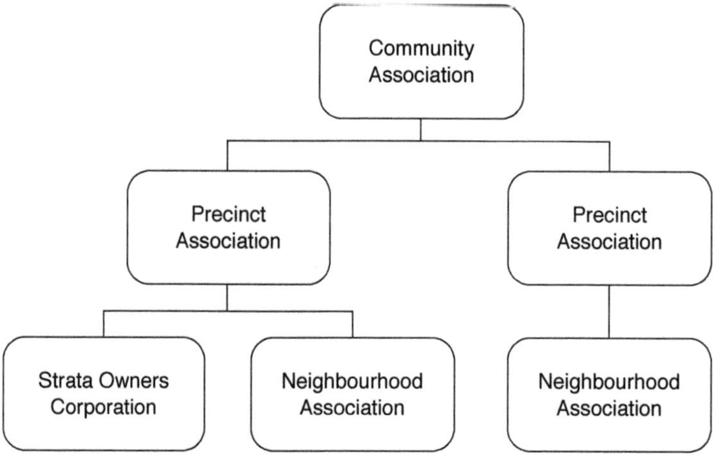

Figure 1.5 Sample governance structure in a tiered community scheme.[126]

neighbourhood association), and those bodies corporate are then members of the body corporate above them. While rational from a planning and financial perspective, tiered schemes create substantial amounts of private governance.

Bodies corporate have the power to enforce the by-laws for their scheme. All owners, mortgagees, lessees and occupiers, as well as the body corporate itself, are bound by by-laws, breach of which can lead to the imposition of fines. In all states, except South Australia, only a tribunal or court can impose fines.

The by-laws for a scheme are land regulation chosen or written by private citizens. Initial by-laws must be lodged for registration with the strata or community plan of subdivision. These may be 'model' by-laws provided in legislation, but developers generally also now have the power to bypass the model by-laws altogether, employing lawyers to draft by-laws specific to a scheme. While some legislation suggests matters that can be covered by by-laws (safety and security, restricting common property, pets, parking, floor coverings, garbage disposal, behaviour, architectural and landscaping guidelines), and some matters that *must* be dealt with in by-laws, (insurance, infrastructure), by-laws are not limited to these. The initial by-law-making power is very broad. Bespoke by-laws, particularly for large schemes, can be very detailed, specifying paint colours, letterbox style, plant type, window washing and lawn mowing, pet type and weight, land management practices and even obligations to attend permaculture courses or to provide the estate manager with a house key. By way of illustration, most private land regulation created with restrictive covenants runs to a few paragraphs, although modern covenants can stretch to 10 pages. In contrast, by-laws or management statements for community title developments are frequently closer to 100 pages, creating all manner of positive and negative restrictions. They are the key to the enduring master plan of the estate.

All states give the body corporate the power to subsequently create new by-laws, typically by special, not unanimous resolution.[127] This means that any single owner, or even a group of owners, will ultimately be powerless to prevent a by-law change, no matter how strongly they object.

Like the initial by-laws, the content of amended by-laws in most states is potentially also extremely broad. In New South Wales, s 136 of the *SSMA 2015* states that by-laws may be made 'in relation to the management, administration, control, use or enjoyment of the lots or the common property and lots of a strata scheme', wording that is similar to s 90(1) of the Tasmanian *Strata Titles Act 1998*, and s 42(1)(c) of the Western Australian *Strata Titles Act 1985*. In Queensland, under the *Body Corporate and Community Management Act 1997 (Qld) (BCCMA)* s 169, by-laws may 'only' be made for the administration, management and control of common property and body corporate assets; the regulation of the use and enjoyment of lots, common property, body corporate assets; services and amenities supplied by the body corporate, as well as other matters permitted under the Act. In the Northern Territory, s 95(2)(a)(i) of the *Unit Title Schemes Act* simply allows by-laws to be made in relation to 'scheme land', which is defined to include common property and individual lots, and s 108(1) of the ACT's *Unit Titles (Management Act) 2011* gives the body corporate the power to 'amend the rules', without any general description of what those rules might relate to, as does s 19 of the South Australian *Strata Titles Act 1988*.[128]

In short, all of these states have given bodies corporate a broad power to create by-laws that regulate not just shared common property, but *also private lot property*. Few by-laws would struggle to meet the description of relating to 'the use or enjoyment' of lots or common property. As summarised by Justice Edelman at [22] in the leading Western Australian case, *Mackie* v. *Henderson*,[129]

> it is trite that by-laws frequently interfere with the property rights of the owner of a lot. They can also interfere with a myriad of personal rights in relation to the lot. The range of possible by-laws can be extremely broad.

Victoria alone has chosen not to give bodies corporate a blanket by-law-making power, or at least there is a possibility Victoria has done this. Under the s 138(1) of the *Owners Corporations Act 2006*, a body corporate has the power to make rules with respect to matters set out in Schedule 1 of the Act. Schedule 1 refers to a limited number of issues, such as health, safety and security, committees, management and administration, design, dispute resolution, and notices and documents. It also includes common property, with the general provision for by-laws relating to its 'use'. This is a broad power. However, in relation to individual lots, Schedule 1 only permits by-laws in relation to the change of use of lots, their external appearance, notice to the body corporate about renovation of lots and times within which work on lots can be carried out. The Schedule empowers by-laws that relate to 'noise and other nuisance control', which might apply to lots, but by-laws regulating the behaviour of owners, occupiers and invitees can only be with reference to common property. The upshot is that unlike all other states, Victoria does not seem to have created a blanket by-law-making power with respect to private lot property.

Unfortunately, the legislation is ambiguous because s 138(2) goes on to say that the body corporate can amend or revoke a rule made under s 1, but does not make it clear whether the amendment must only relate to matters set out in Schedule 1, although logic would suggest that is the case. However, s 138(3) then uses the generic formula used in other states: that a rule must be 'for the purpose of the control, management, administration, use or enjoyment of the common property or of a lot'. This is a much broader power. For example, a blanket ban on pets is a by-law for the use or enjoyment of a lot. However, it is not a by-law that relates to a matter in Schedule 1 of the Act. The Victorian Supreme Court has held that there is an ambiguity in relation to the by-law-making power with respect to lots, which raises a serious question to be tried.[130] The Victorian Civil and Administrative Tribunal has applied the narrower interpretation of the by-law-making power,[131] but unfortunately there is no higher court authority determining the matter.

All states then limit the general by-law-making power with express prohibitions. Most states ban by-laws that restrict transfer, leasing or mortgaging of lots, to prevent the problems that existed in relation to company title, namely banks not wanting to lend on the security of an apartment if they cannot exercise a power of sale quickly and easily. There are some anomalies in relation to community title legislation in this regard that possibly reflect attempts to copy US-style 'themed' communities. This will be discussed in Chapter 5.

Some states' legislation expressly states that by-laws must not be inconsistent with the strata legislation or any other Act, although this is arguably a matter of legislative logic. By-laws are a form of delegated legislation or at least analogous to that, and delegated legislation can never be inconsistent with its parent Act. This concept has not been consistently applied in all states, ironically even those with the express legislative prohibition against inconsistent by-laws. This will be discussed in more detail in Chapter 4.

Most states also prohibit by-laws that restrict guide or assistance animals, evincing some awareness of the harm that private regulation of land can cause others. Some states have a more generic prohibition against 'discriminatory' by-laws, which then relates to difficult questions about the application of discrimination law, which typically only applies in the public sphere, to private strata schemes. This is explored in detail in Chapter 5.

Queensland,[132] Tasmania[133] and now New South Wales[134] have broader prohibitions on by-laws that are oppressive and/or unreasonable, provisions that go to the heart of the argument in this book, that there must be some limit on a private legislative power. Tasmania's Act is the most nuanced, with a prohibition on by-laws that 'adversely affects the health, welfare or safety of any person'. However, the effectiveness of these provisions is entirely dependent on the mindset of reviewing courts and tribunals, in particular their appreciation of the limits of freedom of contract in property law. That issue will be fully explored in the book.

For now, suffice to say that all Australian states have perpetuated the original, relatively simplistic view of by-laws, that they are merely the house-keeping rules chosen by individual communities to manage their common and individually owned property. There seems to be little or no legislative recognition that by-laws are statutorily endorsed positive and negative freehold covenants that might create problems that technical freehold covenant law sought to avoid.

Stratum subdivision

There is one final form of subdivision that must be explained in order to paint a complete picture of the new, legislatively facilitated methods of privately regulating and burdening freehold land. That method is 'stratum' or 'volumetric' subdivision. In New South Wales, stratum subdivisions are created under a relatively brief section of the *Conveyancing Act 1919* (NSW), Pt 23, Div 3B. In Queensland, volumetric subdivisions were created under the *Mixed Use Development Act 1993* (Qld), but are now created under the *Land Title Act 1994* (Qld) (s 48D and Pt 4 Div 4),[135] and in Victoria, while the term 'stratum' is not used, a single building can be divided between multiple owners corporations under the *Subdivision Act 1988* (Vic) (s 27).[136]

Although practised from the early twentieth century by United States railway companies selling the airspace above rail stations,[137] the push for stratum or volumetric subdivision did not begin in Australia until the 1980s when it became evident that the strata titles legislation was not adequate for mixed use developments with residential and commercial owners or multiple commercial owners. The body corporate form was primarily conceived for a group of residential owners to manage their residential community, and not surprisingly, commercial owners did not fit easily into that mix. It is particularly difficult to divide maintenance and service costs equitably between owners with markedly different land uses; for example, ground floor retail owners will never use the lifts and will require commercial garbage disposal; in contrast, residential owners habitually need the lifts, only use residential garbage disposal and will use minimal air-conditioning or other electricity during the day.

The first 'stratum' or 'volumetric' subdivision in Australia was the Paradise Centre in Surfers Paradise, Queensland. It comprised a basement car park, ground floor shopping centre, two large residential towers, and an international hotel. The development was one that generated its own project-specific legislation, the *Registration of Plans (HSP (Nominees) Pty Ltd) Enabling Act 1980* (Qld) and the *Registration of Plans (Stage 2) (HSP (Nominees) Pty Ltd) Enabling Act 1984* (Qld). In New South Wales, the first attempted stratum subdivisions were the Connaught building on Hyde Park, Eastgate at Bondi Junction and Eastpoint at Edgecliff. They were all created with complex easements, covenants and umbrella management agreements, drafted by the developers' lawyers.[138]

To solve the dilemma of mixed use sites, New South Wales, Queensland and Victoria all amended general property legislation. In New South Wales, the *Conveyancing Act 1919* (NSW) was amended to allow the subdivision of a building by deposited plan into separate 'stratum lots', limited by height or depth by reference to the Australian Height Datum. For example, the retail podium of a building might be stratum lot 1, floors 1–5 stratum lot 2, floors 6–11 stratum lot 3, and floors 12–36 stratum lot 4. Those stratum lots are in effect like four adjoining land lots, turned on their side and stacked on top of each other.

Stratum lots can be further subdivided by a strata plan, creating a residential or commercial strata scheme within that stratum lot, which becomes a 'stratum parcel'.[139] For example:

- Stratum lot 1 might be retained by a single registered proprietor of the podium retail section (which is then subdivided by leases);
- Lot 2 might be subdivided by a strata plan creating commercial office space;
- Lot 3 might be bought by a single registered proprietor who runs a motel; and
- Lot 4 subdivided by another strata plan to create a residential scheme.

Although multiple stratum lot/parcel owners do not necessarily want or need to be part of a common community like a strata scheme, they still share a building,

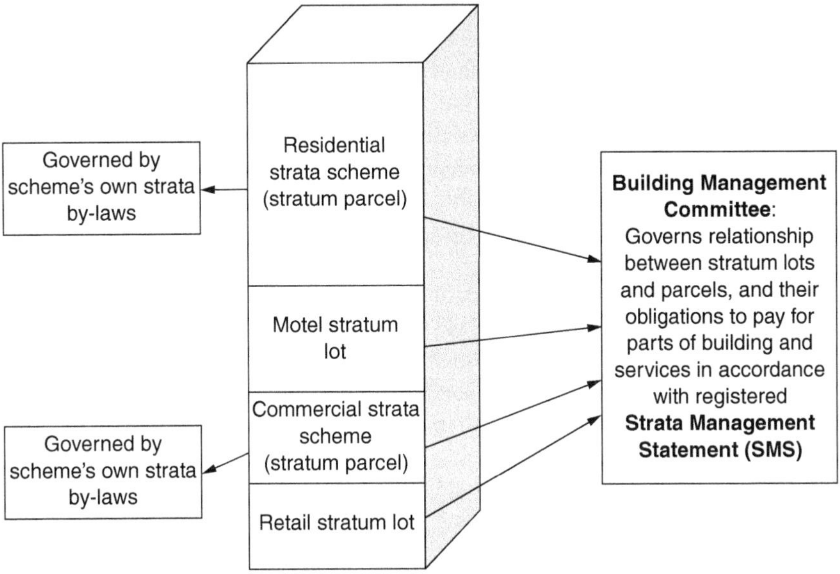

Figure 1.6 Stratum subdivision ('part building strata scheme') under *Conveyancing Act* 1919 (NSW) Pt 23 Div 3B.

and thus rights and obligations between stratum lot/parcel owners, in particular maintenance obligations and insurance, need to be addressed.

Easements are used to grant separate stratum owners access to shared property, which will of necessity be partly or wholly contained in another stratum owner's lot/parcel (for example, lifts, foyer, car parks). However, maintenance costs for the entire building, in particular for the kind of complex plant and equipment that is found in a modern high-rise, need to be imposed on successive owners and as has been noted repeatedly in this chapter, the ban on positive covenants on freehold land means that orthodox property law is not adequate for such a task.

The solution was found in registered Building Management Statements ('BMSs') and Strata Management Statements ('SMSs') in New South Wales and Queensland,[140] although Victoria only requires a brief statement or 'purpose'.[141] BMSs *must* contain provisions on insurance, damage and disputes, and they may also contain provisions on garbage, safety and security measures, the appointment of a managing agent, noise levels, trading activities, service contracts, and an architectural code to preserve the appearance of the building. However, at least in New South Wales, the content of a BMS is specifically not limited to this list.[142] In addition to maintenance provisions, BMSs must establish a 'building management committee', made up of all stratum lot owners and any owners corporations, although owners can be excluded with their consent.

BMSs and SMSs are typically drafted by the original owner(s)' lawyer and are recorded on the Torrens register. Registration turns what began as a contract negotiated by the original parties involved in the development into a statutory property right that will bind all subsequent owners.

Although BMSs were conceived as simply a means to compel the freehold owners of Torrens lots to maintain their shared building, there is no such explicit limit on their content in the legislation. Like strata by-laws, BMSs are the statutory equivalent of positive and restrictive freehold covenants and the power to write them is effectively a quasi-legislative power in the hands of private citizens. This has created the inevitable temptation on the part of original owners to use BMSs to impose novel obligations on successive owners. This will be discussed in more detail in Chapter 4.[143] There is a flash of legislative recognition of the potentially detrimental effects of this power in the Queensland legislation in the provision that invalidates a BMS to the extent that it prevents energy efficient roof colours and window treatments or the installation of solar hot water or solar panels.[144] However, overall, like strata and community title by-laws, there is insufficient legislative and, with respect, judicial appreciation of the full implications of BMSs.

The more citizens who are granted the power to write BMSs, the more significant it becomes. Stratum subdivision is increasingly used in Australian cities for large complex developments, such as Sydney's most significant current development site, Barangaroo.[145] The commercial and residential high-rise development at Barangaroo is a stratum subdivision, and owing to the 'green utilities' built to meet the government's stipulation that the site be carbon-neutral, the BMS is astoundingly complex and will have lasting implications for the thousands of future owners and occupiers of the site.[146]

Conclusion

This chapter has laid the groundwork for the book. It has described the history of low-rise and high-rise development in Australian cities, and the impediment that the traditional doctrine of restrictive covenants presented to the further development of high-rise buildings, as well as suburban estates with collectively owned private facilities. It has described the enactment of strata and community title legislation, which through levying provisions and a broad by-law-making power, overcame the limitation on positive obligations on freehold land. Stratum subdivision then continued that trend.

The chapter has flagged the possibility that the power to regulate freehold land through enduring restrictions and obligations may be more significant than hitherto appreciated. It may in fact be akin to a private legislative power within a defined physical space, raising serious economic, social and political questions that Australian states are yet to address. The next two chapters will delve into the economic, social and political complexity of private regulation of freehold land. Chapter 2 will be a theoretical exploration of property law, which will establish that property law is never simply a matter of individual entitlement, but rather part of the way we define our economic, social and political order. Despite a tendency even on the part of judges,[147] to decline to characterise strata and community by-laws as property rights, they are undoubtedly so, and must be examined within the property rubric. Chapter 3 will examine the very extensive judicial and academic discussion that has already occurred in the United States in relation to freehold covenants. While there was a significant difference between the common law of Australia and the United States in relation to freehold land regulation for most of the twentieth century, strata and community title legislation has effectively removed that difference. As a result, United States debates are now of fundamental importance to Australian land law. Given the propensity of Australian developers and lawyers to follow housing and commercial development trends in the United States, an understanding of these debates becomes more important still. The final chapters of the book will examine the extent to which questions raised in those American debates, as well as fundamental elements of property theory, have been addressed in Australian case law and legislation.

Postscript

In mid-2016, the Victorian Supreme Court, in *Owners Corporation PS 50391P* v. *Balcombe* ('*Balcombe*'),[148] considered the meaning of the rule- (by-law-) making power in s 138 of the *Owners Corporations Act 2006* (Vic), as well as the rule-making power under the previous legislation, regulation 220(1) of the *Subdivision (Body Corporate) Regulations 2001* (Vic). Although on its face, the earlier regulation appeared to give an owners corporation an essentially unlimited power to make rules in relation to lots and common property, Riordan J held that an owners corporation did not have any power to make rules in relation to private lots, beyond enforcing the Standard Rules. This meant that an owners corporation could not ban short-term letting in a strata scheme, the subject of the by-law in question. Riordan J also found that under the current Act, an owners corporation did not have the power to regulate the use of a private lot and that the requirement in s 138(3), that a rule must be 'for the purpose of the control, management, administration, use or enjoyment of the common property or of a lot', operated as an *additional* requirement to rules having to relate to matters set out in Schedule 1. It did not expand the scope of the s 138(1) power in any way.

Riordan J came to this conclusion by applying administrative law principles, drawn from cases that considered the by-law-making powers given to Ministers and their delegates. Riordan J held that the task of the court was to determine the scope of what Parliament intended when enacting the statute that empowered the subordinate authority to make the laws. Riordan J set out a detailed process, which courts should go through to determine the 'Statutory Purpose' and whether the rule had a sufficient and direct connection to this purpose.

Riordan J's judgment stands out as the only state superior court decision that recognizes the potential danger of granting private citizens the power to regulate other people's property; that danger is one of the central foci of this book. Although the current Victorian legislation makes a clear distinction between the power to regulate private lots and common property, the previous legislation, on its face, did not, and it was open to Riordan J to take the same generous approach to the by-law-making power taken by other state courts. However, Riordan J did not, concluding that viewing the Explanatory Memorandum, the Second Reading Speech, and the legislation as a whole, it could not have been the case that the Victorian Parliament intended an owners corporation to regulate private lots; its principal function related to the common property. This decision is consistent with the liberal democratic principle of negative liberty that is discussed in Chapter 5.

The drawback of Riordan J's decision in *Balcombe* is that the administrative law test that his Honour set out is very complicated. It would be fair to say that applying the test would challenge most lawyers, and be well beyond the skills of almost all members of bodies corporate, as well as strata managers. This suggests that the solution to insufficiently limited by-law-making powers may not be found in administrative law, which is typically applied to public officials in their professional capacity, in contrast to volunteer lay members of a body corporate. As this book will argue, the answer is better found in property law, in particular in well-developed United States law, which has taken the essence of complex traditional doctrines of land law, and translated them into modern, comprehensible language for the benefit of owners, residents and managers of private communities.

Notes

1 This phrase is part of the long title of the Act.
2 *CSTA* s 13(2).
3 At this time a special resolution was defined by by-law 35 as 'a majority of not less than three-fourths of the total unit entitlement of the lots, and not less than three-fourths of all members'.
4 'Strata Titles Act is completely novel; there is no precedent. It can fairly be labelled "Made in Australia" ': A F Rath, P J Grimes and J E Moore, *Strata Titles* (Lawbook, 1966) xi.
5 For example, Singapore, Malaysia, Jamaica and even the United Kingdom.
6 D V Cowen, 'The South African Sectional Titles Act in Historical Perspective: An Analysis and Evaluation' (1973) 6(1) *Comparative and International Law Journal of Southern Africa* 1, 3.
7 Kingsley Davis, 'The Urbanization of the Human Population' (1965) 213(3) *Scientific American* 40.
8 It is important to appreciate the role of modern engineering developments in high-rise construction. Without these, there would be no need for the legal complexity discussed in this book. The first engineering development that made high-rise development feasible was the invention of the elevator brake by Elisha Otis (1811–1861). Otis stood on a platform at the 1853–1854 Exhibition of the Industry of All Nations in New York and ordered that the rope holding the platform be cut. The crowd was amazed to see the platform only fall a few inches before being stopped by Otis's specially designed brake. The elevator brake allowed people to be safely transported to higher floors. The development of steel skeletons and non-loadbearing curtain walls subsequently made very high building possible: Edward Glaezer, 'How Skyscrapers Can Save the City', *The Atlantic* (online), 9 February 2011 www.theatlantic.com/magazine/archive/2011/03/how-skyscrapers-can-save-the-city/308387/.

9 *Wohnungseigentumsgesetz* [German Condominium Act] (Germany) 15 March 1951, BGBl I S, 175.
10 C G van der Merwe, 'The Sectional Titles Act and the Wohnungseigentumsgesetz' (1974) 7 *Comparative and International Law Journal of Southern Africa* 165.
11 Hazel Easthope, Bill Randolph and Sarah Judd, 'Governing the Compact City: The Role and Effectiveness of Strata Management' (City Futures Research Centre, 2012) 10.
12 New South Wales Government, 'A New Planning System for New South Wales: White Paper' (2013); New South Wales Government, 'A Plan for Growing Sydney' (2014); Victorian Government, Melbourne '2030: Planning for Sustainable Growth' (2002); Victorian Government, 'Plan Melbourne' (2014); Queensland Government, 'South East Queensland Regional Plan 2009–2031' (2009).
13 Hazel Easthope, Bill Randolph and Sarah Judd, *Managing Major Repairs in Residential Strata Developments in NSW* (City Futures, 2009) 12.
14 Easthope, Randolph and Judd, 'Governing the Compact City', above n 11, 11.
15 NSW Fair Trading, 'Strata & Community Title Law Reform Position Paper' (November 2013) 2.
16 Bureau of Infrastructure, Transport and Regional Economics, *Cities: Population Growth, Job Growth and Commuting Flows in South East Queensland*, Research Report 134 (Australian Government, 2013), 105, 343.
17 Victorian Government, 'Plan Melbourne' (2014), 39.
18 Easthope, Randolph and Judd, 'Governing the Compact City', above n 11, 11.
19 Graeme Davison, 'The Past & Future of the Australian Suburb' (1993) 31(2) *Australian Planner* 63, 64.
20 Ebenezer Howard, *Garden Cities of To-morrow* (Swan Sonnenshein, 1902). Howard advocated for an ideal planned city with an appropriate separation of residential, industrial and rural areas. The governing model suggested by Howard was a trust, with residents being offered long-term leases.
21 Robert Freestone, 'The Great Lever of Social Reform: The Garden Suburb 1900–30' in Max Kelly (ed.), *Sydney: City of Suburbs* (New South Wales University Press, 1987) 53.
22 Ibid. 55.
23 Aidan Davison, 'Stuck in a Cul-de-Sac? Suburban History and Urban Sustainability in Australia' (2006) 24(2) *Urban Policy and Research* 201.
24 Ibid. 203.
25 H J Dyos, *Victorian Suburb: A Study of the Growth of Camberwell* (Leicester University Press, 1961); Donald J Olsen, *Town Planning in London: The Eighteenth and Nineteenth Centuries* (Yale University Press, 1964) 126–43; H J Dyos, 'The Speculative Builders and Developers of Victorian London' in David Cannadine and David Reeder (eds), *Exploring the Urban Past: Essays in Urban History by H J Dyos* (Cambridge University Press, 1982) 154–78.
26 'By 1911, some 50 per cent of the population were already home owners or purchasers – a proportion not matched in Britain for another six decades': Robert Freestone, *Model Communities: The Garden City Movement in Australia* (Nelson, 1989) 43. By 1966 this figure had reached 71.4 per cent and has remained steady ever since: Australian Bureau of Statistics, 'Home Owners and Renters' (ABS Year Book Australia, 2012).
27 Graeme Davison, above n 19, 65.
28 Freestone, *Model Communities*, above n 26, 96.
29 Alex Kondos, 'The Hidden Faces of Power: A Sociological Analysis of Housing Legislation in Australia' in Roman Tomasic (ed.), *Legislation and Society in Australia* (Unwin Hyman, 1980) 338.
30 The best Australian example of this is Canberra, which is an entirely leasehold city.
31 Philip Booth, *Planning by Consent: The Origins and Nature of British Developmental Control* (Routledge, 2003).

32 By consistently enforcing the covenants in initial building leases and subsequent long-term leases, ground landlords like the Dukes of Bedford and Marlborough (or more accurately their estate managers), controlled the development and use of land within their estates: John Summerson, *Georgian London* (Penguin Books, 1978); Dan Cruickshank and Neil Burton, *Life in the Georgian City* (Viking, 1990). The results of this private planning can be seen today in the streets that still make up the Bedford Estate around University College London. See The Bedford Estates www.bedfordestates.com/. The estates of ground landlords who did not enforce covenants in leases often deteriorated into slums.

33 *Tulk* v. *Moxhay* (1848) 2 Ph 774; 41 ER 1143 ('*Tulk*'). For a discussion of the anomalous nature of the decision in *Tulk*, particularly in a Torrens system, see Pamela O'Connor, 'Covenants as Regulation' (2011) 1 *Property Law Review* 145, 147–52 and Sharon Christensen and W D Duncan, 'Is It Time for a National Review of the Torrens' System? – The Eccentric Position of Private Restrictive Covenants' (2005) 12 *Australian Property Law Journal* 104.

34 *Keppell* v. *Bailey* (1834) 2 My & K 517; 39 ER 1042.

35 *Hill* v. *Tupper* (1863) 2 H & C 121; 159 ER 51.

36 Restrictive covenants can come to an end if they have not been enforced for long periods of time or if the character of the neighbourhood has changed so that there is no utility in their enforcement. Equity will not grant a remedy in these circumstances, e.g. *Application of Fox* (1981) 2 BPR 9310. Courts are often expressly empowered to declare that a covenant has become obsolete: see, e.g. *Conveyancing Act 1919* (NSW) s 89, *Property Law Act 1958* (Vic) s 84 and equivalent provisions in other states. Covenants can also be overridden by public planning instruments: see, e.g. *Environmental Planning and Assessment Act 1979* (NSW) s 28(2), although, not all Australian states have legislation to this effect. For a discussion of the difficulties in applying these kinds of provisions, in particular the challenge of balancing public and private interests in the Queensland context, see Michael Weir, 'Public and Private Planning – Implications for Legal Processes' (1994) 11 *Environmental and Planning Law Journal* 137.

37 Outside statute, a covenant cannot exist in gross: *London County Council* v. *Allen* [1914–1915] All ER Rep 1008; [1914] 3 KB 642; *Clem Smith Nominees Pty Ltd* v. *Farrelly* (1978) 20 SASR 227; *Tooth & Co Ltd* v. *Barker* [1960] NSWR 51; (1960) 77 WN (NSW) 231.

38 *Haywood* v. *Brunswick Permanent Benefit Building Society* (1881) 8 QBD 403; *Austerberry* v. *Corporation of Oldham* (1885) 29 Ch D 750 (CA); *Pirie* v. *Registrar General* (1962) 109 CLR 619. The Victorian Law Reform Commission recommended the maintenance of the common law and equitable prohibition on positive covenants: Victorian Law Reform Commission, *Easements and Covenants: Final Report*, Report No 22 (2010) 84–8. Most Australian states now make provision for positive obligations to be imposed on freehold land by public authorities for environmental, heritage or other public purposes: Pamela O'Connor, Sharon Christensen and Bill Duncan, 'Legislating for Sustainability: A Framework for Managing Statutory Rights Obligations and Restrictions Affecting Private Land' (2009) 35 *Monash University Law Review* 233 and O'Connor, above n 33.

39 One of the rare exceptions that proved the rule was the Appian Way, Burwood, Sydney. It was built by industrialist George Hoskins, with Federation houses around a central recreation area for tennis and croquet. The development was leasehold, with tenancies offered to professional people, who then became members of the Appian Way Recreation Club Ltd, by buying shares. The development is now freehold land, protected by a Burwood Council Development Control Plan No 4.

40 See discussion below.

41 Freestone, *Model Communities*, above n 26.

42 Photo by Lola Hubner.

43 *Pirie v. Registrar General* (1962) 109 CLR 619.

44 D N Clarke, 'Commonhold – A Prospect of Promise' (1995) 58(4) *Modern Law Review* 486, 487. Clarke makes it clear that English law has also long recognised 'flying freeholds' but they have always been viewed with suspicion by both conveyancers and lenders and thus underutilised.

45 The United Kingdom now has legislation facilitating the development of freehold apartments: *Commonhold and Leasehold Reform Act 2002* (UK), but it is almost never used by developers. As of 2009, there were only 12 commonhold developments registered in England and one registered in Wales. Developers continue to use the leases, with solicitors suggesting that there is uncertainty about the complexity and efficacy of commonhold: James Brown, 'English Commonhold: A Failed Experiment?' (2013) 3 *Property Law Review* 85.

46 A private London square featured prominently in the popular film *Notting Hill*, in which Julia Roberts, playing the character of an American actress, climbs over the fence of an enclosed square, much to English man, Hugh Grant's horror. At the resolution of the film, the characters are married and she is pictured, pregnant, lying on a bench in the garden. In addition to securing each other's love, they have obviously secured a house bordering the square that entitles them to a key. The garden used in the film is Rosmead Garden, Rosmead Rd, W1, which remains a private garden for residents and is only open to the public as part of an open garden scheme once a year. Notting Hill was developed in the 1850s and 1860s by a retired civil servant Charles Henry Blake. However, the driving force and brains behind his developments was his solicitor: Dyos, 'The Speculative Builders', above n 25, 157–8.

47 Freestone, *Model Communities*, above n 26, 34–8; Evan McKenzie, *Privatopia: Homeowner Associations and the Rise of Residential Private Government* (Yale University Press, 1994) 1–12.

48 Olmsted's work predated Howard's and no doubt influenced the Englishman. Olmsted designed New York's Central Park, the Boston Park system and the Capitol grounds in Washington, DC. See Richard T Richard LeGates and Frederic Stout (eds), *The City Reader* (Routledge, 2003) 302–8.

49 See Chapter 3, n 163.

50 Uriel Reichman, 'Toward a Unified Concept of Servitudes' (1981) 55 *Southern California Law Review* 1177, to be discussed in Chapter 3, 77–8.

51 McKenzie, above n 47. The reasons for this exponential development in the postwar period will be discussed in Chapter 3, 101–2.

52 *Neponsit Property Owners' Assoc Inc v. Emigrant Industrial Savings Bank* 278 NY 248 (1938).

53 This theory frequently failed in practice because of the sequential sale of land in large subdivisions. When a developer sells lots in a subdivision, he or she cannot attach the benefit of promises given by purchasers of later-sold lots to earlier-sold lots because the developer no longer owns the earlier-sold lots; covenants can only attach to the covenantee's land, not land owned by others. Schemes of development (*Elliston* v. *Reacher* [1908] 2 Ch 374) have sometimes been used to overcome this problem, but after the High Court decision in *Westfield Management Ltd* v. *Perpetual Trustee Co Ltd* (2007) 233 CLR 528, it is very unlikely that the doctrine, which depends on proof of matters outside the Register, is applicable to Torrens title land. The problem created by sequential sale was solved in New South Wales by *Conveyancing Act* 1919 (NSW) s 88B, which allows covenants and easements to be created at the point of registration of a plan of subdivision, but the provision was only introduced in 1964. For a discussion of the potentially problematic nature of the mass creation of covenants via registered plans, see O'Connor, above n 33, 152–6.

54 Freestone, 'The Great Lever of Social Reform', above n 21, 53, 66.

55 INHERITage, 'Final Heritage Assessment Report of Rosebery Estate' (2006).

56 Although councils do not directly enforce covenants, not being party to them, they do consider them in granting development consent. For example, in *Brotherton* v. *Sydney City Council* [2004] NSWLEC 475, the applicant had been refused permission to subdivide an 854 sq m lot in an area where the minimum allotment size was 230 sq m. The house was on Ripon Way, Rosebery, which while not a heritage conservation area, was a 'special precinct' under the Development Control Plan (DCP). Murrell C considered the history of the area, as well as the Garden Suburb Movements and concluded at [24] that:

> The proposed subdivision does not warrant approval on the basis that the special character of the area is one that is demonstrated by the consistency in lot sizes. Whilst it is not a conservation area it is a special precinct and the history of the area must be taken into consideration in the assessment of this application. Furthermore, it is relatively intact, in fact the whole of Ripon Way is intact in terms of the presentation of the subdivision pattern which can be seen from the cadastral boundaries located on the aerial photograph.

57 Haberfield is now classified as a heritage conservation area and protected by public planning regulation: Ashfield Council, 'Ashfield Development Control Plan 2007' (21 May 2007).

58 See Cathy Sherry, 'The Legal Fundamentals of High-Rise Buildings and Master Planned Estates: Ownership, Governance and Living in Multi-owned Housing with a Case Study on Children's Play' (2008) 16(1) *Australian Property Law Journal* 1, 1–3.

59 Covenants are more likely to be enforced in newer subdivisions in which the original covenantee is still present: see, e.g. *Levi* v. *Spicer* (2003) 11 BPR 20,927 (enforcement of a covenant preventing building in COLORBOND by original landowner who had gradually subdivided her land on Sydney's western fringe); *Dandaloo Pty Ltd* v. *Darwich* [2006] NSWSC 1305 (enforcement of a 'no commercial vehicles' covenant by the original developer of part of Harrington Park, an 'exclusive' estate in Sydney's south west).

60 For example, *Property Law Act 1958* (Vic) s 84, *Conveyancing Act 1919* (NSW) s 89, and equivalent provisions in other states.

61 Anne Wallace, Michael Weir and Les McCrimmon, *Real Property Law in Queensland* (Lawbook Co., 2015), 831–80; Christensen and Duncan, 'Is it time for a national review of the Torrens' system?', above n 33.

62 For example, the mooted plans for five-storey buildings in the Rocks in 1902, after the outbreak of the Plague, and the completion of the Strickland Flats in 1914, still standing on Cleveland Street, Chippendale: Caroline Butler-Bowdon *et al.*, *Homes in the Sky* (Melbourne University Publishing, 2007) 2–5.

63 Ibid. 2–7.

64 For an evocative depiction of this housing, see Ruth Park, *Harp in the South* (Houghton Mifflin, 1948); Ruth Park, *Poor Man's Orange* (Collins/Angus & Robertson, revised edn, 1992).

65 Butler-Bowdon *et al.*, above n 62, 65–6, 87.

66 Butler-Bowdon *et al.*, above n 62, 8.

67 Louis Nowra, *Kings Cross* (NewSouth, 2013).

68 Quoted in Geoffrey Dutton, *Kenneth Slessor: A Biography* (Viking, 1991) 41, cited in Butler-Bowdon *et al.*, above n 62, 8.

69 'Bohemian' became a marketing slogan for flats of the time, and in some instances, the phrase may have been as hollow as much modern real estate marketing. A satirical piece called 'The Flat', written in 1919, chronicles two young men's failed move from 'Suburbia' to 'Bohemia'. The flat contained 'the inevitable easel and allegedly artistic furniture', but the rent was prohibitively expensive, with the writer concluding that 'if it cost that much to inhale Bohemian atmosphere, there would be very little left over for the midday chop': 'The Flat', *The Age*, 15 February 1919, 6.

70 Butler-Bowdon *et al.*, above n 62, 8.
71 'New Flats Reflect Social Needs of the Age', *Daily News*, 12 March 1938, 9.
72 Although there were some tenants in company title buildings, leasing was often prohibited by the company's articles of association (see below), and thus buildings could be exclusively owner-occupied.
73 Kondos, above n 29, 340.
74 Photo by Lola Hubner.
75 There were also rented apartments marketed to the urban elite which functioned as apartment-hotels. Residents were long-term tenants who took suites in buildings like Melbourne Mansions and Clivedon Mansions in Melbourne, eating in stuffy common dining rooms and frequently mentioned in the social pages of newspapers: Seamus O'Hanlon *Together Apart: Boarding House, Hostel and Flat Life in Pre-war Melbourne* (Australian Scholarly Publishing, 2002).
76 Butler-Bowdon *et al.*, above n 62, 36–40.
77 Jan Roberts (ed.), *The Astor* (Ruskin Rowe, 2003) 41. Some elite apartments were not company title, but functioned as apartment-hotels. Residents were long-term tenants who took suites in buildings like Melbourne Mansions and Clivedon Mansions in Melbourne.
78 In New York co-operatives, a corporation owns the land and building and grants a lease to apartment owners when they purchase shares: Robert M Nelson, 'Examining Cooperative Conversion: An Analysis of Recent New York Legislation' (1982) 11 *Fordham Urban Law Journal* 1089, 1093.
79 These leases were held to be invalid for failure to satisfy the leasehold rule of certainty of duration: *Wilson* v. *Meudon Pty Ltd* [2005] NSWCA 448, [65] (Hodgson JA). A lease which runs as long as someone owns shares in a company title building is uncertain as to duration because it cannot be known when someone might decide to sell their shares/apartment. Ultimately, the invalidity of leases is not particularly problematic, as all shareholders have a vested interest in respecting other shareholders' rights to exclusive occupation of their apartments.
80 Rath, Grimes and Moore, above n 4, xii.
81 There was a third form of apartment ownership whereby people owned buildings as tenants in common. As lending institutions considered this as undesirable a security as company title: Rath, Grimes and Moore, above n 4, xiii, and as it did not create a governing body, tenancy in common buildings were rare. In the process of consulting with the New South Wales Law Reform Commission on *Disputes in Company Title Units*, Report No 115 (2007), the Commission told me that they estimated that there were only six tenancy in common buildings left in Sydney.
82 Kondos, above n 29, 339.
83 Nelson, above n 78, 1093; Matthew Gordon Lasner, *High Life: Condo Living in the Suburban Century* (Yale University Press, 2012), 61–3, 128–30. Co-operative and then condominium conversions continued to be a problem and in the early 1970s, the United States Federal government initiated an inquiry into developer abuse in the process of both condominium and co-operative conversions: Federal Department of Housing and Urban Development, 'Condominium and Cooperative Study' (1975).
84 Kondos, above n 29, 340.
85 The following description relies on Kondos, above n 29.
86 Kondos, above n 29, 351.
87 Ibid. 354.
88 Ibid. 355.
89 See below n 113.
90 Kondos, above n 29, 358.
91 Rath, Grimes and Moore, above n 4, xi.
92 For example, while 'stakeholders' can often have legitimate complaints about the

legislation, they may have variable understanding of the legal genesis of the problem, pushing for legislative change that is unnecessary; e.g. the *Strata Schemes Management Act* 2015 (NSW) ss 58 and 59 require strata managers to provide bodies corporate with details about trust accounts and transactions that relate to body corporate matters, obligations that already clearly existed under the basic common law of agency. Further, some stakeholders are blatantly self-interested and have succeeded in securing legislative provisions that only serve their own interests, e.g. *Strata Schemes Management Act* 2015 (NSW) s 60 authorises strata managers to take commissions so long as they are disclosed to the body corporate.

93 *CSTA* s 15(1)(f).

94 Ibid. s 13(3).

95 Kondos, above n 29, 345.

96 Ibid. 332.

97 Although as Gray points out, the space above land is the lower, useable stratum of air: Kevin Gray, 'Property in Thin Air' (1991) 50 *Cambridge Law Journal* 252.

98 *CSTA* s 4(1)(d).

99 Butler-Bowdon *et al.*, above n 62, 159–67.

100 Harry Seidler, quoted in Robin Boyd, *The Australian Ugliness* (Penguin, 1980) 255, cited in Butler-Bowdon *et al.*, above n 62, 160.

101 Easthope, Randolph and Judd, 'Governing the Compact City', above n 11, 11.

102 DP 270152; see community website at Monarch's Macquarie Links Estate www.macquarielinks.net.

103 DP270613; see community website at Harrington Grove: Grove Life, *Community Documents* http://grovelife.com.au/community-documents.

104 DP270215; see community website at Jackson's Landing www.jacksonslanding.net.au/.

105 DP270347; see community website at Breakfast Point www.breakfastpoint.com.

106 Photo by Lola Hubner.

107 Photo by Lola Hubner.

108 Easthope, Randolph and Judd, 'Governing the Compact City', above n 11, 9.

109 *The Wizard of Oz* (Directed by Victor Fleming and George Cukor, Metro-Goldwyn-Mayer, 1939) 20:51.

110 Gary F Bugden, *Strata Title Management Practice in New South Wales* (CCH Australia 3rd edn, 1981) 2–3.

111 Victoria also enacted the *Cluster Titles Act 1974* but it was rarely used.

112 The Queensland strata title Act at the time was the *Building Units Titles Act 1965* (Qld).

113 Queensland, *Parliamentary Debates*, Legislative Assembly, 17 November 1972, 1839–40 (Burns).

114 E McD Batt, 'Housing Innovation in Brisbane' (1970) 8(3) *Royal Australian Planning Institute Journal* 80, 80.

115 Ibid. 83.

116 Ibid. 81.

117 Ibid. 83.

118 The other Acts were the *Registration of Plans (HSP (Nominees) Pty Ltd) Enabling Act 1980*, the *Registration of Plans (Stage 2) (HSP (Nominees) Pty Ltd) Enabling Act 1984*, the *Integrated Resort Development Act 1987*, the *South Bank Corporation Act 1989* and the *Mixed Use Development Act 1993*.

119 Queensland Parliament, *Parliamentary Debates*, Legislative Assembly, 17 October 1985, 2190 (Hinze).

120 The *Sanctuary Cove Resort Act 1985* operated in conjunction with the old *Building Units and Group Titles Act 1980* (Qld).

121 *Community Titles Act 1996* (SA).

122 *Community Title Act 2001* (ACT).

123 A distinct disadvantage of separate Acts is that, in New South Wales, the community title legislation has been left behind in legislative amendments. Because so many more people live in strata schemes, there is a political incentive to update the strata legislation. While exactly the same concerns often apply to community title, legislative amendments have not been applied to the community Acts. For example, in the 2012–2015 review of New South Wales' legislation, the community title Acts seem to have fallen by the wayside.

124 Wallace *et al.*, 519–28.

125 Cf. In Tasmania, bodies corporate in community schemes, but not strata schemes, have the powers given to them by the constituent documents of the scheme: *Strata Titles Act 1998* (Tas) s 81(3). This is the same as homeowner associations in the United States, where the powers and functions of associations are determined by the documents drafted by the developer's lawyer: see Chapter 3. This legal structure gives private citizens the power to determine multiple management matters that are set by statutory standards in other states.

126 Precinct schemes are in fact rare. The more tiers a community scheme has, the more complicated and expensive it is to run, and so experienced lawyers try to avoid excessive layers of management.

127 Contrary to popular belief a 'special resolution' is not 75 per cent of owners. By way of example, in New South Wales, a special resolution is defined in *SSMA 2015* s 5, as a resolution that is passed at a properly convened general meeting, and not more than 25 per cent of the value of votes cast are against it. A duly convened meeting is one that has had proper notice and which has achieved quorum. Quorum is defined in *SSMA 2015* sch 2 cl 17(2) as either 'if not less than one-quarter of the persons entitled to vote on the motion or election are present either personally or by duly appointed proxy' or

> if not less than one-quarter of the aggregate unit entitlement of the strata scheme is represented by the persons who are present either personally or by duly appointed proxy and who are entitled to vote on the motion or election.

In other words, a special resolution can be passed if a quarter of lot owners turn up in person or give proxies, and fewer than one-quarter of those people vote against the motion. Abstentions do not count as a vote against a motion. So, in a 60-lot strata scheme, if 20 people come to a meeting (or five come and 15 grant proxies), a special resolution is put and if 15 people vote in favour of it, or even five people vote in favour and 10 abstain, it will be passed. In other words, you do not need 45 people ($3/4 \times 60$) to vote in favour of the resolution. The low quorum is necessary to manage the chronic lack of involvement in bodies corporate by most lot owners. If the quorum were higher, many schemes simply could not operate. See Easthope, Randolph and Judd, 'Governing the Compact City', above n 11, 51.

128 The by-law-making power in the *Community Titles Act 1996* (SA) ss 12 and 34, is much more detailed, but ultimately very fulsome. However, it does evidence some awareness of the problematic nature of very broad by-law-making powers, for example, in the s 34(3)(c) limitation of the power to write by-laws for individual lots only to 'prevent interference with the use and enjoyment of other lots'.

129 *Mackie* v. *Henderson* [2011] WASC 197.

130 *Morrish* v. *Republic Tower Body Corporate Strata Plan No 341293D* [2004] VSC 56.

131 *Owners Corporation PS501391P* v. *Balcombe (Owners Corporations)* [2015] VCAT 956.

132 *BCCMA* s 180(7).

133 *Strata Titles Act 1998* s 91 (3)(c).

134 *SSMA* s 139(1).

135 *Mixed Use Development Act 1993* (Qld) Pt 6, as a 'specified Act' is still operative for some older stratum scheme schemes; see *Body Corporate and Community Management Act 1997* (Qld) Ch 8 Pt 1.
136 *Subdivision Act 1988* (Vic) s 27C(4): A subdivision can include limited and unlimited owners corporations. The latter will usually sit 'above' the multiple, limited owners corporations and can carry on the powers and functions of the limited owners corporations.
137 Douglas Harris, 'The Condominium and the City: The Rise of Property in Vancouver' (2011) 36(3) *Law and Social Inquiry* 694, 700.
138 Gary Bugden, *All About Strata and Building Management Statements and Building Management Committees: A Complete Guide to Working with Building Management Committees* (Institute of Strata Management, 2006) www.mystrata.com/doc-store/Building_Management_Statements_Paper_for_ISTM_Nov_06.pdf 2–3.
139 The legislative terminology is confusing. The reader should bear in mind the distinction between a 'stratum *lot*' = single owner and a 'stratum *parcel*' = a stratum lot that has been subdivided by a strata scheme. See Cathy Sherry, 'Building Management Statements and Strata Management Statements: An Unholy Mixing of Property and Contract' (2013) 87 *Australian Law Journal* 393.
140 *Conveyancing Act 1919* (NSW) s 196D; *Land Title Act 1994* (Qld) Pt 4 Div 4.
141 *Subdivision Act 1988* (Vic) ss 27B(2), 27C(2).
142 Ibid. sch 8A cl 5(3); *SSFDA* sch 1C cl 3(3).
143 See discussion of *Italian Forum Ltd* v. *Owners – Strata Plan 60919* [2012] NSWSC 895.
144 *Land Title Act 1994* (Qld) s 54DA; *Building Act 1975* (Qld) Ch 8A Pt 2.
145 Barangaroo is the 22-hectare redevelopment of the industrial wharf area on the western side of Sydney CBD. The underlying freehold is being retained by the state government, which will grant a 99-year lease over the site. The development includes extensive parklands, cultural space, as well as commercial and residential high-rises. See Barangaroo www.barangaroo.com.
146 King & Wood Mallesons, 'Barangaroo: Inside a Deal' (Presentation, 8th Annual ACCAL Conference, 21 March 2013).
147 See discussion of by-law characterisation in Chapter 4, 148–52.
148 [2016] VSC 384.

2 A property theory for strata and community title

> Property concerns things needed for human life. At the most elemental level, property gives us a place to be: a place to work. It gives us the means to thrive: food to eat, clothing to wear. It gives us things that make life enjoyable, meaningful, and fun. But property does not only provide our material needs. It enables us to exercise autonomy, to enjoy our liberties, to shape our destiny, to form relationships with others, to live a human life. We cannot live our lives without the means to do so.[1]

Introduction

This chapter sets out the theoretical framework for the book, discussing the theories of property that might provide us with tools to examine strata and community title. The chapter uses the terms 'property' and 'property law', because they are consistently used in the literature. However, the subject of the chapter, the book itself, and invariably the literature on which it relies, is land, the (arguably most significant) subset of property.

It is crucial to differentiate land from other property for two reasons. First, unusual geological activity aside, land never ceases to exist; it endures in a way that no other property does. Shares, gold, the rights to a song, can all be lost or destroyed, while land remains. The expression, 'you can't take it with you', rings truer for land than anything else. Second, human beings are 'embodied beings'; that is, we all have to be on some piece of land, at all times.[2] Everything we do in life, everything we value – our personal relationships, our jobs, our recreation – must be carried out on land, and thus our relationship to land affects us every minute of every day. In this way, land can be said to be 'elemental'.[3] While many of us no longer live subsistence lifestyles, land remains fundamental to our daily lives and survival; 'without some minimal appropriation – without some minimal taking of the resources necessary to sustain life – we will die'.[4] As Edward J Soja says, '[a]s intrinsically spatial beings from birth, we are at all times engaged and enmeshed in shaping our socialized spatialities and, simultaneously, being shaped by them'.[5]

The combination of these two factors – that land is fundamental to every part of our lives and that it will remain long after we are gone – gives a special significance to land law, absent other areas of property law. When one generation goes, land remains, as needed and valuable for the next generation as it was for the last. Land law, and the theories that support it, must account for this. Of course, most theories of property do, even if their ambit purports to be broader than land. Most property theorists mount their strongest arguments with reference to land, not shares, gold or intellectual property.

There are many theorists whose work could be considered in this section, but for the sake of clarity and brevity (as this book is not intended primarily as a work of theory), the concepts from Joseph William Singer's article 'Democratic Estates: Property Law in a Free and Democratic Society'[6] ('Democratic Estates') will be used to structure the chapter. 'Democratic Estates' draws on Singer's more extensive writing in *Entitlement: The Paradoxes of Property*[7] and *The Edges of the Field: Lessons on the Obligations of Ownership*,[8] as well as the work of fellow property theorists, Laura Underkuffler, Carol Rose, Gregory Alexander and Eduardo Peñalver. This chapter will also explore the work of Hanoch Dagan, Michael Heller, Thomas Merrill and Henry Smith.

Singer's democratic theory of property law is particularly relevant to strata and community schemes for three reasons. First, internally, strata and community schemes have a democratic structure and *ideally* operate as mini-democracies. All owners are automatically members of a governing body,[9] generally exercising equal votes. Owners can elect an executive that exercises functions on their behalf. The larger body corporate operates as a mini-legislature with a power to write and amend the laws that govern the scheme. Arguably, strata schemes are a micro-example of the Hobbesian social contract: when buying into a scheme, people voluntarily give up some of their autonomy to be governed by the rules of their community, thus making their micro-community a better place for everyone to live. Bodies corporate are sometimes referred to as a fourth tier of government in Australia, after Federal, State and local government.[10]

Second, democratic principles are not only relevant to the internal workings of strata schemes, they are relevant to strata schemes' relations with the wider polity. Strata and community schemes provide housing and work places for millions of Australians and are rapidly becoming a dominant form of commercial and residential title. As is the case in the economies of most modern democracies, we rely almost exclusively on the private sector to build new commercial and residential premises. To conceive of this process as developers simply presenting a consumer product to the market place, on a take it or leave it basis, like new televisions, cars or even shares, is to misconceive the reality of the transactions. 'Consumers' who need to live or work within a certain area, for social or financial reasons, do not have unlimited choices. They are limited to the housing or commercial options in that area. For many housing 'consumers' who live in cities like Sydney, strata title is the *only* title choice, because free-standing housing is not financially viable.[11] In the newly developed residential areas around Sanctuary Cove and Hope Island, on the northern end of the Gold Coast, housing is almost exclusively contained in body corporate communities;

it is the only 'product' on offer.[12] Further, housing and work places are not optional consumer items. They are not like televisions or cars, where if you do not like the choices on offer, you can choose nothing. As noted above, we are embodied beings; everyone has to have a physical place to live and work, or as Jeremy Waldron has said, '[e]verything that is done has to be done somewhere'.[13] As a result, when developers build and legally structure strata and community title schemes, they are building the physical and legal landscape in which many of us, and a large proportion of future generations, must live and work. They are building the physical locus of our democracy. The principles that guide that process and regulate the relations between developers and purchasers, owners and their body corporate, must go beyond basic market conceptions. A democratic theory of property law is more nuanced; it embraces a richer array of values and principles than theories based on market efficiency.

Finally, strata and community titles have implications not just for those *inside* schemes, they have implications for those beyond; for people who have never made a voluntary decision to be part of a scheme, but who will nonetheless be affected by them. Strata schemes, by their very bulk and density, have a profound effect on the physical environment, often coming to dominate areas zoned for their construction. Further, as described in the preceding chapter, community title is increasingly being used by councils and state governments to privately fund services that might otherwise have been publicly provided; those who cannot pay for their own private facilities may consequently go without.[14]

When we subdivide large single sites or multiple smaller sites within cities with strata or community schemes, we are allocating large portions of limited physical space to some people, while excluding others. Inside these schemes, we are allocating rights to some people (developers, majority owners), at the expense of others (subsequent purchasers, minority owners). In a democracy that purports to value all citizens equally, we must pay attention to the effects of those allocations.

Property and morality

This chapter assumes that property law is 'the category of legal doctrines concerned with allocating rights to material resources', and that theories of property should attempt to provide 'a normative justification for allocating those rights in a particular way'.[15] This assumption is relatively uncontentious. Most contemporary property theorists would accept that there is no 'inevitable content'[16] to property law and that property theory must always argue for what is inside and outside the sphere of protected property rights. However, the most persuasive theorists accept that those arguments for what is inside and what is outside must go beyond single values like economics or personal liberty and encompass the full range of values relevant to society and individuals (including economics and personal liberty). Further, contrary to lay perceptions of property law, morality – in the broad sense of what is right – plays a central part in discussions of what constitutes a legitimate property right.

For example, Merrill and Smith argue that despite various attempts in our intellectual tradition to separate property and morality, they are inevitably linked. Merrill and Smith assert that no system of property can survive without a connection to basic morality, otherwise it could never secure the widespread acceptance of its terms by large numbers of unconnected people which is necessary for it to be a right in rem, or 'good against the world'.[17] Ultimately, like human or civil rights, what constitutes property is what is widely believed to be morally justifiable. Hanoch Dagon argues that 'no arbitration among the different available conceptions of property is possible without some normative apparatus' and that we are inevitably led back to the question of whether the assertion of property rights, like all private law rights, amounts to 'justificatory practice'.[18] This is determined by reference to the ramifications in people's lives of the assertion of those rights.[19] Carol Rose, in describing what she calls the 'second-best morality' of property (the compromise position of most property rights), memorably said that property's inevitable morality 'does not presume saintliness, but it also is not made for total sinners: she is neither the girl next door nor the woman in red'.[20] In 'A Statement of Progressive Property', Professors Alexander, Peñalver, Singer and Underkuffler state that '[c]hoices about property entitlements are unavoidable, and, *despite the incommensurability of values, rational choice remains possible* through reasoned deliberation'[21] with reference to context, experience and principle. As Singer says:

> Property is not just something we protect or invade, recognize or reject; it is something *we collectively construct*. We must give it its meaning, both social and legal ... [and the] mix of entitlements and obligations we can legitimately claim depends on the kinds of human relationships we can defend, nothing more and nothing less.[22]

Further, when determining what law should be, there is 'no alternative but to make arguments that elaborate fundamental human values and that express our considered commitments to judgments about morality and justice'.[23] There is 'no baseline free from judgment about the fair minimum standards for human relations'.[24] Property law is inescapably normative.

Singer, like Rose, Underkuffler, Alexander and Peñalver,[25] argues that the economic analysis of property law that dominated American scholarship in the late twentieth century,[26] (although not Australian scholarship), is fundamentally impoverished by its excessive focus on satisfaction of individual preference and the free markets this allegedly generates.[27] Singer argues that, unlike political and moral theory, an economic theory of law does not acknowledge that our preferences are not just given, or that in a democracy, not all preferences are legitimate.[28] Further, markets, the ideological bedrock of economic theory, cannot provide answers to what is or is not legitimate. Markets do not operate in a state of nature; they are created and regulated by law. Singer says that:

> It is circular to ask economic analysis to define property rights when one needs to define property rights first to determine what market transactions

> will emerge. There is no escape from the need to use some other form of analysis, such as political and moral theory, to define and shape the basic structure of society.[29]

It is important to note that Singer does not dismiss the importance of markets and market values in his scholarship. A democratic theory of property is inevitably rooted in capitalism and market economies. However, only particular kinds of markets are worth pursuing. As Singer says, 'modern conceptions of property imply *a well-functioning market characterised by fair competition*'.[30]

Singer argues that viewed collectively, the multiple values of contemporary property law specifically promote freedom and democracy. These values include: autonomy (the freedom to determine the shape of one's own life); mobility (people should not be tied to land, unable to move from place to place because they cannot alienate their current land); widespread distribution of property and realistic opportunities for access;[31] freedom of contract with minimum standards for economic and social relationships; and stability and change (property interests that allow for long-term stability but can be modified when necessary). Singer's definition of democracy is not simple majoritarianism, but rather includes the widely accepted, as well as the debated values, of a modern, pluralist, liberal democracy.

Singer's democratic theory is a complex theory insofar as it demands that we look at all of the effects of property law, the good and the bad; that we do not allow ourselves the easy luxury of ignoring the widespread implications of property's choices. Singer is less concerned with formulating a neat, concise theory, than elaborating a complex framework that allows us to identify what are and are not morally justifiable rules of law.

What makes property different? Estates, *numerus clausus* and other 'boundary rules'

That the content of law is a consequence of moral choices, which, in turn, affect and constitute society is true of all law, not just property. However land law has a particular significance, absent more youthful areas of law such as contracts, administrative law, torts and corporate law. Principals of modern land law developed in the same centuries in which our law and political system struggled to wrestle control from the hands of absolute monarchs and vest it in democratically elected parliaments.[32] In this period of history, ownership of land transformed from an institution which bound an individual to multiple feudal relationships of personal obligation and domination, into an institution which allows an individual relative but unprecedented autonomy and freedom. In short, land law played a crucial role in the development of a free land market[33] and liberal democracy. Current crises in capitalism aside, both remain bedrock institutions in our community.

To understand the nature of those institutions, it may be helpful to consider briefly the legal order that emerged from the Norman Conquest in 1066. A feudal system of land ownership was actually or theoretically[34] imposed on all land in England, creating a pyramid of land ownership and obligations with the King at the apex. The most important obligation of landowners who held directly from the King was the provision of knights, but in the absence of an adequate monetary economy, land grants 'paid' for clerical duties, weapons, food, entertainment and agricultural labour.[35] Ownership of land was therefore not a simple, allodial relationship to a piece of earth, but was characterised by multiple relationships with other individuals to whom the owner owed, or from whom the owner could extract, obligations, or 'feudal dues'. The overall effect of the feudal system of tenurial land ownership was *a systemic fragmentation of property rights*, that is, ownership of a single piece of land was split between multiple individuals. Butt cites an example of land that was owned at the turn of the fourteenth century by seven men, the ultimate lord holding of the King of Scotland, who in turn held of the King of England.[36] There were social and economic incentives for owners to further fragment title to pass on their obligations to someone below them.[37]

A fundamental part of the development of a free land market and democratic society was the dismantling of the feudal pyramid by prohibiting tenurial obligations and the *re-aggregation of fragmented interests in land* so that individual owners could freely and thoroughly exploit their resource. Ironic as this may seem, this freedom came from the creation of multiple prohibitive rules. From the thirteenth century onwards judges and legislatures promulgated property rules that prevented excessive fragmentation of property rights, including fragmentation through time; that is, rules that prohibited multiple parties, past, present and future, having interests in the same piece of land.[38] It is hard to find a single box in which to put all of these rules but Michael Heller uses the term 'boundary principle' to describe 'legal doctrines that separate … property categories from each other and help to keep resources well-scaled for productive use'.[39] Examples of boundary rules include the *Statute of Quia Emptores 1290*, which prohibited new layers of feudal obligation. The owner of the fee could not create a new tenurial relationship with a transferee; he could only transfer the land so that the transferee stood in his shoes. Importantly, a landowner could do this without his overlord's consent, even though the overlord would be affected by the quality of person who now stood in the place of their former tenant. The Statute began the gradual shrinking of the feudal pyramid. Feudal tenure was formally abolished in 1660 with the *Tenures Abolition Act*.[40] Other examples of boundary rules include every student's worst nightmare, the rule against perpetuities,[41] the four requirements for easements in *Re Ellenborough Park*,[42] as well as the estates system. The estates system only recognised three freehold estates, the fee simple, life estate and fee tail. The estates system was further simplified by the abolition of fee tails, (discussed in more detail below), and as a result, there are now only two freehold estates in Anglo-Australian land law, constituting a strict boundary rule on the kinds of freehold possessory interests that can exist in land.

Merrill and Smith have focused on the *numerus clausus* principle as the means of rationalising property interests, rather than the estates system or the broader concept of boundary rules.[43] While Singer sets less store in the ultimate simplicity of property law,[44] like Heller, Merrill and Smith, he acknowledges that one of the core attributes of *modern* property law is the stipulation that parties deal in finite, default bundles of property rights that they are not free to reconstruct according to their own preferences. Unlike contract law, parties are not allowed to dream up almost any property entitlements they please. Whether we focus on the terms estate system, boundary rules or *numerus clausus* principle, it is clear that today, as Edgeworth says:

> landowners are not at liberty to customise land rights, in the sense of reworking them in an entirely novel way to suit their individual needs and circumstances. Rather, any new rights must fit within firmly established pigeonholes, of which the law only permits a small and finite number.[45]

In their wide-ranging article on the *numerus clausus* principle, Merrill and Smith note that the significance of a finite number of property interests has, until recent years, gone strangely unnoticed by common law property scholars.[46] However, they say that while they could only find one full article on the *numerus clausus* principle in common law and little explicit judicial recognition, there is consistent judicial adherence to the principle in practice. Whenever courts have been asked to recognise novel interests in land, for example, a covenant to use an existing railway in exchange for payment of a toll,[47] or a life estate with the power to devise the fee simple,[48] they have struck them down. In *Keppell* v. *Bailey*, Lord Brougham famously said that:

> But it must not therefore be supposed that incidents of a novel kind can be devised and attached to property at the fancy or caprice of any owner. It is clearly inconvenient both to the science of the law and to the public weal that such a latitude should be given. There can be no harm in allowing the fullest latitude to men in binding themselves and their representatives, that is, their assets real and personal, to answer in damages for breach of their obligations. This tends to no mischief, and is a reasonable liberty to bestow; but great detriment would arise and much confusion of rights if parties were allowed to invent new modes of holding and enjoying real property, and to impress upon their lands and tenements a peculiar character, which should follow them into all hands, however remote. Every close, every messuage, might thus be held in a several fashion; and it would hardly be possible to know what rights the acquisition of any parcel conferred, or what obligations it imposed. The right of way or of common is of a public as well as of a simple nature, and no one who sees the premises can be ignorant of what all the vicinage knows. But if one man may bind his messuage and land to take lime from a particular kiln, another may bind his to take coals from a certain pit, while a third may load his property with further obligations to employ one blacksmith's forge, or the members of one corporate body, in various operations upon the premises, besides many other restraints as

> infinite in variety as the imagination can conceive; for there can be no reason whatever in support of the covenant in question, which would not extend to every covenant that can be devised.[49]

In *Hill* v. *Tupper*, Pollock CB said:

> A new species of incorporeal hereditament cannot be created at the will and pleasure of the owner of property, but he must be content to accept the estate and the right to dispose of it subject to the law as settled by decisions or controlled by Act of Parliament.[50]

Quoting these passages in *Clos Farming Estates Pty Ltd (Rec and Man Appt)* v. *Easton*,[51] a contemporary easement case, Bryson J noted at [15] that,

> [t]he Common Law and its judges have for centuries favoured land being fully available for its owner to use and sell, and restricted the creation and operation of interests, including new classes of interests, which might have different effects. The enormous complexity of land law from medieval times onwards obscures this policy, but resistance to restrictions on use and alienation of land and on the wishes of property owners to limit the rights of their successors, to restrict alienation and to devise new forms by which to do so is of very long standing.

In later chapters, when deep in the messy complexity of large, modern strata and community title schemes, with extensive by-laws and developer-imposed management structures, it will be worth recalling Lord Brougham's distinction between contract and property law; that is, while there is no mischief in allowing the fullest latitude to contract content,

> great detriment would arise and much confusion of rights if parties were allowed to invent new modes of holding and enjoying real property, and to impress upon their lands and tenements a peculiar character, which should follow them into all hands, however remote.

Further, the problems that plague complex strata schemes also suggest that notice, through modern registration systems and statutorily mandated contractual disclosure, does not eradicate detriment and confusion.

There are clear economic benefits to be gained from limiting the number of interests that can validly be created in land. The *numerus clausus* principle dramatically reduces transaction costs by relieving purchasers of the burden of attempting to discover an unlimited variety of interests that might exist in the land they want to acquire. As Rose says, '[i]n a wide and commercialized property market, property law acts as an axe that purposely chops out nuances and niceties in the things traded. Too many complications spoil the market'.[52] The principle both reflects, and has helped to constitute modern, capitalist land markets. Bernard Rudden notes that 'in all "non-feudal" systems with which I am familiar (whether earlier, as at Rome, or later), the pattern is (in very general terms) similar: there are less than a dozen sorts of property entitlements'.[53]

Today, realistically, most people will be acquiring at law or in equity, a fee simple interest or a leasehold interest and it might be burdened by a mortgage, a lien, an easement, a profit a prendre, a (sub)lease or a covenant, but nothing more. (They might of course just be acquiring the latter interests alone.) Even in the past, when a fee simple might have been affected by a determinable limitation or condition subsequent or it might have been a life estate, possibly with a vested or contingent remainder, there were still few permissible interests in land.[54] As a result, purchasers need to spend very little time and money trying to ascertain the *kind* of interest they might be acquiring. The confusion and detriment referred to by Lord Brougham in *Keppell* v. *Bailey*, above, has been largely avoided.[55]

In addition to reducing transaction costs, the *numerus clausus* principle and the estates system limit the economically stultifying effect of 'dead hand' control. While a current landowner may believe that a myriad of restrictions or conditions would be desirable to attach to his or her land, they may impair or sterilise the uses that the land can be put to by future generations. As Edgeworth argues:

> If parties were free to restrict the usages of land by agreements capable of binding successors in title indefinitely, land could be shackled in ways that might revive all the impediments to economic reform that were endemic in feudal real property law.[56]

As no landowner has a crystal ball, there are clear economic benefits to each generation deciding for itself what it does with its land, rather than having land use predetermined by their great-grandparents.

One of the best ways to demonstrate the importance of the *numerus clausus* principle and estates system is to examine what happens to property regimes when they are absent. Michael Heller did this when he studied new property rights in post-communist Russia (although as we have noted, our own feudal system is an equally obvious example). Heller's research revealed a failure of private property as a result of excessive fragmentation.[57] 'Excessive fragmentation' is the breaking up of property into multiple, idiosyncratic, unpredictable bundles. It leads to too many people having powers of veto in relation to individual parcels of land, which in turn leads to stagnation. Heller examined Moscow storefronts in which multiple government departments, local councils and multiple occupiers all had controlling interests. If one party opposed a use, others would be blocked from exercising their rights. As a consequence, storefronts remained empty, in a constant state of underuse.

Heller labelled this phenomenon of excessively restricted land the 'tragedy of the *anticommons*'.[58] The tragedy of the commons is of course the idea that land to which there is unrestricted access will eventually be depleted and destroyed because no single user has a sufficient disincentive to put another cow on the common, but when everyone does so, the common is destroyed.[59] The tragedy of the anticommons is when

> multiple owners are each endowed with the right to exclude others from a scarce resource, and no one has an effective privilege of use. When there are too many owners holding rights of exclusion, the resource is prone to underuse.[60]

The great risk of an anticommons is that it is much easier for owners to fragment property than to re-aggregate it. Fragmentation 'may operate as a one-way ratchet: Because of high transaction costs, strategic behaviours, and cognitive biases, people may find it easier to divide property than to recombine it'.[61] The feudal pyramid is a good example in point. The creation of multiple feudal relationships was relatively easy and swift; their re-aggregation took centuries.[62] Another example is the mass application of restrictive covenants to large-scale residential subdivisions. Covenants can be created with the stroke of a lawyer's pen;[63] re-aggregation then requires hundreds of owners to negotiate and pay for release of the covenants. For this reason, it is particularly important to guard against unthinking fragmentation with insufficient attention to the consequences.

Importantly, Heller identifies United States condominiums (strata title), and common interest communities (CICs) (another term for homeowner associations and the equivalent of low-rise strata and community title subdivisions), as potential anticommons. He suggests that developers fragmented land and air into individually owned units and common property for economic and social gain – both the developers' and owners' – but that the CIC form was used too enthusiastically before its anticommons risk was fully understood.[64] Heller says that 'like the medieval fee tail, the modern CIC forms may give people too much power to lock resources into low-value uses'.[65] In other words, the CIC form, as we will see in detail below, potentially offends one of the most fundamental principles of a functioning property system. This will be explored further in Chapter 3.

Heller argues that property regimes generally successfully defend themselves from the risk of anticommons by the consistent application of 'boundary rules'. Heller includes the *numerus clausus* principle and the estates system in his concept of 'boundary rules', along with other doctrines like the rule against perpetuities, primogeniture and statutory minimum allotment sizes. Property regimes allow people to divide property, spatially or temporally for economic and social benefit, but 'boundary' rules prevent division into wasteful fragments.[66] Heller says that when dividing property, 'owners may make mistakes, or their self-interest may clash with social welfare'.[67] He concludes that:

> The pervasive presence of boundary rules challenges legal and economic theories that suggest unstructured fluidity to private property. Instead, the overwhelming evidence suggests that the notion of an open-ended bundle of property rights is wrong.[68]

So, the various 'boundary rules', the *numerus clausus* principle and the streamlined estates system make it clear that unlike contract law, property law has never just been about satisfying individual, idiosyncratic preferences, at least in relation to the *creation and transfer* of property interests; of course, it is a very different story when it comes to personal use of property, as we will see below. Property law has always allowed some individual freedom in the creation of property interests like leases (e.g. a range of positive and negative covenants) and easements (e.g. for the passage of water, sewage, humans by foot or by vehicle), but within firmly bounded parameters, (e.g. only an agreement that grants exclusive possession or is for a defined duration, can be a lease;[69] only an agreement that does *not* grant exclusive possession can be an easement[70]).

If property law habitually insists on these prefabricated boxes into which parties must fit their preferences, we need to then ask, how and why does property law come to decide on its boxes?

The norms of a free and democratic society

Singer argues that modern property law determines the parameters of its acceptable boxes by reference to substantive norms, gleaned from the features of a free, democratic society and minimum standards regulation. Logically, the latter could be folded into the former; that is, minimum standards regulation, like our property law, is really a product of a free and democratic society.

Singer points out that a democracy is not a monarchy, it is not a feudal society, a slave society, a caste or corporate society;[71] it is not apartheid, an aristocracy or an oligarchy. We could add to this list that it is not a theocracy. Many property rules that might be acceptable in such societies are never acceptable in a democracy. Singer argues that some of the democratic norms of our property law have been implemented formally, for example, by the provision of free alienation of fees simple without an overlord's consent in the *Statute of Quia Emptores 1290*, or by legislative prohibitions preventing people from excluding others from property on the basis of race. Some norms are protected by social custom, in that today, few people actually think of creating property rights that are inconsistent with freedom and democracy.[72]

While Singer does not do so explicitly, it is suggested that for the purposes of demonstrating the democratic nature of modern property law, the norms promoted by property can be usefully divided into economic norms and personal/political norms. These norms are particularly evident in the historical evolution of property law. Both sets of values were doggedly implemented by property law over the centuries, so much so that property law played a fundamental role in the dismantling of the feudal pyramid and the building of modern democracy.

The economic norm of democratic society

The basic *economic* norm of democratic society is that there should be freely functioning markets, including a freely functioning land market. The discussion of the *numerus clausus* principle and the reformed doctrine of estates, above, has already identified that property law has assiduously applied itself to the task of creating and maintaining such a market. This is not meant to entail support for the free market ideology inherent in law and economic scholarship and which Singer's theory is largely a reaction against. It is simply intended to point out that modern, western land law is fundamentally capitalist. As Rose says,

> Unlike past societies, modern economies – and their laws – tend to treat alienability as one of the principal features of property ... trade is a matter of immense importance in modern economies ... [h]ence when something comes to be treated as property, tradability is presumed ... there are many legal doctrines that simplify property rights, even overriding the wishes of the current owners for more delicate arrangements, so that property will indeed be available for alienation – including alienation to total strangers.[73]

In *A History of The Land Law*, Simpson describes how from as early as 1450 to 1700 the development of property law, in particular doctrines relating to fees tail and perpetuities, was a back and forth, push and pull between the conflicting desires of lawyers and judges to see land on an open market and to retain land in the hands of families.[74] However, over the course of the centuries in which modern democracy and capitalism developed, the sheer weight of property rules facilitating free disposition, pushing property back into the market, tipped the scales in favour of the market. These rules included the impermissibility of restraints on alienation, rules invalidating legal contingent remainders, the rule against perpetuities, mortgagees' powers of sale,[75] registration systems, as well as the *numerus clausus* principle. All simultaneously reflect, and are constitutive of, modern capitalist land markets.

The personal/political norm of a liberal democratic society

The basic *personal/political* norm of liberal democratic society, when viewed from the perspective of property law, is individual autonomy and freedom; freedom from the arbitrary power of government, freedom from family and freedom from religious authorities in relation to our land. Like the economic norm described above, this is *not* meant to connote US-style law and economics or libertarianism,[76] neither of which have ever had much traction in Australian legal debates, particularly the latter. It is simply meant to identify the basic principles that differentiate democratic society from non-democratic societies, including our feudal past. All those committed to living in a democracy may ultimately fail to agree on certain aspects of the principles, but would nonetheless share a consensus about the core. For example, consider John Stuart Mill's 'harm principle', that the only justification for social or government intervention in individual lives, and by extension land, is to prevent harm to others.[77] In classical liberalism, this principle is known as 'negative liberty'. While there has been extensive academic debate about the principle's sufficiency to ensure universal human flourishing,[78] no one argues that we *should* be arbitrarily interfered with by government. The argument is on the edges of what constitutes justifiable interference. The principle's core is accepted; that prima facie and to the greatest extent possible, we should be autonomous and free.

A good illustration of this is Martha Nussbaum's article *Deliberation and Insight: Bloch* v. *Frischholz and the Chicago School of Judicial Behavior Essay*.[79] Nussbaum is one of the most consistent critics of the sufficiency of 'negative liberty'.[80] The article examines the judicial consideration of a condominium association's attempt to remove a mezuzah that an observant Jewish family had placed outside their door, in alleged violation of a condominium rule against objects in hallways. While the bulk of the article is about judicial behaviour, Nussbaum clearly viewed the condominium association's actions as a violation of individual autonomy and liberty, ordinarily protected in a democracy. Similarly, Professors Alexander, Peñalver, Singer and Underkuffler's 'Statement of Progressive Property' refers to property law's 'protection of physical security, the ability to acquire knowledge and make choices, and the freedom to live one's life on one's own terms'.[81] Although a 'progressive' statement of property focuses on the moral demands and obligations that justifiable property law entails, it does not do so at the expense of hard-won, long established, classical liberal/democratic values.

Perhaps the best way to illustrate succinctly both the economic and personal/political norms underpinning modern, democratic property law, is to consider the demise of the fee tail.

Fees tail allowed land to be passed down a line of particular descendants, and in the event that that line failed, the land would revert to the grantor or grantor's estate. In a feudal society where status was assigned by virtue of birth, family ties assumed much greater importance than in modern democracies. Fees tail served a social and financial purpose by protecting family property from dissipation among people unrelated by blood. However, in the process, they seriously limited the current entail holder's ability to economically exploit their land. Leases could not be granted for longer than the entail owner's life,[82] the fee tail could not be mortgaged,[83] nor could it be alienated in its entirety.[84] These economic limitations simultaneously constituted limitations on personal liberty. Irrespective of whether leasing, mortgaging or transfer would have been economically beneficial individually or for markets, the freedom to alienate land, to deal with it as one sees fit regardless of other's views, has value in and of itself. This is perhaps best illustrated by the personal frustration suffered by Mr Bennet in *Pride and Prejudice* and the Earl of Grantham in *Downton Abbey*, as a result of their fictional 'tails male'.[85] Courts and individuals had waged a war of attrition against fees tail from the fifteenth century onwards through ingenious legal devices,[86] but it took many centuries before the fee tail was dealt a fatal legislative blow in the early twentieth century. Acts of fee tail abolition around the common law world[87] were part of a long line of statutory and judicial reform designed to move power 'downward from the feudal lord to the actual possessor of land and from prior generations to current owners'.[88]

In these centuries of rationalisation and consequent democratisation of property law, one anomaly appeared: the *Tulk* v. *Moxhay*[89] restrictive covenant. The preceding chapter explained the significance of limitations inherent in restrictive covenants to the development of strata and community title legislation, and the rationale for those limitations will be explained now.

First, it is very difficult not to conclude that *Tulk* was simply wrong. The basis of the decision was that the purchaser, Moxhay, was bound by a restriction and obligation to maintain Leicester Square as a pleasure garden because he had notice of this. However, that reasoning puts the cart before the horse. 'Notice' in property law relates to notice *of interests in land*, not contracts. The fact that we know that our predecessor in title had an agreement with their neighbour, that the neighbour could come onto our land is only relevant if that agreement amounted to a known interest in land, a lease or an easement. If it is an interest in land, and we had notice of it, we are bound. If it does not amount to an interest in land and is a mere licence, the first hurdle has not been overcome, and notice is neither here nor there. We cannot be bound.

Second, there is rationality to this approach. Contractual agreements of predecessors in title are potentially unlimited. They could include all manner of restrictions and obligations, (e.g. to paint the house pink, to never grow vegetables, to host an annual Christmas party, to hold weekly educational seminars), and if they bind future landowners, they constitute a serious violation of the *numerus clausus* principle. That is why English courts back-tracked so rapidly from the decision in *Tulk*, limiting the newly created doctrine to restrictions and not positive obligations,[90] and to covenants that benefited land, not a person or a business.[91] Courts were trying to contain the potentially economically and socially damaging nature of freehold covenants. As A W B Simpson said:

> The effect of restrictive covenants is to sterilize the use of a parcel of land permanently; in principle it is not at all clear that a private landowner ought to be allowed to do this without public control of his activities. Whatever their merits, restrictive covenants can have a very detrimental effect on the free development of land, which is not in all cases in the public interest.[92]

Merrill and Smith argue that restrictive covenants were not a significant breach of the *numerus clausus* principle, as courts limited the effect of *Tulk* by insisting, not only on notice, but that the covenant touch and concern the land.[93] Further, they suggest that as they are not possessory interests, restrictive covenants amount to a 'fringe area' of property.[94] However, as this book will demonstrate, this might be too optimistic an assessment; the limitations placed on *Tulk* may have been too little, too late. While restrictive covenants are not possessory interests, they attach to the most significant possessory interest, the fee simple estate. When replicated across large geographic areas, their effect is very significant indeed. United States covenant law, which did not further limit *Tulk* to negative obligations as English and Australian law did, allows a vast array of obligations to be attached to the fee, as evidenced in complex HOA communities built from the later nineteenth century onwards.[95] In Australia, *Tulk* was limited to negative obligations, but as we saw in the preceding chapter, strata and community legislation, by permitting both negative and positive by-laws, effectively sweeps this limitation away. In Chapter 3 on United States covenant law and Chapter 4 on strata by-laws, we will see the extraordinarily complicated communities and problems that have been spawned from a tiny chink in the armour of the *numerus clausus* principle, 150 years ago.

Analytical insights

Having argued persuasively that modern property law is informed by the norms of a liberal democracy, Singer goes on to set out four 'analytical insights' that he believes a progressive and democratic approach to property entails.

Relations between people with respect to valued resources

The first insight is that property law concerns not just relations between people and things but relations among persons with respect to valued resources. Singer says:

> Every legal right should be understood not merely by reference to the powers and rights it gives the owner but by reference to the impacts of the exercise of those powers on others and the shape and character of the social relationships engendered by those rights and powers.[96]

Singer acknowledges that this has long been recognised by property theorists,[97] but he suggests that it has not been given appropriate prominence. He argues that it is not enough to recognise rights as relationships; we must focus on the *quality* of those relationships. Libertarian and efficiency scholars set great store in voluntary agreements, assuming that they are inevitably both consensual and efficient.[98]

Singer asserts that while voluntary relationships are indeed good, the assumptions of libertarian and efficiency theorists are too simplistic. He says that we must make judgements not just about whether relationships are voluntary, but whether those relationships are acceptable in a free and democratic society. People cannot agree to master/slave relationships or lord/vassal because

> some relationships are out of bounds … some contract terms are off the table … [and] there are some things you should not ask of others; there are some demands that cannot justly be made in a free and democratic society.[99]

We saw above in the description of the decline of the fee tail, that estate in land created unequal power relations between generations that a liberal democracy ultimately considered unacceptable. In strata and community title, great store is set in the voluntary consent to by-laws that can allegedly be assumed from the act of purchase (consent to both existing by-laws and to any that eventuate from a special majority vote). As will be argued in later chapters, this excessive faith in consent is both at odds with orthodox property doctrines and with the theoretical framework for property advocated by Singer.

Externalities

Singer's second analytical insight flows naturally from the first. It is that property rights involve obligations, externalities and duties of attentiveness. Arguably the most important duty of attentiveness is to recognise that by the very act of owning property, we deny it to another.[100] Laura Underkuffler reminds us, as John Locke did hundreds of years before,[101] that property rights are different from other rights because property is a finite resource.[102] She points out that granting a right to speak to one person does not necessarily take that right from another, but granting property to someone of necessity renders other people non-owners in relation to that thing. Protection of property is inevitably to the 'derogation, exclusion, and often injury of others',[103] or as she evocatively puts it, 'property is rivalrous'.[104] In a liberal democracy, acknowledging this requires us to think about the distributive, and sometimes illiberal, effects of property law.

In addition to rendering other people non-owners, the exercise of property rights can impose a range of externalities on others, and the common law of property, like that of contract and tort, requires us to pay attention to this.[105] We are not free to harm others in the exercise of our property rights. Singer emphasises that property law does not prevent us from *ever affecting* others in the exercise of our rights; it just prevents us affecting them in genuinely harmful ways. Complaints that we are affected by another's use of property can be divided into the legitimate and illegitimate, the latter being ignored by property law. For example, as a general rule, property law places no restriction on the colour someone can paint their house. Neighbours might be deeply annoyed by the shade of pink a property owner has chosen, but property law will not consider their complaint legitimate. A choice of house colour is viewed as a 'self-regarding act' which property law does not regulate.[106] Deciding what is or isn't a self-regarding act comes back to a considered judgment of the legitimate and reasonable expectations of the parties in democratic society.

Considered judgments, in diverse contexts, are messy. Here Singer reminds us that, while central aspects of property law like the *numerus clausus* principle and the reformed estates system promote simplicity and clarity, some property law remains terrifyingly complex. He argues that this complexity is not a historical throwback or a tangled, dated muddle that needs to be remedied; it is complexity that serves an indispensable purpose.[107] Singer argues that much of the complexity of property law comes from the law shielding non-owners from externalities which cannot be justified.[108] What constitutes an unacceptable harm is historically and culturally specific. Absolute property rights, which are blind to these harms are 'contrary to common morality and hence justified expectation'.[109] A good example of the function of property law's complexity can be seen in the discussion in Chapter 3 of the attempts to simplify the United States' notoriously difficult covenant law. Nearly all property scholars involved in the debate concluded that while the rationalisation of property rules was desirable, it would never make the underlying, messy social and political issues disappear.

Multiple individual rights add up to a social and political regime

The third analytical insight Singer identifies is that property is not merely a conglomeration of individual rights; it is an entire social system. He says:

> Property law is part of the way we define a legitimate social order ... [and] we cannot conclude that a particular set of property rules or institutions is acceptable unless we attend to the systemic effects of exercising those property rights.[110]

Singer develops this idea more fully in *Entitlement* where he argues that 'the legal rules we choose may have deep and lasting effects on our social world.... If property shapes social relations, we need to ask ourselves: 'In which world would we rather live?'[111] Every time we make a choice about a legal rule, we are making a choice about the kind of relations we want and thus, the kind of society in which we live. Property is not just an aggregation of individual entitlements; it is a regime.[112]

The most powerful example Singer gives relates to the very subject matter of this book: restrictions attaching to freehold land. Singer cites the widespread United States practice of imposing racially restrictive covenants on land to prevent non-Caucasian families buying into neighbourhoods. The practice developed soon after the Supreme Court ruled in *Buchanan* v. *Warley*[113] that it was unconstitutional for municipalities to racially segregate housing through zoning laws.[114] Although the socially detrimental effects of racially segregated housing had been recognised and outlawed for public institutions, exactly the same practice was allowed to flourish in a private context, on the assumption that it was justified by the ideology of freedom of contract.[115] If individual parties wanted to make such agreements between themselves, then they were free to do so. Ironically, this view was confirmed by the Supreme Court in the very case that effectively outlawed racially restrictive covenants. In *Shelley* v. *Kraemer*,[116] the Supreme Court found that enforcement of a racially restrictive covenant by a state court was discriminatory state action prohibited by the 14th Amendment,[117] but that the covenant itself, as a voluntary agreement between individuals, was prima facie valid.

Of course, what this freedom of contract rationale fails to recognise is that those private agreements affect *everyone* in the community, not just the individuals who make the agreements. Most notably, they hurt the millions of African and Asian Americans who were denied housing in suburbs and whose dignity was affronted when they were denied freedom and rights available to white citizens. The personal preference of individual landowners not to have

African or Asian American neighbours is an example of the kind of preference property law should not legitimate by upholding.

However, the hurt went beyond the individual families who might have wanted to live in residential subdivisions earmarked for Caucasian Americans. Multiplied thousands of times across American cities and towns, they were part of a legal, economic and social system that excluded African Americans from social and political life; a social and political system that caused the Reverend Martin Luther King to proclaim that one hundred years after the abolition of slavery,

> the Negro still is not free. One hundred years later, the life of the Negro is still sadly crippled by the manacles of segregation ... on a lonely island of poverty in the midst of a vast ocean of material prosperity. One hundred years later, the Negro is still languishing in the corners of American society and finds himself an exile in his own land ... [with] basic mobility ... from a smaller ghetto to a larger one.[118]

And because of the nature of property ownership, 50 years later again, the problem remains. Emotional, economic and social attachment to land means that ownership patterns endure; many suburbs that were ‘white’ at the time of *Shelley* v. *Kraemer*, remain ‘white’ today. Original families and their descendants have stayed, while other families do not want to, or cannot afford to, move in.

Further, ownership of land has social and economic ramifications beyond the location of home. One example of my own will suffice: the perennial problem the United States has had with racially segregated schools. While explicit racial segregation of schools was outlawed at the same time as racially discriminatory covenants,[119] suburbs remained racially segregated and children attended their local school. The result is that many American schools are effectively racially segregated. This educational ‘apartheid’ sees many African American and Hispanic children in under-resourced inner urban schools, while white children attend well-resourced, suburban institutions.[120] In the 1970s, court-ordered bussing programs were implemented to move white children into predominantly black schools and vice versa.[121] Bussing was widely criticised and led to ‘white flight’ into the private schools that mushroomed in white neighbourhoods. Racially segregated schooling is a socially, economically and culturally destructive problem that affects the entire nation and which the United States seems unable to solve.[122] It stems squarely from segregated neighbourhoods created by racially restrictive covenants, demonstrating the gross error of evaluating property rules solely on an individual, contractarian basis.

Singer recognises this error and prompts us to think about the exercise of multiple property rights through time. In contrast to more abstract property theories, which focus on hypothetical moments of original acquisition, Singer says:

> the legitimacy of the exercise of a property right depends on the consequences of exercising the right. Those consequences are historical events;

> they take place in time. Thus the time that matters is not the magical moments of acquisition and transfer. Rather, what matters are the continuing moments in which property rights are exercised. The tensions that are a necessary component of the institution of private property affect not only choices about initial acquisition and transfer but also the proper moral and legal response to exercises of rights as they happen over time.[123]

We will pick up this problem of accessing the validity of property rights solely at the point of their creation in Chapter 4, in the discussion of the NSW by-law case *Casuarina Rec Club Pty Ltd* v. *Owners – Strata Plan 77971*.[124]

Conclusion

Singer ends 'Democratic Estates' by asking, 'What difference does [a democratic theory] make?' To answer this, he uses an example of a tenant who wants to post a political sign in the yard of his rented house. Can the landlord prevent this via a covenant in the lease? For the purposes of this book, we could change the characters to a strata or community title lot owner whose body corporate has created a by-law banning all signs.[125] If the owner wants to put a sign in his window or yard, what right does a body corporate have to protect the visual facade of the collectively owned building or the streetscape of a community title subdivision? What right does the body corporate have to regulate activity inside an owner's privately owned lot? If it has such a right, does it extend to curbing the expression of political opinions? Even if someone has apparently voluntarily bargained away their freedom to post signs in their home by purchasing with notice of the ban on signs, should we enforce that bargain? Would it matter if the lot was in a 12-lot strata scheme or in a 1,000-lot community scheme which was effectively an entire suburb?

Singer points out that if we use a democratic theory of property, as opposed to an efficiency or libertarian theory for example, we will be required to:

> make substantive choices about the interests at stake, the values those interests implicate, the relative strength, relevance, and cogency of those values in particular social settings, the social relationships that will result from the choice of legal rule, the opportunities that will be enabled or cut off, and the relation between property rights and political and social life.[126]

In short, when trying to solve a difficult property dilemma like this, we will be obliged to think about 'the full range of human values we care about'.[127] If we accept Singer's argument that property law is creating the social and political regime in which we live, it seems only rational to make our deliberations as far-ranging and deep as possible, in order to ensure that our conclusions remain consistent with liberal democracy.

Notes

1 Joseph William Singer, *The Edges of the Field: Lessons on the Obligations of Ownership* (Beacon Press, 2000) 27.
2 Jeremy Waldron, 'Homelessness and the Issue of Freedom' (1991) 39 *UCLA Law Review* 295.
3 Kevin Gray and Susan Francis Gray, *Elements of Land Law* (Oxford University Press, 5th edn, 2009) 2.
4 Laura S Underkuffler, 'Property: A Special Right' (1996) 71 *Notre Dame Law Review* 1033.
5 Edward W Soja, *Seeking Spatial Justice* (University of Minnesota Press, 2010) 18.
6 Joseph William Singer, 'Democratic Estates: Property Law in a Free and Democratic Society Essay' (2008) 94 *Cornell Law Review* 1009.
7 Joseph William Singer, *Entitlement: The Paradoxes of Property* (Yale University Press, 2000).
8 Singer, *Edges of the Field*, above n 1.
9 In Australia there has been little discussion of the property-based enfranchisement in strata and community schemes. Tenants, including long-term tenants, are not members of the body corporate and as a result, have no vote. Their landlords, including absentee landlords, have the vote associated with the lots. For a discussion of alternative bases for enfranchisement, see Robert C Ellickson, 'Cities and Homeowners Associations' (1981) 130 *University of Pennsylvania Law Review* 1519 and Frank Michelman, 'Universal Resident Suffrage: A Liberal Defense' (1981) 130 *University of Pennsylvania Law Review* 1581. In New South Wales, if at least half of the lots in a scheme are tenanted, the tenants can elect a representative to serve on the strata committee but the tenant representative does not have a vote: *SSMA 2015*, s 33.
10 H Easthope, B Randolph and S Judd, 'Governing the Compact City: The Role and Effectiveness of Strata Management' (City Futures Research Centre, 2012) 1.
11 Judith Yates, 'The Rhetoric and Reality of Housing Choice: The Role of Urban Consolidation' (2001) 19(4) *Urban Policy and Research* 491.
12 This is the case in some areas of the United States, like California and the Sunbelt states, where almost all new housing built in the last 40 years has been in homeowner associations: Evan McKenzie, *Privatopia: Homeowner Associations and the Rise of Residential Private Government* (Yale University Press, 1994) 11–12; Evan McKenzie, *Beyond Privatopia: Rethinking Residential Private Government* (Urban Institute Press, 2011) 75–7; Casey Perkins, 'Privatopia in Distress: The Impact of the Foreclosure Crisis on Homeowners' Associations' (2010) 10 *Nevada Law Journal* 561.
13 'Everything that is done has to be done somewhere. No one is free to perform an action unless there is somewhere he is free to perform it. Since we are embodied beings, we always have a location': Waldron, above n 2, 296.
14 B Gleeson, *Australian Heartlands: Making Space for Hope in the Suburbs* (Allen and Unwin, 2006) 80–102.
15 Gregory S Alexander and Eduardo M Peñalver, *An Introduction to Property Theory* (Cambridge University Press, 2012) 6.
16 Hanoch Dagan, 'The Limited Autonomy of Private Law' (2008) 56 *American Journal of Comparative Law* 809, 814.
17 Thomas W Merrill and Henry E Smith, 'The Morality of Property' (2007) 48 *William and Mary Law Review* 1849.
18 Dagan, 'The Limited Autonomy of Private Law', above n 16, 814–15.
19 Ibid. 817.
20 Carol M Rose, 'The Moral Subject of Property' (2007) 48 *William and Mary Law Review* 1897, 1900.

21 Gregory S Alexander *et al.*, 'A Statement of Progressive Property' (2008) 94 *Cornell Law Review* 743 (emphasis added).
22 Singer, *Entitlement*, above n 7, 215–16, emphasis added.
23 Joseph William Singer, 'Normative Methods for Lawyers' (2008) 56 *UCLA Law Review* 899, 904.
24 Singer, *Entitlement*, above n 7, 215.
25 Alexander and Peñalver are at the forefront of a new 'virtue ethics' school of property thought, which focuses on Aristotelian notions of human flourishing: Alexander and Peñalver, *An Introduction to Property Theory*, above n 15.
26 R H Coase, 'The Problem of Social Cost' (1960) 3 *Law and Economics* 1; R H Coase, 'Coase Theorem and the Empty Core: A Comment' (1981) 24 *Journal of Law and Economics* 18; Harold Demsetz, 'Toward a Theory of Property Rights II: The Competition between Private and Collective Ownership' (2002) 31 *Journal of Legal Studies* 65; Guido Melamed Calabresi and A Douglas, 'Property Rules, Liability Rules, and Inalienability: One View of the Cathedral' (1971) 85 *Harvard Law Review* 108; Thomas Merrill and Henry E W Smith, 'What Happened to Property in Law and Economics?' (2001) 111 *Yale Law Journal* 357.
27 Singer, *Entitlement*, above n 7, 117–30.
28 For further critique of preference theories, see Cass R Sunstein, 'Social Norms and Social Roles' (1996) 96 *Columbia Law Review* 903.
29 Singer, 'Democratic Estates', above n 6, 1054.
30 Singer, *Entitlement*, above n 7, 143 (emphasis added).
31 Ibid. 140–3, 167–71.
32 Douglass C North and Barry R Weingast, 'Constitutions and Commitment: The Evolution of Institutions Governing Public Choice in Seventeenth-Century England' (1989) 49(4) *Journal of Economic History* 803; A W B Simpson, *A History of the Land Law* (Clarendon Press, 2nd edn, 1986).
33 References to a free land market are not intended as references to an unregulated market. My argument is not underpinned with the economic rationalism or libertarianism so common in United States academic debates. By free land market I mean the kind of basic free market that we would expect to find in all democratic societies and which is absent in undemocratic countries like China or the Middle East: see, e.g. Katherine Wilhelm, 'Rethinking Property Rights in Urban China' (2004) 9 *UCLA Journal of International Law and Foreign Affairs* 227. The Dubai International Financial Centre has strata title laws (Strata Title DIFC Law No 5 or 2007), written by Emeritus Professor Peter Butt, and based on the New South Wales legislation. This law is only operative inside the Dubai International Financial Centre, and was needed because western citizens and business are less likely to invest in land markets that do not have the basic stability and protections to which they are accustomed.
34 In return for aiding the conquest of England, William made grants of land directly to the lords who came with him across the Channel and the local lords who agreed to accede to his authority. However, many land grants were fictional, that is, the Anglo-Saxon landowners colluded in the pretence that they owned their land by virtue of grants from William: Simpson, above n 32; Peter Butt, *Land Law* (Thomson, 5th edn, 2006) 66–7. This accounted for the almost universal inclusion of land in England within the feudal pyramid.
35 Butt, above n 34, 67–73. Most agriculture labour was performed by villeins within the manorial land system rather than the feudal system of free tenure. The obligations that villeins owed their lord were often characterised by uncertainty, being at the lord's discretion.
36 Peter Butt, *Land Law* (LBC, 3rd edn, 1996) 53.
37 Simpson, above n 32, 5; Michael A Heller, 'The Boundaries of Private Property' (1999) 108 *Yale Law Journal* 1163, 1171–2.

38 Simpson, above n 32, 208–9.
39 Heller, 'The Boundaries of Private Property, above n 3, 1166.
40 Charles Harpum, *Megarry and Wade: The Law of Real Property* (Sweet and Maxwell, 6th edn, 2000) 27–30.
41 The rule against perpetuities, in both its common law and statutory forms, essentially requires interests in land to vest within a generation, preventing landowners from determining ownership too long after their own deaths. While there have been calls for the abolition of the rule in recent times, it remains important to prevent 'dead hand control' and intergenerational inequity: see Brendan Edgeworth *et al.*, *Sackville & Neave Australian Property Law* (LexisNexis Butterworths, 9th edn, 2013) 657–62.
42 *Re Ellenborough Park* [1956] Ch 131; [1955] 3 All ER 667.
43 Thomas W Merrill and Henry E Smith, 'Optimal Standardization in the Law of Property: The Numerus Clausus Principle' (2000) 110 *Yale Law Journal* 1.
44 See discussion below, n 107.
45 Brendan Edgeworth, 'The *Numerus Clausus* Principle in Contemporary Australian Property Law' (2006) 32 *Monash University Law Review* 387.
46 Merrill and Smith, 'Optimal Standardization in the Law of Property', above n 43, 3–7.
47 *Keppell v. Bailey* (1834) 2 My & K 517; 39 ER 1042, 1048.
48 Merrill and Smith, 'Optimal Standardization in the Law of Property', above n 43, 13–14.
49 *Keppell v. Bailey* (1834) 2 My & K 517; 39 ER 1042, 1048, 1049.
50 *Hill v. Tupper* (1863) 2 H & C 121; 159 ER 51, 53. Rudden provides a list of other judicial pronouncements on the impermissibility of novel rights or 'fancies': Bernard Rudden, 'Economic Theory v. Property Law: The *Numerus Clausus* Problem' in John Eekelaar and John Bell (eds), *Oxford Essays on Jurisprudence* (Clarendon Press, 1987) 239, 244.
51 *Clos Farming Estates Pty Ltd (Rec and Man Appt)* v. *Easton* [2001] NSWSC 525.
52 Rose, 'The Moral Subject of Property', above n 20, 1917.
53 Rudden, above n 50, 241.
54 These interests are still of course permissible; they are just much less commonly used in liberal democracies where personal autonomy and freedom to deal with land as one pleases, are widely held values.
55 Singer questions the simplicity of property law that Merrill and Smith claim is the result of the *numerus clausus* principle. He points to the fact that the widespread use of homeowners associations (HOAs) means that buyers of land must search the voluminous covenants, conditions, and restrictions contained in the recorded declaration, as well as the governing rules of the association, to find out what their rights will be if they buy the property. Further, even if they discover all this detail, they are subject to the possibility that rules will change: Singer, 'Democratic Estates', above n 6, 1025–6. However, as this book demonstrates, HOA servitudes are an anomaly in general property law. They are the problematic exception that proves the general rule.
56 Edgeworth, 'The Numerus Clausus principle', above n 45, 394. See also Uriel Reichman, 'Toward a Unified Concept of Servitudes' (1981) 55 *Southern California Law Review* 1177, 1233, Chapter 3, 78.
57 Michael A Heller, 'Tragedy of the Anticommons: Property in the Transition from Marx to Markets' (1998) 111 *Harvard Law Review* 621.
58 Ibid.; Heller, 'The Boundaries of Private Property', above n 37.
59 Garrett Hardin, 'The Tragedy of the Commons' (1968) 168 *Science* 1243.
60 Heller, 'Tragedy of the Anticommons', above n 57, 624.
61 Heller, 'The Boundaries of Private Property', above n 37, 1165–6.
62 See Harpum, above n 40, 27–30.

63 For example, in New South Wales, a single s 88B instrument under the *Conveyancing Act 1919* (NSW), registered with plan of subdivision, will create hundreds of covenants benefiting and burdening individual lots in the subdivision.
64 Heller, 'The Boundaries of Private Property', above n 37, 1184. See also Ben W F Depoorter and Francesco Parisi, 'Fragmentation of Property Rights: A Functional Interpretation of the Law of Servitudes' (Working Paper No 284, John M Olin Center for Studies in Law, Economics and Public Policy, 22 May 2003) http://digitalcommons.law.yale.edu/lepp_papers/284/.
65 Heller, 'The Boundaries of Private Property, above n 37, 1184.
66 Ibid.
67 Ibid. 1165.
68 Ibid. 1192.
69 *Radaich* v. *Smith* (1959) 101 CLR 209 (exclusive possession); *Wilson* v. *Meudon Pty Ltd* [2005] NSWCA 448 (certainty of duration).
70 *Re Ellenborough Park* [1956] Ch 131; [1955] 3 All ER 667; *Clos Farming Estates Pty Ltd* v. *Easton* [2002] NSWCA 389; *Copeland* v. *Greenhalf* [1952] Ch 488; *Moncrieff* v. *Jamieson* [2007] UKHL 42; [2008] 4 All ER 752. See Michael Weir, 'Easements for Storage and Parking: A Time for a Rethink' in Lyria Bennett Moses, Brendan Edgeworth and Cathy Sherry (eds), *Property and Security: Selected Essays* (Lawbook Co, 2010) 249 for a critical analysis of these rules.
71 In *People ex rel Moloney* v. *Pullman's Palace-Car Co*, 175 Ill 125 (1898), the Supreme Court of Illinois ordered that a town built by the Pullman Palace Car Company be sold off, because a privately owned company town was 'incompatible with the theory and spirit of our institutions': cited in Singer, *Edges of the Field*, above n 1, 18. Other company towns were built in the United States, but courts were ever wary of the concept: *Marsh* v. *Alabama*, 326 US 501 (1945).
72 For example, although there is no doubt that bodies corporate in New South Wales can ban smoking inside people's homes, irrespective of whether smoke is affecting others (*Salerno* v. *Proprietors of Strata Plan No 42724* (1997) 8 BPR 15,457), the frustration and dispute around smoke drift in buildings suggests that many schemes do not seem to understand that they have this blanket power or are reluctant to use it. Bodies corporate seem to assume that, consistent with the principles of a liberal democracy (but in fact contrary to the strata legislation), they do not have the power to regulate activities inside other people's homes.
73 Rose, 'The Moral Subject of Property', above n 20, 1917.
74 Simpson, above n 32, 208–9.
75 The good faith standard that equity traditionally imposed on a mortgagee selling property was not onerous, facilitating a genuine but swift sale, moving the property back into the market: *Kennedy* v. *De Trafford* [1897] AC 180.
76 Richard Allen Epstein, *Simple Rules for a Complex World* (Harvard University Press, 1995).
77 John Stuart Mill, *On Liberty* (Longman, Roberts & Green, 4th edn, 1869) 1–15. See also Isaiah Berlin, 'Two Concepts of Liberty' in Henry Hardy (ed.), *Liberty* (Oxford University Press, 2002) 167.
78 It is impossible in this context to do any justice to the extensive scholarship critiquing classical liberalism. The concept of negative liberty has been rightly challenged as insufficient by Charles Taylor, John Rawls, Amartya Sen and Martha Nussbaum, to name a few. Taylor, for example, asks would we consider Albania a freer country than Britain because it has fewer traffic lights and citizens are thus routinely less interfered with as they negotiate traffic each day: C Taylor, 'What's Wrong with Negative Liberty?' in A Ryan (ed.), *The Idea of Freedom* (Oxford University Press, 1979) 175. Sen and Nussbaum have focused on 'capabilities' and the fact that without government intervention and social assistance, negative liberty simply means that many people will be free to starve. For a general account of critiques of

liberalism and references, see Gerald Gaus and Shane D Courtland, 'Liberalism' in Edward N Zalta (ed.), *Stanford Encyclopedia of Philosophy* (2011) http://plato.stanford.edu/entries/liberalism/. For more specific legal critiques of negative liberty, see, e.g. Cass R Sunstein, 'Legal Interference with Private Preferences' (1986) 53(4) *University of Chicago Law Review* 1129.

79 Martha C Nussbaum, 'Deliberation and Insight: Bloch v. Frischholz and the Chicago School of Judicial Behavior Essay' (2010) 77 *University of Chicago Law Review* 1139278 NY 248, 259–60 (1938).

80 Martha Nussbaum, 'Capabilities as Fundamental Entitlements: Sen and Social Justice' (2003) 9(2–3) *Feminist Economics* 33; Martha C Nussbaum, *Frontiers of Justice: Disability, Nationality, Species Membership (The Tanner Lectures on Human Values)* (Harvard University Press, 2006).

81 Alexander *et al.*, 'A Statement of Progressive Property', above n 21, 743.

82 Simpson, above n 32, 90–1.

83 Ibid. 90.

84 The statute *De Donis Conditionalibus 1285*, 13 Edw 1, c 1; see Butt, *Land Law* (5th edn), above n 34, 125–6.

85 Jane Austen, *Pride and Prejudice* (T Egerton, 1813) and ITV, *Downton Abbey*, 26 September 2010. In *Pride and Prejudice*, Mr Bennet wants to provide financially for his five daughters, but cannot because the estate is entailed to the 'odious Mr Collins'. In *Downton Abbey*, although the Earl of Grantham is happy to accept that his estate is entailed, only considering himself its custodian for his lifetime, the rest of his family, in particular the Dowager Countess, are appalled by the prospect of the estate passing to distant cousin Matthew, a Manchester solicitor.

86 Harpum, above n 40, 73–5.

87 For example, in New South Wales, fees tail were converted to fees simple by *Conveyancing Act 1919* (NSW) ss 19, 19A. *Law of Property Act 1925* (UK) s 1 allowed only fees simple or terms of years to exist at law.

88 Singer, *Entitlement*, above n 7, 147.

89 *Tulk* v. *Moxhay* (1848) 2 Ph 774; 41 ER 1143.

90 *Haywood* v. *Brunswick Permanent Benefit Building Society* [1881] 2 QB 403; *Austerberry* v. *Corporation of Oldham* (1885) 29 Ch D 750 (CA); *Rhone* v. *Stephens* [1994] 2 AC 310; [1994] 2 All ER 65; cf. *Neponsit Property Owners' Assoc Inc* v. *Emigrant Industrial Savings Bank*, 278 NY 248, 259–60 (1938).

91 *London & South Western Railway Co* v. *Gomm* (1882) 20 Ch D 562.

92 Simpson, above n 32, 257. As O'Connor has argued, a jurisprudential justification for restrictive covenants is hard to pin down: Pamela O'Connor, 'Covenants as Regulation' (2011) 1 *Property Law Review* 145, 147–52. In relation to the public/private tensions produced by restrictive covenants, see also Michael Weir, 'Public and Private Planning – Implications for Legal Processes' (1994) 11 *Environmental and Planning Law Journal* 137.

93 Merrill and Smith, 'Optimal Standardization in the Law of Property', above n 43, 17. See *Haywood* v. *Brunswick Permanent Benefit Building Society* (1881) 8 QBD 403; *Austerberry* v. *Corporation of Oldham* (1885) 29 Ch D 750 (CA); *Pirie* v. *Registrar General* (1962) 109 CLR 619. See also Victorian Law Reform Commission, 'Easements and Covenants: Final Report', Report No 22 (2010) 74, 84.

94 Merrill and Smith, 'Optimal Standardization in the Law of Property', above n 43, 23.

95 McKenzie, *Privatopia*, above n 12.

96 Singer, 'Democratic Estates', above n 6, 1047.

97 See Singer's discussion of Wesley Hohfeld's 1913 and 1917 articles, 'Some Fundamental Legal Conceptions as Applied in Judicial Reasoning' (1913) 23 *Yale Law Journal* 16; (1917) 26 *Yale Law Journal* 710, which argued that rights constituted jural relations: Singer, *Entitlement*, above n 7, 131–4.

98 See discussion of Richard A Epstein, 'Notice and Freedom of Contract in the Law of Servitudes Comments' (1981) 55 *Southern California Law Review* 1353 in Chapter 3, 80–2.
99 Singer, 'Democratic Estates', above n 6, 1048.
100 Singer, *Edges of the Field*, above n 1, 25–30; Singer, *Entitlement*, above n 7, 140–78.
101 Locke said that man could have property in whatever he mixed his own labour with, provided he left 'enough and as good in common … to others': John Locke, 'Second Treatise of Government' in Thomas Hobbes and John Locke (eds), *On Sovereignty* (Collector's Library of Essential Thinkers, 2005) 396. This is known as the 'Lockean proviso'.
102 Underkuffler, 'Property: A Special Right', above n 4.
103 Ibid. 1043.
104 Laura Underkuffler, 'A Theoretical Approach: The Lens of Progressive Property' in Susan Bright and Sarah Blandy (eds), *Researching Property Law* (Palgrave, 2016) 11–23, 14.
105 Joseph William Singer, 'Corporate Responsibility in a Free and Democratic Society' (2008) 58 *Case Western Reserve Law Review* 1031, 1036–8.
106 Singer develops this argument more fully in Joseph William Singer, 'How Property Norms Construct the Externalities of Ownership' in Gregory Alexander and Eduardo Peñalver (eds), *Property and Community* (Oxford University Press, 2010) 57. Of course, strata and community title associations do regulate things like house colour, which is precisely why their by-laws sometimes trouble us. This will be explored more fully in Chapter 5.
107 Singer relies on the scholarship of Laura Underkuffler here. Underkuffler argues that we have a 'common' conception of property as absolute rights and an 'operative' conception that involves complexity: Laura S Underkuffler, *The Idea of Property: Its Meaning and Power* (Oxford University Press, 2003), cited in Singer, 'Democratic Estates', above n 6, 1027.
108 Singer, *Entitlement*, above n 7, 30–1. Singer defines externalities as 'effects on others not directly involved in a transaction or act': Singer, 'How Property Norms Construct the Externalities of Ownership', above n 106, 61.
109 Singer, 'Democratic Estates', above n 6, 1028. A good example of this is the *in personam* exception in Torrens registration systems. This is the judicially crafted exception to indefeasibility of title, which avoids the guillotine effect of registration in appropriate circumstances. The seminal expression of the exception is found in *Frazer* v. *Walker* [1967] 1 AC 569, 585, where Lord Wilberforce said that an indefeasibility provision 'in no way denies the right of a plaintiff to bring against a registered proprietor a claim *in personam*, founded in law or equity for such relief as a court acting *in personam* may grant'. See Lyria Bennett Moses and Brendan Edgeworth, 'Taking it Personally: Ebb and Flow in the Torrens System's in Personam Exception to Indefeasibility' (2013) 35 *Sydney Law Review* 1.
110 Singer, 'Democratic Estates', above n 6, 1049–50.
111 Singer, *Entitlement*, above n 7, 137–8; see also, Singer, *Edges of the Field*, above n 1, 20.
112 Singer, *Entitlement*, above n 7, 143.
113 *Buchanan* v. *Warley* 245 US 60 (1917).
114 McKenzie, *Privatopia*, above n 12, 56–78; James A Kushner, 'Apartheid in America: An Historical and Legal Analysis of Contemporary Racial Residential Segregation in the United States' (1979) 22 *Howard Law Journal* 547; Stephanie M Stern, 'The Dark Side of Town: The Social Capital Revolution in Residential Property' (Working Paper, Chicago-Kent College of Law, April 2013) http://papers.ssrn.com/sol3/papers.cfm?abstract_id=2243134 or http://dx.doi.org/10.2139/ssrn.2243134.
115 McKenzie, *Privatopia*, above n 12, 68–74.

116 *Shelley* v. *Kraemer* 334 US 1 (1948).

117 The equal protection clause in the 14th Amendment to the *United States Constitution* provides that no state shall deny persons within its jurisdiction equal protection of the law.

118 Martin Luther King, Jr, *I Have a Dream* (1963) Archives www.archives.gov/press/exhibits/dream-speech.pdf.

119 Explicit racial segregation in American schools was outlawed by the Supreme Court in the landmark case *Brown v. Board of Education of Topeka*, 347 US 483 (1954).

120 Gary Orfield and Chungmei Lee, 'Why Segregation Matters: Poverty and Educational Inequality' (Civil Rights Project, Harvard University, 2005). The Civil Rights Project, originally at Harvard University and now at UCLA, has been researching racial segregation in American schools for almost two decades. There are multiple reports available at http://civilrightsproject.ucla.edu/research/k-12-education/integration-and-diversity. Problems in American schools are exacerbated by the fact that schools are typically funded by local property taxes, in contrast to Australia, where they are funded by state and Federal governments.

121 Stephen E Asher, 'Interdistrict Remedies for Segregated Schools' (1979) 79 *Columbia Law Review* 1170.

122 Jonathan Kozol, *The Shame of the Nation: The Restoration of Apartheid Schooling in America* (Crown Publishers, 2005).

123 Singer, *Entitlement*, above n 7, 174. Singer contrasts his view with that of Robert Nozick, *Anarchy, State, and Utopia* (Basic Books, 1974) and Epstein, *Simple Rules for a Complex World*, above n 76, who judge the legitimacy of private property by reference to initial acquisition and subsequent transfer: Singer, *Entitlement*, above n 7, 171–3.

124 *Casuarina Rec Club Pty Ltd* v. *Owners – Strata Plan 77971* (2011) 80 NSWLR 711.

125 This question has been the subject of litigation in the United States in *Committee for a Better Twin Rivers* v. *Twin Rivers Homeowners' Assoc*, 890 A 2d 947 (NJ Sp Ct App Div, 2006), revd 192 NJ 344 (2007).

126 Singer, 'Democratic Estates', above n 6, 1058.

127 Ibid. 1059.

3 United States law and homeowner association practice

Introduction

Chapter 1 flagged the significance of United States' property law and academic debate to this book. A divergence in Australian and American property law in the mid-nineteenth century meant that United States developers were able to impose positive obligations on freehold land, which facilitated the provision of private facilities and the rise of private residential associations. Despite physical similarities in suburban development in both countries, the continued limitation of freehold covenants to restrictions and not positive obligations meant that Australia did not have a corresponding development of private facilities and residential associations. However, the passage of the strata titles legislation created the statutory equivalent of positive obligations on freehold land in levies and by-laws, as well as the statutory equivalent of private residential associations in bodies corporate. The enactment of the group or community titles legislation extended both of these legal developments to horizontal subdivisions. All Australian states now have the equivalent of United States' homeowner associations (HOAs) in both high- and low-rise forms.

This radical legislative change came about with limited judicial and academic consideration. As noted in Chapter 1, it is not entirely surprising that the legislative empowerment of private citizens to write laws for their neighbours raised few questions when the power related to single, relatively small buildings and groups of residents, rather than large swathes of land. Further, the initial legislation, as well as subsequent enactments, were often driven by developers and their legal advisers, who were justifiably more concerned with creating a marketable product than engaging in sociological, political or jurisprudential analysis. Also, it would be fair to say that parliamentarians, like the 'Minister for Everything', the Honourable Russell Hinze, may not have fully appreciated the finer points of property law they were changing with new legislation.[1] Finally, because the amendments to the law were made by legislation and not the gradual development of the common law, there has been less opportunity for judicial analysis of the changes. Higher courts have the ultimate responsibility for the maintenance and progressive amendment of the common law,[2] unlike legislation which remains the responsibility of the legislature and which judges must apply according to its terms.

In contrast, in the United States, HOAs, while now regulated by legislation, are not the product of it. They depend on common law and equitable doctrines for their existence and have been the subject of over 150 years of judicial consideration. HOAs were also used extensively in low-rise residential property development from the 1960s onwards,[3] so that by the early 1980s, the academics who turned their attention to them had considerable material, both judicial and physical, to study. This material is invaluable for a proper appreciation of the legal, social and political complexity of strata and community title. Current academic work on strata and community title in Australia is limited, and owing to the differences in relative populations of citizens, judges and academics, we are unlikely to ever have an equivalent body of Australian work. As a result, a significant part of this book's attempt to fill our research gap is to document *the breadth and depth* of United States' research. A familiarity with debates between great property scholars like Robert Ellickson, Carol Rose, Richard Epstein and Gregory Alexander on the United States' equivalent of strata and community title, will give future Australian deliberations, (academic, policy and ideally judicial), a depth that they have hitherto lacked. The work of HOA scholars like Susan French and Evan McKenzie, who have observed the operation of thousands of private residential communities over decades, has produced insights from which Australian researchers and policymakers could learn. There is no need to reinvent the wheel. While there are social and political differences between the United States and Australia, we share a common democratic tradition, and as a result, United States discussions of the limits of small group autonomy in the liberal state, for example, are equally applicable here. Inter-generational tensions in relation to land use are universal. As noted in Chapter 1, Australian developments, in particular community title, are often copied from the United States. For these reasons, close attention will be paid to the United States law and literature; this chapter is intentionally substantial.

For the sake of reader understanding, the following should be noted.[4] In the United States, HOAs are also called common interest developments (CIDs), common interest communities, (CICs) or planned unit developments (PUDs). They are low-rise master planned estates and are the equivalent of group or community titles. Condominiums are generally apartments, the equivalent of strata title.[5]

The legal underpinnings of United States HOAs and condominiums are servitude regimes, commonly referred to as 'C, C & Rs' (covenants, conditions and restrictions). By a 'declaration' or other appropriate instrument, a developer imposes a range of servitudes on subdivided land. Servitudes include:

- Easements, which will give all individual owners the right to use common areas and facilities;
- Restrictive covenants, which limit use of individual homes; and
- Positive or 'real' covenants, which, inter alia, require all owners to pay assessments for the upkeep of the community and facilities.

Servitudes run with the land and bind all future owners. The declaration will also set the number of votes allocated to each unit and the assessments (levies) to be paid. Assessments are usually calculated on lot size rather than market value. Declarations are registered.

The developer also creates a mandatory-membership association, which may be an incorporated association, a not-for-profit, mutual benefit corporation or even a trust. The powers of, and law governing the association will vary depending on the legal structure chosen by the developer. A set of by-laws will also be written to deal with association governance. Hyatt and French say that

> [o]ne way to think about the relationship between the declaration and the bylaws is to regard the declaration as the constitution of the community and the bylaws as the statutes. Fundamental provisions dealing with property rights and voting rights belong in the declaration; operational provisions belong in the bylaws.[6]

Surprisingly, by-laws may or may not be registered. Hyatt and French note that while recording will notify all purchasers of the by-laws' content, it will make it more difficult to alter them.[7]

A 'plat map' is included in the declaration, showing the boundaries of the development and the location of the buildings, as well as the roads and easements. Condominium developments will also have floor plans showing the horizontal and vertical boundaries of the units.

Because United States developments are so heavily dependent on the common law, legal regulation and review is affected by the *Restatement (Third) of Property, Servitudes* (2000), as well as state condominium and/or HOA or CIC legislation. United States *Restatements* of various areas of private law are produced by the American Law Institute (ALI), an independent organisation made up of judges, academics and practitioners. The *Restatements* seek to clarify and rationalise the law that has emerged from the application of already complex common law in multiple state jurisdictions. Professor Tarlock said that the

> ALI Restatements have been extremely influential because they bring order and uniformity to a welter of cases in the 50 states.... Modern judges have less time and patience to synthesize long lines of cases. General principles, embodied in Restatements, allow them to decide cases.[8]

The Uniform Law Commission, another independent organisation of legal experts, attempts to do the same in the area of statute law by drafting uniform legislation available for state enactment. The Commission has drafted a *Common Interest Ownership Act* (*1982*, *1994*, *2008*), a *Condominium Act* and a *Planned Community Act*, that have been taken up inconsistently throughout the Union. Some states like California[9] have their own legislation. Neither the *Restatements* nor Uniform Statutes actually make law unless they are adopted by legislatures or judges, but they are an authoritative guide for legal concepts and policy issues that judges, academics and respected practitioners believe are central to an area of law in question.

In comparison to Australia, the legal structures and regulation of apartments and planned communities in the United States are extraordinarily varied; this is because many of the titling issues that have to be addressed by the C, C & Rs and other documents are automatically taken care of by our legislation, which applies uniformly to all multi-owned communities.[10] However, the question of what kind of land-use and behavioural regulation can and should be included in C, C & Rs and rules of associations is exactly the same as the question of what kinds of regulation can and should be included in strata and community by-laws. Whether in the United States or Australia, those questions address the complex and potentially conflicting claims of multiple present owners and their multiple predecessors in title, in relation to the individually and socially valuable limited resource that is land. For that reason, when moving through this Chapter, readers are asked to remember that the terms 'C, C & Rs' or 'servitudes' are synonymous with the Australian term 'by-laws'.

United States debates on covenant law reform

The United States law on covenants, servitudes and easements was notoriously complex. A 1974 general property law text described it thus:

> The law in this area is an unspeakable quagmire. The intrepid soul who ventures into this formidable wilderness never emerges unscarred. Some, the smarter ones, quickly turn back and take up something easier like the income taxation of trusts and estates. Others, having lost their way, plunge on and after weeks of effort emerge not far from where they began, clearly the worse for wear. On looking back they see the trail they thought they broke obscured with foul smelling waters and noxious weeds. Few willingly take up the challenge again.[11]

Despite this legal complexity, for a range of social and economic reasons discussed later in this chapter, covenants, servitudes and easements had increasingly been used in modern residential and commercial developments. As a result, United States academics began to consider the case for simplification. They started by attempting to identify the social and economic policy underlying the law to determine if there was any functional importance to the technical rules. In a 1982 Symposium published in the *Southern California Law Review*, Professors Reichman and French were the 'intrepid souls' who ventured into the covenant wilderness, and from the safety of the close, Professors Epstein and Rose commented on their findings.

Reichman's article 'Toward a Unified Concept of Servitudes' provided a detailed history of the development of real covenants, equitable servitudes and easements in American law.[12] He emphasised that tracing the development of servitude law provided a 'vivid picture of dynamic economic and behavioural patterns in American society'[13] and that the bulk of servitude law was associated with residential planning, dating from the first half of the nineteenth century. It was at this time that developers began to explore the possibilities of private communities, with private rules and common property, which inspired the common interest communities in which so many Americans (and increasing number of Australians) live today.

Reichman identified the critical difference between English and American law on freehold covenants, noted in Chapter 1. From the early nineteenth century, American law did not share England's reticence to allow either restrictions *or positive obligations* on land to run with the fee. Reichman attributes this to the greater availability of land and thus a lesser concern with the cluttering of titles with encumbrances, as well as an early recording system, which allowed American purchasers to discover negative or affirmative burdens that were not always obvious from an inspection of land.[14]

By the mid-nineteenth century, America had developed a concept of 'real covenants' which ran with the fee and were enforceable at law so long as they 'touched and concerned' the land. Reichman explained that in an 1852 commentary on *Spencer's Case*,[15] American Judge Hare anomalously decided that privity of estate existed in the grantor/grantee relationship, not merely the landlord/tenant relationship, as was and still is the case in Anglo-Australian law. As a result, in America, agreements between a grantor and a grantee of the fee, whether positive or negative in nature, would run with the land so long as they touched and concerned it.[16] The phrase 'touch and concern' has proved extremely difficult to define, but one judicial definition suggests that the

> greater the degree the covenant imposes obligations unique to the covenanters, which cannot exist independently of them, the less likely the covenant ... touches and concerns the land. Conversely, the greater the effect of the covenant on the land itself, without regard for who owns it, the more likely it will be binding on successor owners.[17]

Reichman argued that Judge Hare's development of the law dovetailed with the nineteenth-century enthusiasm for freedom of contract. Real covenants 'offered a potential mechanism whereby the notion of freedom of contract might be injected into the law of real estate' and the 'right to mold legal relations freely and to fully vindicate bargained-for expectations became imprinted throughout the law'.[18] A century and a half later, the wisdom of this application of freedom of contract theory to land obligations remains a serious question, and one of the primary subjects of this book.

Following the decision of *Tulk* v. *Moxhay*,[19] which demonstrated a similar enthusiasm for blurring contract and property law by holding a purchaser to a contractual promise of a predecessor in title of which he had notice, the United States adopted the doctrine of 'equitable servitudes'. As Reichman noted, English courts later restricted the doctrine to negative covenants[20] and promises that affected the land.[21] The United States eventually adopted the latter but not the former, presumably because the existence of real covenants, which allowed affirmative obligations to attach to land, made any restriction to negative obligations in equitable servitudes otiose.

Reichman concluded that by the twentieth century, real covenants, equitable servitudes and easements could be used interchangeably to achieve almost identical results, binding successors in title to the original covenantor to do something on their own land, refrain from doing something on their land or allow another to do something on their land. Further, with the fusion of law and equity, real covenants could be specifically enforced. The only difference between equitable servitudes and real covenants was the privity of estate rule. Reichman argued that the unification of all three servitude rules would allow easier application of the law and clarify the law's underlying policy considerations.

Reichman identified the central underlying policy consideration of servitudes to be the efficient use of land.[22] Rather than needing to acquire complete possession of another's land to prevent or compel activity on it, a grantee of a servitude could acquire certain sticks in the servitude grantor's 'bundle of rights' that satisfied their own land-use needs. The grantor permanently gave up certain rights which would be replaced with an obligation to do or not do certain acts and he or she could not change this by transferring title.

Reichman argued that the theoretical underpinning of servitudes in the efficient use of land had normative implications. Only those promises that actually related to land, doctrinally defined as 'touch and concern', were traditionally enforced. Personal promises that did not 'touch and concern' the land did not run with the land binding assignees. What might seem a fussy technicality was actually an important limitation on parties' freedom to negotiate agreements that might frustrate efficient land use.[23] Reichman considered it a limitation that remained useful. Idiosyncratic personal agreements were potentially inefficient. Further, he noted that consensual termination of them could be hindered by prohibitive transactions costs: for example, a benefited owner may 'hold out' and refuse to release a servitude for anything other than an exorbitant fee. While it might be easy to agree on restrictions and obligations, getting rid of them is not so simple.

Reichman acknowledged a further significance to the touch and concern doctrine:

> Private property is sanctioned by society not only to promote efficiency, but also to safeguard individual freedom. Servitudes are a kind of private legislation affecting a line of future owners. *Limiting such 'legislative powers' to an objective purpose of land planning eliminates the possibility of creating modern variations of feudal serfdom.* There might be nothing objectionable in personal agreements concerning personal labor, adherence to ideologically prescribed modes of behaviour, or promises to buy from a certain supplier. When such obligations, however, become permanently enforced against *an ever-changing group of owners*, the matter acquires different dimensions.[24]

In 'Democratic Estates', Singer makes a similar analogy between HOAs and feudalism, arguing that as 'covenants and condominium rules can change over time as the homeowners association chooses ... [o]wners of such homes are subject to the whim of their neighbors, just as feudal tenants were subject to the whim of their lords'.[25]

Reichman's identification of the economic and social function of the 'touch and concern' doctrine supports the argument made in Chapter 2, that modern property law promotes democratic norms in the economic and social sphere. While the doctrine of real covenants constituted a breach of the *numerus clausus* principle by potentially allowing idiosyncratic contractual agreements to attach to land, the 'touch and concern' doctrine sprang up as a new 'boundary rule', limiting that freedom of contract to a (fuzzily) defined box acceptable to property law.

After surveying the exceptional technicality of the sometimes subtle differences between easements, real covenants and equitable servitudes in American law, Reichman concluded that for the sake of the thousands of Americans who lived in low- and high-rise developments governed by servitude regimes, all three should be fused into one concept. As long as the *functional purpose and limitations* of servitudes were borne in mind, there were no insurmountable doctrinal obstacles to the achievement of that end.

Susan French, in the article immediately following Reichman's, 'Towards a Modern Law of Servitudes: Reweaving the Ancient Strands'[26] reached the same conclusion, but like Reichman, acknowledged the important policy reasons behind the terrifying technicality of the law. Easements, real covenants and servitudes are all characterised by one critical feature: permanence. Their ability to run with the land and bind successors in title to the original agreement is their value, but with this value comes danger. Servitudes freeze land use, prevent economically productive development, impose unreasonable burdens that depress land values and impose 'significant dead hand controls over land'.[27] French argued that it was their simultaneous benefits and danger that led to judicial ambivalence, which in turn produced such complexity. French noted that while it would be tempting to discard the entire messy edifice of servitudes, this would lead to a loss of valuable lessons from the past.

French argued that when a functional analysis was made of the technical rules of servitudes, two underlying concerns were revealed: economic productivity and fairness. For example, the technical rule that the benefit of an easement or equitable servitude be 'attached' to the dominant tenement promoted fairness and productivity. Servitudes that were 'attached' to neighbouring land were discoverable by inquiry in the neighbourhood, while burdens in gross were far more time-consuming and difficult to discover as the benefited party may not be obvious.[28] It was both unfair and costly for purchasers to be bound by interests they could not reasonably ascertain. Similarly, the 'touch and concern' doctrine ensured that 'parties [would] be bound only to the obligations which a reasonable purchaser would expect to have incurred, and [would] acquire only the benefits which a reasonable purchaser would expect to have gotten',[29] as opposed to the potentially limitless idiosyncratic benefits and burdens to which the original grantor and grantee might have agreed. Finally, the 'changed condition doctrine' allowed courts to refuse to enforce equitable servitudes that had become obsolete or unreasonably burdensome. French argued that courts struck a balance between the stability that servitudes could promote and the need for development and change of use in land. French concluded that the focus on fairness and productivity provided a reasonable structure for the operation of private land agreements, but that overlapping doctrines and obsolete terminology needed to change.[30]

French argued that a primary reason early law restricted the creation of servitudes was its lack of ability to terminate them. As modern American law now had the potential to terminate servitudes, restrictions on their creation were no longer necessary. She concluded that agreements creating servitudes should be treated like other agreements and the law should give effect to the parties' intentions.[31]

French outlined two possible methods for termination or modification of servitudes. The first drew on already common subdivision practice and the *Uniform Planned Community Act*,[32] both of which provided for the termination of servitudes by the less than unanimous agreement of the benefited parties, that is, by vote. French argued that a statute providing for the modification or termination of servitudes by the majority allowed stability in land-use arrangements, while avoiding high negotiation costs potentially incurred to remove obsolete servitudes. While the point French makes is important, particularly as the provisions she is discussing are equivalent to the power Australian bodies corporate have to vote to change by-laws, Heller's observations on fragmentation in Chapter 2 must be borne in mind. He said that fragmentation 'may operate as a one-way ratchet: Because of high transaction costs, strategic behaviours, and cognitive biases, people may find it easier to divide property than to recombine it'.[33] Heller argued that CICs (HOAs) had done exactly that, dividing property between hundreds of benefited parties who were then reluctant to give up any rights. In effect, even with modification and termination provisions which are more efficient than market buy-outs, servitude regimes can give people 'too much power to lock resources into low-value uses'.[34] While in theory restrictions and obligations can be removed by majority vote, in practice, owners are often inclined to cling to the status quo.[35]

The second method French posited was judicial termination through an expanded changed condition doctrine. This would allow for the termination of servitudes that were obsolete, unduly burdensome or economically wasteful or unreasonable. Judicial termination, along with judicial declaration of servitude invalidity ab initio, became central elements of the *Restatement (Third) of Property: Servitudes* (2000), of which French was the ultimate author. This will be discussed in more detail below.

Reichman and French's pieces were followed by comments from Professors Epstein and Rose. Professor Epstein has since become an extremely prolific proponent of minimal legal regulation,[36] and his approach to servitude law was consistent with that trajectory. Epstein argued that while he agreed with Reichman and French's conclusion that the law of servitudes could be radically simplified, he would go further. Epstein asserted that as long as public recording of covenants occurred giving notice to third parties acquiring interests in the land, there was no need for any other regulation, either judicial or legislative. Epstein said, '[m]y thesis is simple: with notice secured by recordation, freedom of contract should control'.[37] Epstein's 'thesis' is particularly relevant to Australia, where by-laws are always recorded on the publicly accessible Torrens registers, providing purchasers and tenants with notice of by-law content.[38] In contrast, the United States does not have a Torrens system – a system of *title by registration*, not registration of title[39] – and thus registration is more haphazard.

With recordation and notice as his assumed starting point, Epstein took issue with Reichman's desire to maintain the 'touch and concern' doctrine. Epstein said that insistence that personal obligations may frustrate the objectives of a private land-holding system, presupposed that we had a 'collective vision' of what that system was supposed to do. Epstein argued that:

> one traditional argument for both freedom of contract and private property is that they define domains in which individuals may establish both the means and the ends for themselves, to pursue as they see fit (*so long as they do not infringe upon the rights of third parties*) [emphasis added]. Private property is an institution that fosters individualized, if not eccentric, preferences; it does not stamp them out. We may not understand why property owners want certain obligations to run with the land, but as it is *their* land, not ours, some very strong reason should be advanced before our intentions are allowed to control [emphasis in original].[40]

Further, Epstein saw no reason for courts to eradicate servitudes at a later date with the changed condition doctrine, saying that 'when disputes arise there is no apparent reason to consider judicial coercion superior to consensual renegotiation'.[41]

That covenants should be treated in the same way as contracts with individual freedom as the guiding principle has been one of the central strands of covenant scholarship in the past 25 years and remains deeply influential, both in the United States, and with respect to by-laws, in Australia.[42] Epstein's argument has some validity; voluntary agreements are good and property law's enforcement of them has led to the individual autonomy and freedom that we now associate with private property.[43] But, as we saw in the last chapter, voluntary agreements are not always benign and they do not only affect the parties to the transaction; they affect the wider community, particularly if agreements relate to the limited resource of land, and many other individuals make that same agreement. As a result, 'we cannot conclude that a particular set of property rules or institutions is acceptable unless we attend to the systemic effects [as opposed to merely the individual effects] of exercising those property rights'.[44] The most obvious example is that of racially restrictive covenants; agreements that were voluntarily made but which harmed millions of other people.

Despite the libertarian stance he purports to take, Epstein recognises the necessary limits on individual control of land with the phrases, 'so long as they do not infringe upon the rights of third parties' and 'some very strong reason should be advanced before *our* intentions are allowed to control'. Epstein accepts that while freedom of contract may be an essential norm, it has limits. At some point we must draw a line in the sand. Epstein's caveat reveals that the *real* debate and the truly difficult question, is not *whether* we draw a line, but *where* we draw it; that is, what constitutes an 'infringement' on the rights of third parties and what are the 'very strong reasons' that would justify 'our' (that is, the wider community's) desires prevailing over individuals'. In reality, 'freedom of contract' is a furphy, designed to push the line in the sand as far away as possible, but no one, including Epstein, seriously argues that it should not exist at all. He may disagree that the touch and concern doctrine is the appropriate place to draw the line (that is, a line between land-related and personal obligations, the former being permissible, the latter not), but there is some point at which he would draw it. Epstein's thesis is not as 'simple' as he claims. Similarly, while Australian legislatures give private citizens extremely expansive powers to write by-laws, they have all enacted express prohibitions on particular by-laws, evidencing their 'line in the sand'. Despite this, as we will see in later chapters, like Epstein, Australian legislatures and the judiciary continually default to the simplistic freedom of contract position, which is ultimately untenable. As we saw in Chapter 2, at some point, collective decisions about the legitimacy of property entitlements is unavoidable, and messy as this may be, rational choice is only possible through 'reasoned deliberation'.[45] There is 'no baseline free from judgment'[46] in property law, and we have 'no alternative but to make arguments that elaborate fundamental human values and that express our considered commitments to judgments about morality and justice'.[47] We cannot, (and as a matter of fact, do not), abdicate all responsibility for property law by simply allowing individuals to determine its content through 'voluntary' agreements.

Professor Rose's commentary followed Epstein's. Now an equally well-known and respected property scholar, albeit of a more progressive persuasion,[48] Rose made some important comments about the Symposium's aim to instigate the modernisation and simplification of property law, that are directly relevant to novel, modernising legislation like strata and community title Acts.

Rose acknowledged that it was sensible to root out the *function* of old rules, as Reichman and French had done, because the old doctrines balanced the competing expectations of successive generations. However, she warned that in 'devising new strategies in an old war among generations of land owners … *we should not be overly surprised if we wind up in some of the same old trenches*'.[49] This relates to the argument made by Singer and Underkuffler, noted in the preceding chapter: the technicality of property law serves an important function of mediating externalities produced by the exercise of property rights. In the context of servitudes, the potential externalities are imposed on future generations by predecessors in title, and by individual vendors and purchasers on the wider community. Simplifying the language of the common law doctrines cannot eradicate the inevitable tensions that will exist between those people in relation to the limited and essential resource that is land. Exactly the same point can be made about strata and community title. Amending the common law doctrines that limited regulation of freehold land, by the enactment of modern statutes, will not make the age-old tensions and difficult *choices* about land and people go away. This will be developed as a theme of the book in Chapters 4 to 6, which focus on Australian states' by-law-making powers.

Rose noted that the voluntary acceptance of servitudes by new owners was a central element of old doctrines, as assent justified binding later generations with the agreements of their predecessors. She suggested that voluntary agreement was likely to be even more important in contemporary society, which aspires 'to conduct its affairs on the basis of free choice'.[50] She acknowledged that recording systems were a significant solution to the problem of voluntary assent. However, Rose raised problems that were already becoming a source of significant debate in the United States among academics, practitioners and judges, and which are becoming a source of debate in Australia. In HOAs, a purchaser may consent to abide by all of the rules in force from time to time, but can this include consent to *any* rule the association dreams up in the future? Surely, Rose said, only rules that are 'reasonable'.[51] Rose suggested that 'reasonable' was what a person could have 'expected'; for example, a rule for the maintenance of the community. She pointed out that this could just as easily be stated as a rule that 'touches and concerns' the land; that is, although the rule has a different label, we have wound up in the same old trench. Further, she noted that in interpreting the word 'reasonable', a court may have to resort to old distinctions between servitudes that (1) let someone else do something on one's land (an easement), (2) restrict someone from doing something on their own land (an equitable servitude), or (3) compel someone to do something on their own land (a real covenant).

So, (1) having to allow someone to come on to land to repair an electricity system and (2) not being allowed to put in a wading pool without architectural committee approval might both seem reasonable, while (3) compelling someone to keep a Doberman dog may not. Even if they are given one simplified label, 'servitude', that will not obliterate the distinctions between them or make the most problematic of them – the positive obligation – any less so. In short, Rose is arguing that we can attempt to dismantle the technicality but we will still find the hard questions about people and land, freedom and compulsion lurking underneath.

The 1982 Symposium brought out some critical issues raised by enduring regulation of land through property law by private citizens. These included that:

1 Enduring restrictions and obligations affecting multiple land titles are 'a kind of private legislation affecting a future line of owners'.
2 We could say that so long as new owners have notice of those restrictions and obligations, this gives no cause for concern because purchasers have voluntarily consented.

3 Alternatively, we can acknowledge that as owners are self-interested and can agree to restrictions and obligations that,

   a actively harm others ab initio (for example, racially restrictive covenants);
   b harm others after the passage of time (economically and socially outdated covenants or personal covenants that only ever related to the original parties and not the land); or
   c allow for subsequent *change* of rules to which purchasers have not consented;

   property law should control both the creation and ongoing operation of restrictions and obligations on land.
4 Property law has always chosen option 3, not option 2, even though that function of control might have been obscured by the technicality of traditional rules.
5 Even if those rules are simplified, the underlying function of determining what are and are not acceptable land-use rules between private citizens will remain just as difficult.

All of these issues are raised just as squarely by strata and community by-laws as they are by United States servitudes, by-laws simply being the statutory equivalent of covenants on freehold land.

The limits of group autonomy in the democratic state

In the same period that Reichman and French were wading through the 'unspeakable quagmire' of United States covenant law, other scholars were beginning to address the broader political dimensions of communities created with C, C & Rs, (although the technical rules were of course themselves political). These scholarly discussions all bordered on or actively engaged with much larger questions relating to democracy, communitarianism, liberalism and libertarianism. In one sense, the scholarship discussed in this section is merely a subset of the debate on the adequacy of classical liberalism. However, discussion will be restricted to writing with a central HOA focus.

The first group of articles, written in the 1980s, focused on the question of how far small groups should be permitted to secede from the wider polity in liberal democracies. This is a fascinating question raised by alternative communities like the Amish, the Shakers, Hasidic Jewish and other religious groups, communes and eco-communities. Owing to our divergent history of settlement, the United States offers much richer fodder for exploration of the issues at stake than Australia. The initial European 'settlers' in America were the Pilgrim Fathers, who were fuelled by a desire to escape mainstream Europe and establish a society ordered by their own religious values; many United States communities continued this separatist tradition. In contrast, arriving at his Majesty's pleasure, the original European 'settlers' in Australia were somewhat less intentional in their plans. While Australia has had its share of communes and utopian visionaries, the desire to secede into small, autonomous groups free from central government control and ruled by their own values, does not permeate the Australian psyche and society in the same way as it does in the United States. Nonetheless, questions about small group autonomy still arise in Australia. For example, should private schools be required to teach sex education? Can Islamic schools limit girls' participation in sport? Can clubs exclude people on the basis of ethnicity or ban all entrants from wearing anything on their head? And in the context of this book, can strata and community title schemes ban sustainable building practices, children or pets? In other words, are their limits to the freedom we grant groups to operate inconsistently with wider community values?

Owing to their nation's history, academics in the United States were quick to identify the separatist potential of HOAs and the implications for democratic society. As this is a legal text, the work of two prominent property theorists, Robert Ellickson and Gregory Alexander, will be the focus of discussion in this section.[52] Ellickson and Alexander were particularly interested in determining the limits that could be placed on HOA autonomy by judicial review (and potential invalidation) of their founding documents and subsequent actions.

In response to an article by Gerald Frug which identified cities and business corporations as the only significant candidates for decentralised power,[53] Robert Ellickson argued that HOAs, while 'relatively exotic' 20 years earlier in the 1960s, now outnumbered cities and were an ideal model for decentralised power.[54] Ellickson argued that HOAs had much in common with cities: they had a territorially defined area, they regulated and 'taxed' (through levies or assessments), they had majoritarian and representative procedures and they even had a quasi-power of eminent domain in their ability to expel members. However, the one crucial difference between cities and HOAs was the presence of involuntary members in the former, but not the latter. Ellickson said that 'the presence of involuntary members is both a necessary condition of the use of the adjective "public" in ordinary language, and also a powerful explanation for the different legal treatment currently accorded to public and private organizations'.[55] In short, voluntary membership was a justification for fewer constraints on private, as opposed to public, action. The characterisation of HOAs as public, private or a hybrid somewhere in between, has become a theme of HOAs literature, discussed below. It is also central to the difficult question of whether bodies corporate in Australia are purely private actors, or partly public, and thus are, or should be, captured by discrimination legislation, or administrative law. These questions will be discussed in Chapter 5.

Ellickson explored the growing United States case law on judicial review of HOA founding documents, (the C, C & Rs and initial rules), and subsequent HOA actions. Was there a place for public, judicial oversight of private communities and their rules, or should private communities be left to their own voluntary regulation?

In relation to the founding documents, Ellickson argued for minimal judicial oversight. He said that by voluntarily joining the association, original members unanimously consented to the founding documents, giving them the status of a private constitution. He stated that '[t]he original documents – which today typically include a declaration of covenants, articles of association (or incorporation), and by-laws – are a true social contract'.[56] Significantly, Ellickson provided this caveat: 'I assume throughout that an association is created through a fair contracting process – that is, one not tainted by fraud, duress, gross inequality of bargaining power, and so on'.[57] This is a questionable assumption in the context of property markets. Founding documents, just like initial by-laws and management statements, are more the product of profit-driven decisions of developers, than any social contract between members; and property development, particularly in booms, is frequently tainted by fraud, inequality of bargaining power, and purchaser abuse.[58]

Nonetheless, Ellickson argued for strict judicial enforcement of founding documents to protect members' reliance interests and to foster diverse types of residential developments, giving people a genuine choice of communities.[59] He stated that 'proper legal nourishment could enable private associations to supply quasi-public goods, such as parks and security services, which in the past had been more frequently supplied by cities'.[60] Here, Ellickson was alluding to James Buchanan's 'club goods theory' and Charles Tiebout's theory of 'residential shopping', both of which have been influential in United States housing debates, as well as other areas of public policy, particularly as public choice theory gained traction in the 1980s. Buchanan's 'club goods theory' aimed to identify the optimal size of a club to efficiently provide goods that might otherwise be publicly provided.[61] Tiebout's theory of 'residential shopping' posits that people 'shop' between residential areas, choosing the place that best suits their needs and desires.[62] Put together, the two theories amount to an argument that private HOAs can supply services more efficiently than public municipalities, and that if there are sufficient HOAs, people can 'shop' between them.[63] While there are flashes of 'residential choice theory' in Australian housing debates, state governments' enduring preference for metropolitan plans and urban consolidation, often in the face of intense community opposition, indicates an unwillingness to leave residential planning to the market. However, Australian state governments have been attracted to privatisation of public goods and services in a range of areas. Although obscured by marketing slogans like 'exclusive' and 'resort-style living', the privately owned infrastructure inside strata and community title schemes (sporting facilities, parks, water, sewerage, telecommunications, transport, roads and bushland) is privatisation through 'club goods'. If the developer builds it and purchasers pay to maintain it, government does not have to.

While insisting on strong protection for founding documents, Ellickson conceded that external legal norms should strike down 'offensive' provisions, such as those based on race. What, in addition to race, constituted an 'offensive' provision for Ellickson was unclear.[64] He argued that if a group of orthodox Jews set up a condominium and stipulated that on holy days all men were required to wear yarmulkes, courts should enforce this. He made reference to the fact that the Supreme Court had recently declared age not to be a 'suspect classification', (although Congress subsequently disagreed[65]). Overall, Ellickson argued that the 'pattern of judicial decisions tends to honor the suggested principle of greater private associational autonomy'.[66] What would be impermissible for public actors was acceptable by private HOAs in their founding documents. Ellickson supported a Florida Court's finding, (in litigation that would go on to become seminal decisions in HOA law), that original restrictions were:

> clothed with a very strong presumption of validity which arises from the fact that each individual unit owner purchases his unit knowing of and accepting the restrictions to be imposed ... a use restriction in a declaration of condominium may have a certain degree of unreasonableness to it, and yet withstand attack in the courts.[67]

Like Epstein's argument, there is validity in Ellickson's stance. It can seem indefensible for someone to voluntarily join a group and then complain about the application of its existing rules to them. However, like Epstein, Ellickson acknowledged that there were limits to the freedom of individuals to create rules. Some things should not be contractually agreed and enforced, such as restrictions based on race. There was a line in the sand that could not be crossed, but Ellickson, like Epstein, did not seem particularly concerned with how we find that line.

Ellickson then considered judicial review of HOA *subsequent* actions to implement their constitution and govern their community. Here Ellickson agreed that the 'reasonableness' standard which had been developing in HOA case law should apply to judicial review of association actions. The reasonableness rule had been stated by the District Court of Appeals of Florida in *Hidden Harbour Estates* v. *Norman* as follows:

> the association is not at liberty to adopt arbitrary or capricious rules bearing no relationship to the health, happiness and enjoyment of life of the various unit owners. On the contrary, we believe the test is reasonableness. If a rule is reasonable the association can adopt it; if not, if cannot. It is not necessary that conduct be so offensive as to constitute a nuisance in order to justify regulation thereof. Of course, this means that each case much be considered upon the peculiar facts and circumstances thereto appertaining.[68]

The 'reasonableness test', along with the business judgement rule,[69] has since been applied many times by United States courts to assess the validity of association decisions. It is routinely accepted that there are limits to HOA freedom of action, and that those limits should be determined by the concept of 'reasonableness'. However, Ellickson argued that 'reasonableness' should not be defined by *external* community norms, but rather by the association's *internal* collective values as evidenced in original documents.[70] The private group should be allowed to create its own voluntary culture and values, and henceforth have its actions judged against those values and not those of wider society.[71]

Finally, Ellickson considered how courts should approach changes to founding documents. Like Rose, he was recognising that while a person can consent to the existing rules of a community by joining, this is qualitatively different to consenting to *any* subsequent rules the community might enact. Ellickson addressed this dilemma by arguing that 'wealth-shifting' amendments, that is, those that alter members' voting rights, shares of ownership of the common property and shares of common expenses, should require unanimity, (generally speaking, this is the case in Australia, where these issues require a unanimous vote, or they are set by the legislation and cannot be altered by any vote). In contrast, 'wealth-creating' amendments should only require an extraordinary majority, but significantly, drafters should explicitly include, or reviewing courts should imply, a 'taking clause'. Here, Ellickson recognised that the power to vote to amend founding documents comes with the power to strip minority owners of valuable rights, whether the right to keep a pet or have a child, retain a view or access to facilities. He argued that while the procedure for granting and methods for calculating compensation might be complicated, if the amendment would in fact be wealth-creating for the proposers, they would still favour it, even if required to compensate losers. Ellickson's takings clause for HOAs has never been adopted by courts, drafters of founding documents or regulating legislation,[72] but it remains an important suggestion, (despite the obvious fact that some losses, such as not being able to have a child or even a pet in one's existing home, may not be monetarily compensable.) Like Rose,[73] Ellickson was acknowledging the double-edged sword of private communities: while HOAs allow members a level of control over their homes and communities that might not be possible in public cities and municipalities, they simultaneously render members vulnerable to their neighbours' collective control, in a way that they never would be in public cities or municipalities.[74]

There is a practical difficulty with Ellickson's differential judicial review for founding documents and subsequent association actions – the 'founding documents' of a community will be different for different members, depending on the point in time at which they joined. Although governing documents might change, if someone joins the community after that point, they can be said to have assented to the changed documents by their act of voluntarily joining. It would seem unworkable for a court to treat one version of governing documents as founding for earlier-joining members and other versions as founding for later-joining members.[75]

Ellickson's discussion of whether internal or external norms should be used to judge the validity of private rules was taken up by Gregory Alexander in 'Dilemmas of Group Autonomy: Residential Associations and Community'.[76] Alexander, now a leading proponent of an Aristotelian view of property,[77] asked 'what degree of autonomy ought the polity, in our "segmented society", grant to social groups, "each presuming autonomy in its own domain, each requiring homogeneity of its membership, and each demanding the right to fulfil its destiny without interference"?'[78] While many scholars have addressed this question generally,[79] Alexander explored it with specific reference to HOAs, which interestingly, he thought 'may represent the dominant aspect of the late twentieth-century contribution to American residential group life'.[80]

Alexander located his discussion in the context of two visions of American residential life (which hold equally true for Australia).[81] In the nineteenth century, the idealised residential life was outside teeming industrialised cities, with their congestion and lack of privacy. Ownership of the single family dwelling, segregated from the world of work and the private space of others, became the goal of middle-class families. The middle class wanted 'to escape from the social heterogeneity of urban life ... [to] an environment that excluded all of the dissonant elements of the city'.[82] Exclusion became a central element of residential housing.

By the second half of the twentieth century, the vision of secluded residential life was not so rosy. Quasi-rural suburbia came to be associated with isolation and a lack of community. HOAs and condominiums, with their commonly owned recreational facilities and resident associations, represented an opportunity for homeowners to become members of an instant community, to have that connectedness to their neighbours for which they yearned.[83] As Alexander said, 'it is no coincidence that developers advertise their projects as "communities" and "villages."'[84] Australian developers are equally fond of neo-traditional imagery, with ads replete with references to borrowed cups of sugar, kids on bikes and 'days gone by'. Developers are even known to refer to huge, urban high-rises as 'vertical villages'.

Alexander identified two approaches to the external regulation of groups: first, a public choice/contractarian/interest-group pluralism perspective or second, communitarianism. The first perspective saw the individual as the basic unit of society and the formation of groups was merely the expression of self-interest by a collection of individuals. The more groups there are, the more possibility individuals have to freely choose between groups that suit them.[85] Individuals are best placed to determine what is in their own welfare and if they have voluntarily chosen to join a group, there was no need for others, via the judicial process, to interfere with the character or quality of relationships within the group.[86]

In contrast, communitarians do not see autonomous individuals as the sole source of obligation or value, nor that 'good' can be reduced to the satisfaction of individual preference. Rather, they believe that 'the main source of obligation and value are collective social entities'[87] and that far from being autonomous, a person's identity is constructed by their community. Communitarians define 'community' as 'the well-ordered society, in which individuals are enabled fully to realise their own identities'.[88] Alexander argued that communitarians make a crucial distinction between voluntary associations and communities, the former being 'mere aggregations of self-seeking individuals', while the latter are made up of members 'drawn together by shared visions that constituted for each of them their personal identity'.[89] Alexander conceded that HOAs could be a combination of both; groups that were initially voluntary associations formed through the process of purchase can flower into true communities based on shared values.[90] Alexander identified the paradoxical position of both contractarian and communitarian theorists in the context of HOAs. The former, who favour individual autonomy, find themselves defending the relinquishment of autonomy by individuals to a group, while the latter, who generally favour community, find themselves casting aspersions on its virtues.

Alexander noted that, unlike contractarians, communitarians do not accept that groups should be accorded almost unfettered freedom to determine their own rules. If rules are dominating or antithetical to human flourishing, the state, as ultimate regulator of groups, must step in. Alexander stated that:

> Interpreting community as an ethic of inclusion, communitarians regard social intervention as sometimes necessary to prevent groups from practicing unacceptable forms of exclusion. Just what exclusionary social practices should be considered unacceptable is, of course, an enormously controversial question. It is just because that question is open-ended and intensely political, however, that we cannot permit groups to isolate themselves from dialogue with the rest of society. Strong group autonomy would enable groups to end all public dialogue, in effect imposing their answers to questions of exclusion and privileging their interpretation of 'community'. In the end, such an interpretation transforms community into a strategy for domination.[91]

Alexander singled out Ellickson's argument that 'reasonableness' in HOA decisions should be internally defined as an example of contractarian 'strong group autonomy'. Contractarians argue that as there are different conceptions of good and they are all subjective, the state should remain neutral and support all contracts, so long as they were freely agreed. This view has been critiqued nicely with the phrase, 'there is no view from nowhere; the "neutral" view will be as partisan as any other'.[92] Alexander also pointed out that this position was internally flawed because there is always a point at which contractarians will concede that the state needs to step in and limit freedom of contract – when it harms other people. We have seen this in both Epstein's and Ellickson's articles.

Alexander usefully reminded his readers that group existence depends on exclusion, which is not necessarily good. That is clearly the case when exclusion is based on race, gender or ethnic differences. Alexander referred to racially restrictive covenants to demonstrate that strong group autonomy can lead to domination, and that neutrality can only exist when the state intervenes.[93] Alexander pointed out that despite this obvious limit, contractarians continue to argue that those outside groups, 'but who are connected with the group by virtue of our common humanity and our common participation in society', should not be allowed to make collective judgements about group practices.[94] Alexander argued that this position threatens our collective welfare. This is the same point that Singer makes, discussed in the preceding chapter: the exercise of property rights does not only affect those inside the transaction, but those outside as well; thus, those outside have a legitimate interest in property entitlements.[95]

Alexander's conclusion was that HOAs should be fostered,[96] but that we must recognise the deep ambivalence we feel about communal autonomy, based as it is on members' 'desire to isolate themselves from others who they perceive as different in some significant way'.[97] A 'deep ambivalence' might be triggered in readers when they learn that HOAs in the United States routinely use 'single family' and 'residential only' covenants to exclude group homes for the intellectually or physically disabled or even for the elderly.[98] I once attended a session at an HOA lawyers' conference that was only slightly ironically called 'Keeping Out the Riff Raff'. To maintain a sense of 'groupness', two potentially undesirable activities have to occur: others have to be excluded from the group by rules, and those within the group must be controlled by rules. Alexander appropriately quoted Unger's observation that, '[b]y its very nature, community is always on the verge of becoming oppression'.[99]

Alexander's solution was 'open-ended group autonomy', that is, autonomy limited by state intervention in the form of judicial review. He said that:

> No single group within a pluralistic communitarian society can be permitted to arrogate to itself unilateral control over the inside-outside problem. There must be some agency that is responsible for maintaining the framework within which a variety of groups can engage in a dynamic process of formulating and reformulating the good ... Courts are an appropriate institution to maintain that framework. Through the dialogic process of adjudication, courts can serve as a bridge between communities and society, preventing the former from lapsing into a solipsistic perspective that would pervert the ideal of community ... the role of the legal system is to connect nomic groups with the rest of society even while recognising their separateness. Disputes over the enforcement of group restrictions present opportunities for a dialogue about the group's shared values and the extent to which they advance the ideal of community or impede it by creating relations of domination.[100]

Alexander argued that courts should use standards rather than rules to review groups' internal actions, because standards 'create opportunities for those inside and those outside to engage each other in dialogue'.[101] The 'reasonableness' standard was appropriate, as long as reasonableness was judged against the culture of the wider community.[102] Alexander closed by stating that:

> To realize their potential for creating community, groups must be held to an obligation more demanding than mere compliance with their own internal aspirations. Groups must accept a civic obligation to maintain community within our society. Meeting that obligation requires a praxis of openness and dialogue.[103]

Professors Ellickson and Alexander's articles described two seemingly conflicting theories for HOA association governance, which mirror conflicting views within wider political and philosophical debates. Ellickson's view is the classic liberal position that individual freedom and autonomy are overriding values that democratic communities and their judiciaries should promote. There is no justification for the judiciary to second-guess private decisions on what constitutes a 'good' community or life, unless those private decisions clearly harm others. Alexander's view is that such a limited role for the judiciary and the wider community inevitably leads to a range of harms that need to be addressed, not ignored.

These ideological positions remain consistently present in HOA debates in the United States, and while rarely explicitly acknowledged in Australia, lurk beneath the surface of our strata and community title law. But are the positions really so polarised? Like Epstein's argument on covenants above, Ellickson is possibly engaging in a sleight of hand. While promoting individual freedom and autonomy, Ellickson accepts that there is a point at which freedom and autonomy must be controlled. He accepts that there is a line in the sand over which individuals and their groups must not step. As argued above, the real debate then becomes not *whether* to draw a line in the sand, but *where* to draw the line. The property theorists discussed in Chapter 2 tell us that while there is no simple way to define that line, definition is possible through 'reasoned deliberation'.[104] When determining property entitlements, we have 'no alternative but to make arguments that elaborate fundamental human values and that express our considered commitments to judgments about morality and justice'.[105] As a result, we need legal space to make those arguments. As Alexander argued, a standard of 'reasonableness' provides space for courts, individuals and communities to engage in those arguments, and to explicitly acknowledge and assess the values that are at stake. We will return to this point in the discussion of the *Restatement (Third) of Property: Servitudes* (2000), below.

From theory to practice

The 1982 dialogue between French, Reichman, Epstein and Rose stimulated intense debates about the legitimate role of servitudes in American law which carried on through the 1980s and 1990s. In the mid-1980s, the American Law Institute began a project on a third *Restatement of Property: Servitudes* with Professor Susan French as the Reporter.

One of the foci for debates was the question of coercion and consent. As noted by Rose, above, the orthodox theory of servitudes is that while third parties cannot be bound by others' contractual agreements, if those agreements are intended to bind land and a third party purchases with notice of them, the third party has consented to the servitude and thus enforcement is legitimate. Exactly the same argument applies to the by-laws of strata and community schemes, (although it is technically prior registration and not notice that binds subsequent purchasers). The argument is based on freedom of contract: parties are free to negotiate the contracts they desire, but once they have been concluded, they should be bound by them. Similarly, purchasers are free to buy or not to buy land, but having made a decision to do so knowing the land is affected by covenants (or by-laws), they should be held to them. Freedom of contract goes hand in hand with liberal convictions about wealth-maximising behaviour of individuals. People act in ways that maximise the value of their assets, including land. Thus, left to their own devices, individuals will negotiate agreements that produce the most efficient use of land. As we have been discussing, liberalism also assumes that individuals are better placed than governments or courts to know what is in an individual's personal interest. As a result, privately agreed land regulation such as servitudes is generally preferable to public regulation such as zoning.[106]

Not surprisingly, a number of academics challenged this faith in consent from both a doctrinal and empirical perspective. Professor Stewart Sterk addressed the question of individual foresight and whether, as Epstein continued to argue,[107] individuals really do act with sufficient foresight by negotiating only wealth-maximising servitudes.[108] Sterk argued that a plethora of legislative and judicial actions limiting personal autonomy, or compensating for lack of foresight in a wide range of areas, suggests that neither courts nor legislatures share Professor Epstein's faith in individual capacity to consistently make good long-term decisions. Nor, according to Sterk, do judges share Epstein's scepticism about the incapacity of public decision-makers to make superior decisions. Judges seem to believe, as does Sterk, that hindsight gives public officials an undeniable advantage over private actors.

Sterk argued that judicial intervention was inevitable. He accepted that individual autonomy is central to our legal system and judges tend to respect private agreements, but there was always a point where judges would step in. This intervention could be minimised by ensuring that formal doctrine is as close to the shared norms of decision-makers as possible;[109] however, 'no doctrine is sure to command universal respect or to account for all contingencies that might arise'.[110] Sterk argued that it was vital for intervention to occur in an open, explicit way, rather than under the covert guise of judicial interpretation of the parties' alleged intentions. He said that:

> when the real grounds for intervention are obscured, the common law process will not refine the principles that underlie the intervention. Lawyers, parties, and even judges may fail to understand the basis for intervention, resulting in uncertainty and its attendant costs.[111]

Openness about the limits of freedom of contract would produce a more stable system of servitudes.

Elsewhere, Sterk had identified that one of the triggers for judicial or legislative intervention was the point at which private land agreements imposed unacceptable externalities on third parties or the public at large.[112] Sterk used the obvious illustration of racially restrictive covenants which have an adverse effect on land prices for minorities, and more fundamentally, are an affront to human dignity. Sterk also cited covenants in restraint of trade as private agreements which produced 'external diseconomies' that constitutions, statutes or the common law (through the 'touch and concern' doctrine) have struck down. Sterk rightly concluded that, '[d]etermining when external diseconomies justify restricting contractual freedom may be difficult, but the argument that freedom of contract ought to be absolute is hardly tenable'.[113] Again, the real question is not *whether* to draw a line in the sand, but *where* to draw it.

In the same article, Sterk explicitly addressed the issue of intergenerational equity in land-use transactions, a concern of property law for centuries. Sterk argued that although land cannot be physically destroyed, like the Grand Canyon or a historic building, nor utterly exhausted like the world's oil supply, its use can be so restricted that it becomes economically useless. If the legal system allowed current landowners to impose significant restrictions, which could not be removed by future landowners, or at least not without considerable expense,[114] then current landowners would be permitted to use up future generations' land.[115] This is the point that was made at the beginning of Chapter 2: land is different to other forms of property. Land always remains, after those who have used or owned it have died or moved on, and all human beings need to use land. It is not an optional resource, as we all have to be somewhere, all of the time.

Sterk pointed to the existence of traditional rules of property – the rule against perpetuities, the abolition of fees tail and statutory limits on defeasible fees – which routinely limit the freedom of current landowners to control land into the future. The doctrines define the boundaries of acceptable land control. Sterk's overall conclusion on the technical doctrine of servitudes was similar to both French and Reichman. Mindless, archaic technicality serves no purpose; or '[i]t is revolting to have no better reason for a rule of law than that so it was laid down in the time of Henry IV'.[116] However, like Australian freehold covenant law, the technical rules of servitudes are not mindless, except perhaps when recited by students or lawyers who, quite understandably, owing to their insufficiently clear formulation, may not have fully understood them. Foreshadowing Singer's arguments on the limited ability of free market principles to create functional land law,[117] Sterk said:

> However quaint in formulation and muddled in application, the touch and concern requirement, for instance, has at last some potential to serve as a modest check against externalities, inadequate foresight, and intergenerational imposition. The market itself cannot adequately account for these problems, and throughout property law a variety of restrictions on freedom of contract have been developed to deal with them.[118]

Sterk, like Rose before him,[119] issued a warning that is central to this book: when modernising property law, either by a Restatement or legislation, it is important not to toss the baby out with the bathwater. We can remove the archaic, technical language of property law, but that will not remove the complexities inherent in people's use of land, which neither individual preference nor the market are capable of adequately addressing.

Debates of this nature tended to be conducted at a relatively abstract level, divorced from the lived and litigated reality of HOA residents. From the 1970s onwards, one-third of all new housing was governed by C, C & Rs and in some areas, like California, this figure was two-thirds.[120] By 1990, 30 million Americans lived in 130 000 HOAs across the country.[121] As a result, there was a growing body of empirical research, as well as reported cases, that allowed servitude regimes to be viewed in situ, so to speak, to investigate the validity of certain assumptions about human behaviour that were inherent in theoretical debates.

James Winokur's article 'The Mixed Blessings of Promissory Servitudes: Towards Optimizing Economic Utility, Individual Liberty, and Personal Identity'[122] was a good example of the scholarship that tried to grapple with servitudes at both a doctrinal and practical level.[123] Like many others, Winokur identified the role servitude regimes play in relation to zoning in the United States,[124] where zoning is considerably more contested than in Australia.[125] There is a common belief in the United States that zoning is an inefficient land-use allocator, has excessive administration costs, can amount to an expropriation of private property,[126] and is associated with government corruption. Zoning is also considered too general to achieve the detailed regulation that many people desire, such as mandating house colours and fencing styles and too liable to change to safeguard against non-residential development. Finally, in some liberal thinking, zoning represents state actions that can be done more effectively, efficiently and most importantly, more legitimately through private action.

As the title 'Mixed Blessings' suggests, Winokur noted that while all these criticisms of zoning had some validity, and servitude regimes had some clear benefits, they were a double-edged sword. For example, servitude regimes may have created uniform, green suburbs of great beauty, but they had also created large swathes of dull, homogenous, low-density residential development with a complete absence of services. He argued that servitude-mandated minimum allotment size and residential-only use were partly responsible for the endless suburban sprawl that blights so much of America, where it is impossible for residents to access doctors, entertainment or even buy a carton of milk without a car. Winokur referred to the work of Jane Jacobs, the champion of organic, high-density, mixed-use cities. Jacobs condemned insular, single-use suburban development as 'silly, and even harmful',[127] preferring the diverse, intense and continuous use of space in New York City.[128] Winokur highlighted the fact that in the process of negotiating servitudes, individual people do not have foresight in relation to large-scale, long-lasting urban and suburban planning. The results of their private 'planning' efforts often spoke for themselves.

Winokur challenged the classical liberal assumption that servitudes represent the efficient allocation of resources by freely negotiated market agreements, which can be equally freely released when obsolete. Examining the kinds of restrictions that are found in servitude regimes, such as restrictions on age and childbearing, religious practice, financial and social compatibility, family and marital status, commercial and political speech, plant type, housing colour and all alterations without architectural committee approval, Winokur questioned the true motives of purchasers who bind themselves to such 'long-term, anti-autonomous arrangements'[129] or in Reichman's words, 'modern variations of feudal serfdom'.[130] Winokur argued that the idea that all or any of these restrictions are freely negotiated is more imagined than real. C, C & Rs are drawn up by developers prior to the sale of lots and individual purchasers are never given the opportunity to alter the terms.[131] This is not just because they do not have the bargaining power to do so, but because of the very nature of servitude regimes. The restrictions are part of a large-scale common plan achieved through the reciprocal application of servitudes to all lots, and by definition, later purchasers cannot negotiate a different regime for their own lot. Negotiation becomes even more unrealistic with the widespread use of boilerplate contracts,[132] which compromise the supposed ability of purchasers to 'shop around' for different servitude regimes.[133] As has increasingly become the case since Winokur wrote his article, in some areas of the United States, typically in the Sun Belt states, buyers do not even have the option of buying land with no C, C & Rs; almost all housing available is part of an HOA.[134] In some places like the city of Las Vegas, all new housing construction is effectively mandated to be inside an HOA.[135]

In relation to the all-important question of notice, Winokur argued empirical research suggests that buyers do not read or understand the servitude regime and they frequently do not employ lawyers to advise them on the purchase. One study indicated that as many as 85 per cent of homeowners were not aware of the servitudes or even the existence of an HOA at the time of purchase and only became aware when they attempted to do something prohibited.[136] Further, the United States recording system is not nearly as strict as a Torrens system and buyers have found themselves bound by unwritten, unrecorded covenants of which the only notice they had was the uniform physical appearance of the neighbourhood.[137] Even if the covenants were written and recorded, another academic writing on the same question concluded that, '[i]n many parts of the country, the recording system is so poorly organized, so overloaded, and so incompetently operated as to be of limited utility'.[138] As noted at the beginning of this chapter, in some areas of the United States, even if C, C & Rs are required to be registered, association rules or by-laws are not.[139] Winokur concluded that '[a]s standardized servitude regimes increasingly dominate housing markets – with buyers rarely reviewing the incomprehensible, non-negotiable, and sometimes non-recorded servitude documents – buyer consent to servitudes is reduced to a purely theoretical premise'.[140] It should be noted that even in Australia, where by-laws are all clearly recorded on the Torrens register, and are available for public inspection at minimal cost, it seems that many purchasers do not read their by-laws.[141] As by-laws are often initially disclosed to purchasers in contracts that exceed 100 pages, and are non-negotiable, this is not surprising.

Also not surprising, is that lack of understanding of servitudes led to extreme resident dissatisfaction with their enforcement. Empirical research relied on by Winokur suggested that more than association management of common areas or payment of financial dues, regulation of perceived 'private' behaviour through servitude enforcement generated the greatest resentment, sometimes spilling over to violence. Winokur cited examples of fistfights between members, shotgun threats and an association president having his leg bitten.[142] Today, firearm threats and fatal shootings of HOA board members are not uncommon.[143]

Winokur argued that dissatisfaction with servitude regimes and their implementation by inexperienced association boards led to state appellate court decisions on the issue multiplying sevenfold from 1977 to 1982 and then doubling again from 1982 to 1987.[144] Winokur suggested that the small participatory democracies that HOAs were feted to be have never been realised in many communities. Residents were overwhelmingly uninvolved in their association and one study reported that 84.3 per cent of respondents agreed that owner apathy was the greatest problem facing HOAs.[145] When associations attempted to regulate member action through servitude enforcement, apathy often turned to litigation. Winokur argued that regardless of the theories of consent, on the ground, many residents simply did not accept the legitimacy of servitudes to regulate their behaviour in their own homes.

Winokur logically raised Margaret Jane Radin's well-known thesis that special provisions must be made for property that is justifiably personal to its owner.[146] Winokur cites Radin's assertion that:

> [legal protection of the] sanctity of the home ... contains a strand of property for personhood. It is not just that liberty needs some sanctuary and the home is a logical one to choose because of social consensus ... [One's home is] the scene of one's history and future, one's life and growth. In other words, one embodies or constitutes oneself there. The home is affirmatively part of oneself – property for personhood – and not just the agreed-on locale for protection from outside interference.[147]

Winokur argued that the growing predominance of increasingly intrusive servitude regimes 'frustrates the notion of property as protecting personal identity, and home as a favoured context for individual flourishing'.[148] He said that as the home is of central importance to an individual's sense of self and to 'the shaping of that individual's unique history and future by the expression of his or her own distinctive personality' each resident should 'presumptively ... control the details of his or her own residential uses'.[149] These details include when to take out the rubbish, the colour of the family swing set or interior curtains, whether to fly the United States flag, childbearing, ages of ownership and religious practices. Persons seeking to regulate any of these areas through servitudes should bear a heavy burden of persuasion to justify interference with 'activities integral to one's self in one's own home'.[150]

Winokur's solution to the 'mixed blessings' of servitudes was to recommend that their large-scale application be legislatively restricted to 20 years' duration. Beyond 20 years, servitudes should be enforceable only within smaller 'pods' of 12 homes or less. Within these pods, both enforcement and buy-outs are likely to remain meaningful in a way that they are not in large-scale developments.

The change of circumstance doctrine would become less significant because smaller groups of homes would have a realistic chance of being able to renegotiate the servitude regime. Their decision on what is beneficial for them is likely to be more accurate than a judicial decision.[151]

One final contribution to the consent and coercion debate that should be considered was that of contract professor, Clayton P Gillette, in 'Courts, Covenants, and Communtities'.[152] Like Alexander above, Gillette recognised the ironic position of communitarians in relation to private communities. Communitarians are traditionally in favour of protection of minorities like the Amish or orthodox Jews, but generally opposed to protection of groups who chose to separate themselves from the wider community because they do not like satellite dishes or fear crime. Gillette said, 'it seems odd to afford substantial protection to communities furthest from the majority culture while affording little protection for those only marginally different from the majority'.[153] Gillette makes an important point that needs to be acknowledged: it is not defensible to criticise to private communities that centre on white shoes and 18 holes of golf a day, while promoting communities based on permaculture. Any objection to private communities must be based on *genuine* harm they may cause and not just that they are made up of people 'not like us'. Some commentary on private communities risks falling into this trap.[154]

Privatopia

Perhaps the most significant contribution to the debate on HOAs in the 1990s, and which deserves particular attention, is Evan McKenzie's book *Privatopia: Homeowner Associations and the Rise of Residential Private Government*.[155] McKenzie's research is particularly useful for two reasons. First, as an academic and practising lawyer, McKenzie understands the technical detail of servitudes and the role they have played in the development of HOAs. Second, while McKenzie appreciated the significance of technical servitude law, his work predominantly addressed the social, historical and political significance of servitude regimes and the effect they have had on America's urban and suburban landscape.

McKenzie traced the rise of 'privatopia' back to Ebenezer Howard's visionary book *Garden Cities of To-morrow*,[156] which described a utopian city with agriculture, industry, housing, schools and services all rationally planned in concentric circles around public buildings and a central park. The legal structure for the city was a land trust with leasehold residents and it aimed to create 'a new civilisation based on service to the community and not self-interest'.[157] As noted in Chapter 1, Howard's work was hugely influential in suburban planning in the United Kingdom, Australia and in the United States, although it was not embraced by all. Jane Jacobs, who was delightfully dismissive of great men of planning, including legendary high-rise architect Le Corbusier,[158] cuttingly described Howard's aim as 'the creation of self-sufficient, small towns, really very nice towns if you were docile and had no plans of your own and did not mind spending your life among others with no plans of their own'.[159]

McKenzie identified two crucial differences between the adoption of Howard's ideas in the United Kingdom and America. First, in the United Kingdom, government played a central role in the implementation of planning, while in the United States, private developers and businessmen were the 'dominant forces'.[160] In the 1920s in the United States, large corporate builders emerged, constructing 'hundreds, then thousands of houses at a time for short-term profit, not for long-term social transformation'.[161] Australia could reasonably be positioned somewhere between the United Kingdom and the United States in this regard; private developers, driven by profit, have built the majority of our housing, but local and state governments have regulated the process, and state governments have traditionally provided the bulk of the infrastructure.[162]

Second, as an Englishman, Howard envisaged residents only having leasehold interests. As discussed in Chapter 1, while long-term leasehold was the norm in the United Kingdom, it was not favoured in the New World, with courts and governments disliking land arrangements that smacked of feudalism or aristocratic oppression.[163] American governments and the real estate industry liked the 'finished product' of the Garden Suburb Movement, believing that careful planning promoted stable property values, but they needed a legal mechanism other than the long-term lease to implement the private system of planning. Servitudes, in particular 'real' covenants, became the key. Judge Hare's anomalous mid-nineteenth-century decision that privity of estate existed between vendor and purchaser[164] allowed detailed restrictions and obligations to attach to land without the need for a politically offensive leasehold structure. In *Neponsit Property Owners' Association, Inc* v. *Emigrant Industrial Savings Bank*,[165] the New York Supreme Court confirmed that a restriction in a deed requiring all lot owners to pay annual assessments for the maintenance of roads, parks, beaches and sewers touched and concerned the land and ran with the fee. It also held that the Property Owners' Association, which did not actually own any benefited land itself, could enforce the restriction as agent for the property owners. These two factors – the ability to make monetary payments run with the land and have a collective body to enforce payment – were critical to the rise of private communities in the United States. As noted in Chapter 1, they account for the physical differences in suburban development in Australia and America in the twentieth century.

There is a deep irony in this history. As Reichman noted, servitudes are 'a kind of private legislation affecting a line of future owners', which carries the risk of 'creating modern variations of feudal serfdom'.[166] McKenzie's descriptions of HOAs regulating signs, newspaper distribution, political gatherings, parking of commercial vehicles, children's swing sets, fencing height and colour, roofs, patios, dog walking and dog weight, certainly do not sound like the freedom and personal autonomy that we have come to associate with modern fee simple ownership. McKenzie cited the example of one HOA posting notices around the development accusing a resident of 'parking in a circular driveway kissing and doing bad things for over 1 hour' and threatening her if it happened again.[167] McKenzie argued that HOAs have created an 'ideology of hostile privatism … [in which] rational planning is distorted into an emphasis on conformity for its own sake'.[168]

One of the most important contributions that McKenzie's book made to scholarship on HOAs was to describe the history of racially restrictive covenants in the United States and connect this to the ideology of private communities. While the existence of racially restrictive covenants was acknowledged by academics in favour of private regulation like Epstein and Ellickson, above, they failed to address properly the endemic nature of this abusive land-use practice, and its reliance on the doctrine of freedom of contract that they so strongly promoted. This is a serious flaw in the liberal and libertarian scholarship. As we saw in Chapter 2, Singer, although not discussing HOAs specifically, takes full account of the racially discriminatory nature of private property practice in the United States in formulating his theory of property.[169]

McKenzie documented the real estate industry and Federal government's response to one of the greatest internal migrations in history, the movement of 6.5 million African Americans from the rural South to the industrial North and West between 1910 and 1970. Racially restrictive zoning by government had been declared unconstitutional in 1917,[170] but private agreements were not 'state action' regulated by the Constitution. The National Association of Real Estate Boards (NAREB), an alliance of real estate agents and builders, in conjunction with the Federal Housing Administration (FHA), pursued a policy of strict racial segregation in housing, for the alleged purpose of maintaining real estate values. Until 1950, article 34 of the NAREB Code of Ethics contained the provision that, '[a] Realtor should never be instrumental in introducing into a neighbourhood a character of occupancy, members of any race or nationality, or any individuals whose presence will clearly be detrimental to property values in that neighbourhood'.[171] The FHA, which was created during the Depression to stimulate building by insuring residential mortgages against loss, promoted the use of racially restrictive covenants in its *Underwriting Manual* for staff. It recommended that restrictions should include 'prohibition of the occupancy of properties except by the race for which they are intended'.[172] McKenzie argued that the dominant ethos of housing development at both private and government level was homogeneity. Consequently, thousands of communities all across the country included covenants restricting occupation to Caucasians. By 1960, one large-scale builder could boast a community in Long Island with 82 000 residents, not one of whom was black.[173] In this context, Unger's statement, '[b]y its very nature, community is always on the verge of becoming oppression', quoted by Alexander above,[174] rings very true.

As noted in Chapter 2, in 1948 the Supreme Court in *Shelley* v. *Kraemer*[175] declared racially restrictive covenants unconstitutional, or more accurately, the enforcement of racially restrictive covenants by courts unconstitutional.[176] Convinced that homogeneity was still the key to successful sales and communities, real estate operators immediately began to create less explicit racially segregating covenants such as single family, minimum allotment size, minimum cost of dwelling, maximum number of occupants per room and owner-occupied covenants, all of which militated against African American families buying or renting in developments.[177] These were supplemented by association or club fees to maintain private facilities, many of which put developments out of reach of African American families.

In addition to illustrating the dangers of excessive reliance on individual preference in property law, McKenzie also called into question the very existence of a widespread individual preference for covenants. He argued that the rise of HOAs had more to do with land economics than purchaser demand for covenant-restricted land. Like Alexander,[178] McKenzie identified the historical importance of suburbia as an escape from nineteenth- and twentieth-century high-density, heterogeneous city living: '[the] suburban prescription for social health was a single-family house surrounded by a large yard planted as a vast expanse of green lawn. It promised privacy, autonomy, and self-determination for the middle-class family'.[179] In the post-war period, rising land costs made it increasingly difficult for builders to profitably construct these middle-class suburban utopias. In 1948, developed land for housing cost a builder on average 11 per cent of the price of the housing unit; by the late 1970s, it had risen to 30 per cent.[180]

McKenzie described how 'cluster subdivisions' with HOAs were seen as the solution to this problem by the real estate industry and the Federal government. Individual plots could be reduced in size and open space and amenities provided collectively on co-owned land. Mandatory membership associations, separate bodies entrusted with the task of maintaining and regulating common property, were seen as the key to success. After studying an HOA development, fittingly in destined-to-be-iconic Orange County, California,[181] the FHA began to insure these developments in 1963, paving the way for their nationwide distribution. While servitude-regulated communities had been the preserve of a wealthy minority for most of the twentieth century, the 1960s would see them trickle down into the bulk of new housing. Communities with an HOA increased from less than 500 in 1962 to 20 000 in 1975,[182] and 130 000 in 1990.[183] There are now over 300 000 HOAs in the United States, with 60 million residents.[184]

As HOAs were being promoted for low-rise suburbia, condominiums were being promoted for high-rise, high-density, urban development. While apartment buildings had existed in large United States cities for over a century, most had a single owner with individual leaseholds. Co-operative ownership had enjoyed fitful popularity in New York and Chicago,[185] but like Australian company title, it only allowed individual ownership of shares, not land. Condominiums, like strata title, allowed for individual ownership of apartments, which was significantly more profitable for builders. The FHA extended insurance to condominiums in 1961,[186] reducing much of the risk for builders moving into this new form of housing. Although condominiums still use orthodox C, C & Rs and association rules to regulate properties and residents, condominium subdivision is assisted by legislation, and by 1967, all 50 states had legislation facilitating condominium ownership.[187]

McKenzie's research demonstrated the way in which large-scale private housing development can occur with insufficient consideration of its social, political or wider economic consequences. In particular, title structures or 'servitude regimes' were chosen with no reference to their long-term implications for micro and macro communities. There is a risk that exactly the same process is occurring in Australia with strata and community title; the legal structures are being enthusiastically employed by developers and local councils for short-term financial gain and insufficient concern for their long-term implications.

Private builders are (legitimately) concerned with their own profit and if their activities are not critically overseen by government, profit is the only end that will be served. McKenzie argued that in the United States:

> Public policy regarding housing is made in an increasingly privatized political environment ... in which economic considerations are central and social issues are viewed as externalities.... Given a large degree of [government] latitude, the real estate industry followed the path of greatest profit.[188]

Another way of stating this is to acknowledge that the private process of negotiating the profitable sale of houses, in combination with the private process of choosing an individual family home, will not necessarily produce large-scale urban and social planning that meets the needs of everyone outside, or even inside, those negotiations.

The final crucial point that McKenzie makes in *Privatopia* is that while land economics and forces in favour of privatisation drove the creation of private residential associations, few stakeholders anticipated the problems that would arise when groups of private citizens were given significant power over their neighbours' lives and homes. Like Australian bodies corporate, associations administer rules that regulate behaviour on common property and in individual homes; they set monthly assessments or levies and decide how they are spent; they have the power to make new rules and decide if rules have been breached. Their boards are elected but may contain members who have no experience of decision-making, accounting, property management or architecture. Unlike most Australian bodies corporate, United States associations can also fine members or exclude them from amenities or even the community,[189] and they can also acquire liens over homes and foreclose. McKenzie argues that few people involved in promoting these associations seemed to appreciate the complexity of these roles, their potential for abuse and to foster dispute. However, as HOAs multiplied, contrary to the arguments of Ellickson above,[190] some people began to ask whether HOAs were a form of quasi-government whose powers should be restrained as such.

McKenzie describes how proponents of HOAs were initially eager to characterise associations as 'mini-governments', private alternatives to local government. In 1976, Wayne Hyatt, professor, lawyer and former president of the Community Association Institute, wrote that '[u]pon analysis of the association's functions, one clearly sees the association as a quasi-government entity paralleling in almost every case the powers, duties, and responsibilities of a municipal government'.[191] Garbage collection and park maintenance are equivalent to municipal services, assessments equivalent to taxes and private security guards equivalent to a police force.[192] However, a decade later, HOA advocates were retreating from this stance because of the threat that constitutional limitations on municipal government, in particular freedom of speech, would apply to HOAs.[193] As far back as 1945, the Supreme Court in *Marsh* v. *Alabama*[194] had held that the refusal of a company town to grant Marsh, a Jehovah's Witness, a licence to distribute religious leaflets and her subsequent charge of trespass by the State were a violation of her First and Fourteenth Amendment rights to free speech. The Supreme Court held that while the town's pavements might be private property, property rights had to be balanced against other rights such as freedom of religion and speech.

At the time McKenzie was writing *Privatopia*, litigation had started to challenge HOAs' power to ban signs and the distribution of pamphlets. Bans were largely designed to prevent visual 'pollution', but the application of these rules to political signs and pamphlets, particularly in large-scale communities the size of suburbs, raised questions about conflict with free speech provisions in the United States' and individual states' constitutions. Bans also captured the display of flags, causing community outrage when people were instructed to remove the national flag, or in Texas, the Lone Star; sentiment ran particularly high after September 11.[195] In *Committee for a Better Twin Rivers (CFBTR)* v. *Twin Rivers Homeowners' Association (TRHA)*,[196] homeowners in HOAs with 10000 residents alleged that the Association rules, as well as its decisions on posting signs, access to a community room, and access to a community newsletter were violating their rights to free speech and free association. They claimed that the President of the Board used the community newsletter as his 'personal political trumpet' and that there was a discriminatory two-tiered charging system for use of the community room, clubs approved by the Association being given a lower rate. The plaintiffs argued that, as the Association effectively replaced the role of a municipality, it should be subjected to the free speech and association clauses of the New Jersey constitution. Despite the authority of *Marsh*, the Court disagreed. It held that Twin Rivers was a private residential community whose residents, including the plaintiffs, had contractually agreed to abide by the rules and regulations of the Association. Those rules were subject to the business judgment rule, legislative provisions on HOAs, and traditional principles of property law, but not constitutional protections.[197] It is hard not to see the irony in citizens of a nation with such strong traditional limits on government power, being subjected to a proliferation of largely unlimited mini-governments. This is another illustration of the liberal paradox of HOAs, identified by Alexander earlier in the chapter.[198]

In the mid-1990s, part way through the project of rewriting the *Restatement on Property (Servitudes)*,[199] this irony was not lost on Professor Susan French. Just prior to the publication of McKenzie's book *Privatopia*, French argued that developers should include a bill of rights in their governing documents for HOAs.[200] French suggested that purchasers would inevitably become wise to the excessive risks to their autonomy and freedom posed by C, C & Rs and by-laws, and not want to buy in HOAs. She said that:

> As the resemblance [of HOAs] to government grows stronger, the more people will fear the lack of controls that limit the powers of public governments. People will feel comfortable with this most local form of government only if it is subjected to the same kinds of restrictions we, as a people, have imposed on our other levels of government.[201]

French's article begins with a Hobbesian quote from the seminal HOA case *Hidden Harbour Estates, Inc* v. *Norman*, in which the Court said:

> Inherent in the condominium concept is the principle that to promote the health, happiness, and peace of mind of the majority of the unit owners ... each unit owner must give up a certain degree of freedom of choice which he might otherwise enjoy in separate, privately owned property.[202]

French accepted the basic validity of this statement, but questioned whether there needed to be a limit on the amount of freedom a developer, in drafting documents, could ask people to relinquish; that is, are there limits on how far people are free to agree to be less free? More importantly, French highlighted the fact that if the developer left the association with broad discretion to write new restrictions, the value of people's property and their personal well-being could be adversely affected by rules they did not anticipate. French said:

> Dreams of homeownership can turn sour for people whose building or landscaping plans are not approved and for people who learn too late that they will not be permitted to put up political signs, for sale signs, or holiday decorations. Dreams can turn to nightmares for homeowners who are forced to a choice of moving or giving up beloved pets, lovers, or even children.[203]

French was referring to the notorious fact that at this time United States developments routinely used covenants to determine family composition, including provisions that excluded unmarried couples and all children.[204]

French's suggested bill of rights included protections that are relevant to anyone living in collectively owned property, where their neighbours have the power to legally regulate their lives. The bill of rights included:

- Equal treatment: all owners and residents should be treated similarly.
- Speech: the right to display political signs, religious symbols,[205] or holiday displays should not be unreasonably restricted.
- Household composition should not be restricted, except to a single housekeeping unit and to limit the number of occupants.
- Activities within the individually owned property should not be restricted except to the extent they interfered with other people.
- Pets: unless a rule prohibiting pets was in place from its inception, an HOA must not adopt a rule prohibiting pets, unless a particular pet is a danger or nuisance.

- The initial allocation of financial burdens and rights to use common facilities should not be changed to the detriment of a particular lot owner, unless they consent in writing.
- Alienation: the association must not adopt rules that prohibit the transfer of lots nor impose any fee on transfer greater than an amount reasonably based on costs incurred to the association by the transfer.

To my knowledge, French's bill of rights has never been taken up by developers, however her prediction that purchasers would avoid HOAs because of perceived invasion of privacy and basic rights associated with homeownership has proved true. Many United States real estate agents now advertise homes as 'No HOAs' or 'No HOAs fees'. However, even if purchasers would prefer to avoid a homeowner association, a survey of buyers in California HOAs found that 84 per cent had only bought by default, having been unable to find a non-HOA alternative.[206] This is an example of the finite nature of property requiring us to be attentive to the affect of property rights; contractual notice and consent are not sufficient protections because there may be limited or no alternative property to buy or rent.

Twenty-five years later, with the benefit of hindsight and a knowledge of the exponential growth of HOAs, McKenzie reaffirmed the argument he had made in *Privatopia* about the potentially dangerous power of voluntary associations and their tendency to foster dispute. He said:

> [one] point cannot be overemphasized: the entire institution of common interest housing rests on the volunteer directors, yet they are unpaid, untrained, often unqualified, and almost entirely unsupported by the governments whose work they are often doing.[207]

He noted that the only limit on their activity has been civil litigation, which is not sufficient. He argues that:

> The most basic principle of liberal democratic and constitutional government is the requirement that it must include enforceable limits on the power of government … Liberal democrats of the Founders' era believed that people cannot be trusted with unlimited power, because we are naturally selfish creatures and our emotions override our intellect. We are easily convinced that, by an amazing coincidence, the very course of action that suits our own self-interest just happens to be the morally correct and wise rule for the entire society.[208]

Perhaps the best way to conclude this section, and to validate McKenzie's point about the dangers of private governments, would be to remember the case of Trayvon Martin, the 17-year-old African American boy who was fatally shot by a neighbourhood watch volunteer, George Zimmerman, inside a Florida gated community, The Retreat at Twin Lakes. Zimmerman, who was not meant to be carrying a gun, had been appointed by his neighbours in the HOA to patrol the community.[209] While Zimmerman was acquitted of criminal charges, his HOA settled a wrongful death suit with Martin's family for an undisclosed sum.[210]

The *Restatement (Third) of Property: Servitudes* (2000)

To end this chapter, we come back to where we began, to efforts by legal experts to rationalise the United States' extraordinarily complex covenant law in light of its enthusiastic use by the real estate industry. In 2000, the American Law Institute's *Restatement (Third) of Property: Servitudes* was released, with the aim of making it 'easier to use servitudes for the important roles they play in modern land development'.[211]

In keeping with the initial thinking of French and Reichman back in 1981, the *Restatement* focused on the functional operation of covenants, easements and profits, applying uniform rules whenever appropriate. While accepting the basic principle of freedom of contract by creating a presumption of validity for servitudes, the *Restatement* stopped well short of the libertarian thesis of Epstein that freedom of contract should invariably prevail. The *Restatement* noted that courts' traditional reluctance to enforce affirmative burdens in particular, 'may be explained by the long struggle to free land from the feudal incidents, which burdened landowners with a wide variety of affirmative obligations'.[212] Covenants potentially threaten centuries of hard-won gains achieved through the consistent application of modern property doctrines and statutes discussed in Chapter 2. The *numerus clausus* principle, the estates system, and 'boundary rules' all simplified interests in land by eradicating novel interests and attempts by predecessors in title to control future land use. Covenants run directly counter to this simplifying trajectory.

The *Restatement* abandoned the 'touch and concern' doctrine, which was considered to be confusing in its application, as well as ineffectual in combating racially or religiously discriminatory covenants or those that threatened free speech.[213] However, the role the doctrine had played, albeit confusingly, as a mini-boundary rule which allowed courts to invalidate covenants that were socially and economically undesirable, was recreated and expanded in a provision that covenants could be declared invalid if they were illegal, unconstitutional or against public policy,[214] were unreasonable restraints on alienation, undue restraints on trade, or were unconscionable.[215] The *Restatement* said that:

> it is not intended to remove courts from their historic role of safeguarding the public interest in maintaining the social utility of land resources. By reformulating the inquiry to ask the question directly – whether the servitude in issue poses such a threat to the public welfare that the rights or obligations it creates should not be allowed to run with land – the law will encourage clearer identification of the issues and clearer explanations of the reasons why servitudes may not be used to implement particular arrangements than was ever possible using the touch-or-concern doctrine.[216]

The *Restatement*'s stipulation, that servitudes that constituted an unreasonable restraint on trade were invalid, is worth noting for future reference to Australian case law.[217] Utility-tying arrangements are sometimes used in residential development, requiring all homeowners to pay for a water or sewerage service, or even recreation services such as a gym. United States case law in the area is inconsistent, with courts sometimes upholding arrangements,[218] and other times striking them down, for example, if they require landowners to continue to pay for a service they do not need, such as continuing to purchase water from a particular supplier after the homeowner had sunk their own well.[219] The *Restatement* envisaged that such covenants could continue to be invalidated on the grounds they were an unreasonable restraint on trade or alienation, or that they were unconscionable.

The inclusion of an express provision that servitudes were invalid if they were unconscionable[220] was a result of increasingly frequent appeals to the contractual concept of unconscionability in servitudes cases and that the old doctrine of 'touch and concern', as well as resistance to affirmative burdens, had always functioned to invalidate unconscionable servitudes. Unconscionability was described as 'overreaching, unfairness, surprise, or harshness that leads to the conclusion that the servitude should not be enforced, even though the disadvantaged party could have protected him- or herself through the exercise of proper precautions'.[221]

Comment on the section expressly recognised the risk of unconscionable arrangements put in place by developers of HOAs and condominiums. It noted that:

> In common-interest-ownership developments, the developer has the opportunity to create long-term obligations for the property-owners association that will fall on purchasers through the assessments they become obligated to pay. Such obligations can be imposed in the original declaration of covenants governing the project, or *they can be created by contracts entered into while the developer retains control of the homeowners association*. The servitudes imposing the obligation to pay may be unconscionable with respect to these obligations, *particularly if the circumstances include self-dealing by the developer*.[222]

The commentary cited arrangements whereby the developer extracted obligations never to sue the developer as an illustration of an unconscionable servitude. It suggested that a requirement that all property owners pay for the services of an adjoining recreation club owned by the developer would not be unconscionable, so long as the price was comparable to other clubs in the area.

The *Restatement* contained a whole chapter on *residential* common interest developments because they had become such a substantial proportion of the American housing market and because of the significance of property, which is also a home. Echoing the reasoning of Margaret Jane Radin,[223] the *Restatement* acknowledged that a home is not only a 'haven of personal autonomy, liberty, and security, but, for many, it is also a major financial investment' and 'community associations often have substantial power to affect both the quality of life and financial health of their members'.[224] Obviously accepting the reasoning of commentators like Winokur and Sterk,[225] the introductory commentary explicitly recognised that most residential purchasers have no opportunity to negotiate servitudes and they are unlikely to have understood their significance. It notes that even if they have read the servitudes they may not appreciate or be able to anticipate the kinds of rules that the servitude regime empowers the association to make in the future. It recognised the public interest in HOAs which had taken over the role of local government in many areas, a role that would inevitably go back to the public should developments fail, along with all of their infrastructure and expenses.[226] The introductory note to the chapter encapsulated the concerns that had emerged in the literature of the preceding 20 years, highlighted in this chapter:

> The law of residential common-interest communities reflects these tensions between protecting freedom of contract, protecting private and public interests in security of the home both as a personal base and as a financial asset, and protecting the public interest in the ongoing financial stability of common-interest communities. It also reflects the tensions between protecting the democratic process at work in common-interest communities and protecting the interests of individual community members from imposition by those who control the association.

In short, in the development of common interest communities, there is much more at stake than freedom of contract.

Section 6.7 of the *Restatement* deals with an association's power to create rules. Except as limited by statute or governing documents (the C, C & Rs), it states that a community has the implied power to create *reasonable* rules. These can govern use of the common property or the use of individual property 'to protect the common property'. In the absence of specific authorisation in the declaration, the common-interest community does not have the power to adopt rules that 'restrict the use or occupancy of, or behavior within, individually owned lots or units'.

The *Restatement*'s provision on rules is based on the clear distinction between servitude regimes and subsequent rules created by the community. The former are only invalid if illegal, unconstitutional, unconscionable or against public policy, or they are unreasonable restraints on alienation or trade; the latter are invalid if they are unreasonable.[227] The requirement of reasonableness, as well as an assumption that rules that cannot regulate behaviour within lots, was written to ensure that the rule-making power did not conflict with 'traditional expectations of property owners that they are free to use their property for uses that are not prohibited and do not unreasonably interfere with the neighbors' use and enjoyment of their property'.[228]

Despite the *Restatement*'s description of a broad, existing judicial power to invalidate servitudes and association rules, it would be misleading to suggest that

United States courts are in fact quick to invalidate. The case law cited by the commentary in the *Restatement*, as well as that decided in its wake,[229] is characterised by a level of unpredictability. Courts are often torn between holding purchasers to their bargains, particularly when servitudes were clearly disclosed, and preventing undesirable or exploitative practices. Holding purchasers to bargains often wins out. For example, in the leading case of *Hidden Harbour Estates, Inc* v. *Norman*[230] the homeowner who challenged the association rule to ban the sale of alcohol in the clubhouse lost. However, whether litigants are successful in challenges to covenants or rules is not the essential point. The essential point is that the discretionary power that United States courts have to invalidate servitudes and rules allows courts and landowners *to conduct a dialogue about the desirability of the entitlements created by private citizens, and for the courts to adjudicate on those entitlements via a process of reasoned deliberation.*

This is essential to contain the potentially damaging effects inherent in privately negotiated covenants. As we saw in Chapter 2, property law does not operate like a contract, allowing parties to create almost any entitlement they please. Property law operates through default bundles of rights that have been constructed to serve free market and democratic ideals, and which create the kind of human relationships that we are happy to defend.[231]

Covenants are a profound anomaly in this system of default bundles. If permitted to be both positive as well as negative, their content is potentially as varied as a contract. They threaten to obliterate the *numerus clausus* principle, allowing all manner of 'incidents of a novel kind [to] be devised and attached to property at the fancy and caprice of any owner',[232] risking the creation of 'modern variations of feudal serfdom'.[233] The only limits on this are common law or statutorily designed boundary rules that force these contractual agreements back into a predefined, property box. Property boxes (ideally) only include entitlements that do not compromise democratic or efficient land use.

The traditional boundary rule for covenants in the United States was the 'touch and concern' doctrine, which the *Restatement* replaced with a prohibition on servitudes that are illegal, unconstitutional, unconscionable or against public policy, or which are unreasonable restraints on trade or alienation. For association-made rules, the *Restatement* continued to use the judicially constructed boundary rule of 'reasonableness'. None of these provisions can necessarily be applied easily, but that is not fatal. In fact, the rules' greyness is their very function: it allows space for a dialogue about legitimate entitlements between people with respect to land. As the theorists in Chapter 2 rightly insist, '[c]hoices about property entitlements are *unavoidable*, and, despite the incommensurability of values, rational choice remains possible through reasoned deliberation'.[234] When determining property rights, there is 'no alternative but to make arguments that elaborate fundamental human values and that express our considered commitments to judgments about morality and justice'.[235] When making those arguments we must consider the 'the full range of human values we care about'.[236] United States covenant law, whether through the old 'touch and concern' doctrine, its reformulation in the *Restatement*, or the requirement of 'reasonableness', allows citizens and courts to do exactly that. In Part II of the book, we will turn back to Australian strata and community title law and investigate the extent to which these ideas have been implemented or even recognised in the context of by-laws, the statutory equivalent of United States servitudes and association rules.

Notes

1 See Chapter 1, n 119.
2 *Mabo* v. *Queensland (No 2)* (1992) 175 CLR 1, 29 (Brennan J).
3 Susan F French, 'Design Proposal for the New Restatement of the Law of Property – Servitudes' (1987) 21 *UC Davis Law Review*; Evan McKenzie, *Privatopia: Homeowner Associations and the Rise of Residential Private Government* (Yale University Press, 1994).
4 The following description is from the leading United States textbook on HOAs and condominiums: Wayne S Hyatt and Susan F French, *Community Association Law: Cases and Materials on Common Interest Communities* (Carolina Academic Press, 2nd edn, 2008) 19–41.
5 For the sake of completeness, as explained in Chapter 1, United States co-operatives are the equivalent of Australian company title.
6 Hyatt and French, above n 4, 32.
7 Ibid.
8 Dan Tarlock, 'Touch and Concern is Dead, Long Live the Doctrine' (1998) 77 *Nebraska Law Review* 804, 821.
9 *Davis–Stirling Common Interest Development Act* Pt 5, Cal Civ Code is the section of the *California Civil Code* that governs condominiums, co-operatives and common interest developments.
10 For example, in Australia, there is no need for lawyers to create a governing body or easements giving owners access to common property. On registration of a strata or community plan of subdivision, a body corporate is automatically created and title to common property vests in it for the benefit of all owners as tenants in common. As tenants in common, owners are entitled to 'occupy the whole', and do not need easements over common property created in their favour.
11 E Rabin, *Fundamentals of Modern Real Property Law* (Foundation Press, 1974) 489, cited in Susan F French, 'Toward a Modern Law of Servitudes: Reweaving the Ancient Strands' (1981) 55 *Southern California Law Review* 1261, 1261.
12 Uriel Reichman, 'Toward a Unified Concept of Servitudes' (1981) 55 *Southern California Law Review* 1177.
13 Ibid. 1183.
14 United States discussions of servitudes set great store by their recording system, which from an Australian perspective seems a little ironic. The prevalence of title insurance in the United States is a consequence of the unreliability of their recording systems; see n 138. Further, as noted above, by-laws affecting HOAs are often not recorded as this makes them more difficult to alter: above n 7. In contrast, in Australia, title insurance is almost unknown, as the Torrens system of state guaranteed title is so rigid that failing to register is a risk few owners will take. As a result, our registers are genuine 'mirrors' of title; cf. the questionable practice of including provisions in community title registered management statements which purport to grant the body corporate power to create unregistered rules: see discussion in Chapter 6, n 85.
15 *Spencer's Case* (1583) 5 Co Rep 16a; 77 ER 72. *Spencer's Case* created the doctrine that leasehold covenants must 'touch and concern' the land to run with the lease.

16 Reichman, above n 12, 1213.
17 *City of New York* v. *Delafield 246 Corp*, 662 NYS 2d 286, 294 (1997), cited in Tarlock, above n 8, 830.
18 Reichman, above n 12, 1217.
19 *Tulk v. Moxhay* (1848) 2 Ph 774; 41 ER 1143.
20 *Haywood* v. *Brunswick Permanent Benefit Building Society* (1881) 8 QBD 403.
21 *London & South Western Railway Co* v. *Gomm* (1882) 20 Ch D 562.
22 Reichman, above n 12, 1231.
23 Ibid. 1233.
24 Ibid. (emphasis added).
25 Joseph William Singer, 'Democratic Estates: Property Law in a Free and Democratic Society Essay' (2008) 94 *Cornell Law Review* 1009, 1025–6.
26 French, 'Towards a Modern Law of Servitudes', above n 11.
27 Ibid. 1265.
28 Burdens in gross are for the benefit of specific people, as opposed to a piece of land, the benefit of which then flows to whoever owns that land from time to time. Burdens in gross are useful for public authorities, like local councils, which might benefit from the right to run pipes along, walk across, take water from or prevent building on someone else's land, although the authority has no particular land of its own to which to attach the benefit. Rights in gross are now often created by legislation, for example, *Conveyancing Act 1919* (NSW) s 88A.
29 French, 'Towards a Modern Law of Servitudes', above n 11, 1290.
30 Ibid. 1304.
31 Ibid. 1305.
32 §2–117, 7A ULA (1981).
33 Michael A Heller, 'The Boundaries of Private Property' (1999) 108 *Yale Law Journal* 1163, 1165–6.
34 Ibid. 1184.
35 For example, see *Taunton Enterprises Pty Ltd Appeal Against Order No C579394 Strata Title Act* [2005] TASRMPAT 76, discussed in Chapter 5.
36 Richard Allen Epstein, *Simple Rules for a Complex World* (Harvard University Press, 1995).
37 Richard A Epstein, 'Notice and Freedom of Contract in the Law of Servitudes Comments' (1981) 55 *Southern California Law Review* 1353, 1358.
38 Although the use of unregistered Community Handbooks, referred to in the by-laws of some developments, undermines this principle: see Chapter 6, n 85.
39 *Breskvar* v. *Wall* [1971] HCA 70; (1971) 126 CLR 376, per Barwick CJ, [15].
40 Epstein, 'Notice and Freedom of Contract', above n 37, 1359.
41 Ibid. 1366.
42 For example, 'optional' 'model' by-laws in Australian states are evidence of a strong adherence to freedom of contract principles. For example, in New South Wales, important provisions on pets and child safety have been put into model by-laws on the assumption that the provisions are desirable. However, their placement in alterable by-laws, as opposed to the body of the legislation, demonstrates a tendency on the part of the legislature to default to freedom of contract principles; that is, if citizens voluntarily decide they do not want these provisions, no matter how ideal the legislature believes them to be, they are free to exclude them.
43 Gregory S Alexander *et al.*, 'A Statement of Progressive Property' (2008) 94 *Cornell Law Review* 743.
44 Singer, 'Democratic Estates', above n 25, 1049–50.
45 Alexander *et al.*, 'A Statement of Progressive Property' above n 43, 744.
46 Joseph William Singer, *Entitlement: The Paradoxes of Property* (Yale University Press, 2000) 215.

47 Joseph William Singer, 'Normative Methods for Lawyers' (2008) 56 *UCLA Law Review* 899, 904.
48 Carol M Rose, *Property & Persuasion: Essays on the History, Theory and Rhetoric of Ownership* (Westview Press, 1994).
49 Carol M Rose, 'Servitudes, Security, and Assent: Some Comments on Professors French and Reichman' (1981) 55 *Southern California Law Review* 1403, 1404, emphasis added.
50 Ibid. 1405.
51 Ibid. 1408. Rose referenced the emerging United States case law on 'reasonableness' in association rules.
52 Sociological analysis of private communities can be found in books like Edward J Blakely and Mary Gail Snyder, *Fortress America: Gated Communities in the United States* (Brookings Institution Press, 1997) and Setha M Low, *Behind the Gates: Life, Security, and the Pursuit of Happiness in Fortress America* (Routledge, 2003).
53 Gerald E Frug, 'The City As a Legal Concept' (1979) 93 *Harvard Law Review* 1057.
54 Robert C Ellickson, 'Cities and Homeowners Associations' (1981) 130 *University of Pennsylvania Law Review* 1519. Ellickson had previously argued for private, as opposed to public regulation of land in the United States, in particular covenants in preference to zoning: Robert C Ellickson, 'Alternatives to Zoning: Covenants, Nuisance Rules, and Fines As Land Use Controls' (1972) 40 *University of Chicago Law Review* 681.
55 Ellickson, 'Cities and Homeowners Associations', above n 54, 1523.
56 Ibid. 1527.
57 Ibid. 1527, note 28.
58 The United States condominium and co-operative industry was so badly affected by exploitative and fraudulent developer practice that a Federal inquiry was initiated in the early 1970s: Federal Department of Housing and Urban Development, 'Condominium and Cooperative Study' (1975). See also W E Shipley, 'Self-Dealing by Developers of Condominium Project as Affecting Contracts or Leases with Condominium Association' (1976) 73 *American Law Reports* 613; G Batsel, 'Florida Condominiums – Developer Abuses and Securities Law Implications Create a Need for a State Regulatory Agency' (1972–1973) 25 *University of Florida Law Review* 350; Notes, 'Condominium Regulation: Beyond Disclosure' (1975) 123(3) *University of Pennsylvania Law Review* 639; J Young, 'Areas of Dispute in Condominium Law' (1976) 12 *Wake Forest Law Review* 983. Developer abuse led to the creation of the Community Association Institute, the leading United States professional body for private communities, which, in an effort to save the industry, began to publish advisory manuals on optimal developer practice: see, e.g. Urban Land Institute and Community Association Institute, *Creating a Community Association: The Developer's Role in Condominium and Home Owner Associations* (Urban Land Institute, 1977). Developer abuse in modern developments is common in the UK, Australia, Hong Kong and New Zealand: Sarah Blandy, Ann Dupuis and Jennifer Dixon, *Multi-Owned Housing: Law, Power and Practice* (Ashgate, 2010), and has been a feature of booming property markets for centuries: Philip Booth, 'Speculative Housing and the Land Market in London 1660–1730: Four Case Studies' (1980) 51(4) *Town Planning Review* 379.
59 Ellickson, 'Cities and Homeowners Associations', above n 54, 1527.
60 Ibid.
61 James M Buchanan, 'An Economic Theory of Clubs' (1965) 32(125) *Economica* 1.
62 Charles M Tiebout, 'A Pure Theory of Local Expenditures' (1956) 64(5) *Journal of Political Economy* 416.
63 Fred Foldvary, 'Proprietary Communities and Homeowner Associations' in David T Beito, Peter Gordon Tabarrok and Alexander Tabarrok (eds), *The Voluntary City: Choice, Community, and Civil Society* (University of Michigan Press, 2002) 258.

64 The same can be said for United States courts which have not made any clear ruling on whether religiously or ethnically based restrictions would be struck down on the *Shelley* logic: Nomi Maya Stolzenberg, 'Return of the Repressed: Illiberal Groups in a Liberal State, The Liberalism and Illiberalism' (2001) 12 *Journal of Contemporary Legal Issues* 897.

65 Although Congress subsequently disagreed in relation to housing, enacting amendments to the *Fair Housing Act*, 42 USCA §3601 (1968) to prevent discrimination against families with children: Nicole Napolitano, 'Fair Housing Act Amendments and Age Restrictive Covenants in Condominiums and Cooperatives, The Note' (1999) 73 *St John's Law Review* 273.

66 Ellickson, 'Cities and Homeowners Associations', above n 54, 1528.

67 *Hidden Harbour Estates, Inc* v. *Basso*, 393 So 2d 637 (Fla Dist Ct App, 1981) (emphasis in original); ibid. 1529.

68 309 So 2d 180, 182 [15] (Fla Dist Ct App, 1975). In this case the Court had upheld a decision by the association to ban the sale of alcohol in the condominium clubhouse. The association had canvassed the views of residents and 70 per cent were in favour of the ban: Hyatt and French, above n 4, 96.

69 The business judgment rule was borrowed from corporate law. The rule is that the judiciary does not interfere with a decision made by company directors, if the decision was made in good faith and legitimately furthering corporate purposes. It is more commonly used for review of financial decisions, while the reasonableness test is used for review of governance decisions. See Hyatt and French, above n 4, 211–66; Wayne S Hyatt, 'Common Interest Communities: Evolution and Reinvention' (1997) 31 *John Marshall Law Review* 303.

70 Ellickson, 'Cities and Homeowners Associations', above n 54, 1530.

71 One of the best ways to illustrate the use of internal norms to judge private rules is to consider the case of *Wisconsin* v. *Yoder*, 406 US 205 (1972). This was an attempt by the State of Wisconsin to force Amish communities to educate their children beyond 8th grade. The Supreme Court ruled in favour of the Amish parents on the basis that the First Amendment of the United States Constitution protected the practice of genuine religious beliefs. In essence, the internal norms of the Amish community were used to test the validity of their rules in relation to schooling, in preference to the external norms and rules of the wider community.

72 Courts and drafters of founding documents and legislation have instead opted for an all or nothing approach; that is, the action is entirely prohibited without unanimous consent or it is entirely permissible with majority or supermajority consent and the minority just has to wear the consequences. The intermediate position suggested by Ellickson, that it might be permissible but compensation must be paid, has never been implemented.

73 Above n 49.

74 In the New South Wales context, see Cathy Sherry, 'How Indefeasible is Your Strata Title?' (2009) 21(2) *Bond Law Review* 159.

75 A distinction between original and amended by-laws has rarely been considered in Australian case law, but in *Casuarina Rec Club Pty Ltd* v. *Owners – Strata Plan 77971* (2011) 80 NSWLR 711, Young JA said in *obiter* comments that some

> very different considerations arise when one is considering whether an original by-law is valid as opposed to an amended by-law. In the case of an original by-law, people have vested rights which are not lightly to be diminished by an amendment at the behest of the majority:

at 718 [50]. Counsel did not seem to have been eager to pursue the matter, possibly because the distinction is arguably impractical to apply.

76 Gregory S Alexander, 'Dilemmas of Group Autonomy: Residential Associations and Community' (1989) 75 *Cornell Law Review* 1.

77 Gregory S Alexander and Eduardo M Peñalver, 'Properties of Community' (2009) 10 *Theoretical Inquiries in Law* 127; Gregory S Alexander and Eduardo Moisés Peñalver, *Property and Community* (Oxford University Press, 2010); Eduardo M Peñalver, 'Property's Memories' (2011) 80 *Fordham Law Review* 1071; Alexander *et al.*, 'A Statement of Progressive Property', above n 43; Gregory S Alexander, 'The Social-Obligation Norm in American Property Law' (2008) 94 *Cornell Law Review* 745; Gregory S Alexander, 'Pluralism and Property' (2011) 80 *Fordham Law Review* 1017.

78 Gregory S Alexander, 'Dilemmas of Group Autonomy', above n 76, citing Robert H Wiebe, *The Segmented Society: An Introduction to the Meaning of America* (Oxford University Press, 1975) 2.

79 See, e.g. Stolzenberg, above n 64; Larry Alexander, 'Illiberalism All the Way Down: Illiberal Groups and Two Conceptions of Liberalism' (2001–2002) 12(2) *Journal of Contemporary Legal Issues* 625.

80 Gregory S Alexander, 'Dilemmas of Group Autonomy', above n 76, 5.

81 Caroline Butler-Bowdon *et al.*, *Homes in the Sky: Apartment Living in Australia* (Melbourne University Publishing, 2007); Robert Freestone, *Model Communities: The Garden City Movement in Australia* (Nelson, 1989). See discussion in Chapter 1, 10–26.

82 Gregory S Alexander, 'Dilemmas of Group Autonomy', above n 76, 9.

83 See also McKenzie, *Privatopia*, above n 3; Robert D Putnam, *Bowling Alone: The Collapse and Revival of American Community* (Simon & Schuster, 2000). After an exhaustive review of empirical material, Putman concludes that commuting time to suburban developments is the single biggest cause of Americans' dramatic drop in civic participation. This would suggest that, ironically, HOAs and community title developments built on the urban fringe and marketed as community-enhancing, potentially reduce community engagement.

84 Gregory S Alexander, 'Dilemmas of Group Autonomy', above n 76, 10.

85 See the theory of 'residential shopping', above n 62.

86 Gregory S Alexander, 'Dilemmas of Group Autonomy', above n 76, 21.

87 Ibid. 23.

88 Ibid.

89 Ibid. 26.

90 Ibid. 39–43.

91 Ibid. 4.

92 Larry Alexander, above n 79, 629.

93 Gregory S Alexander, 'Dilemmas of Group Autonomy', above n 76, 38. See also Cass R Sunstein, 'Legal Interference with Private Preferences' (1986) 53(4) *University of Chicago Law Review* 1129.

94 Gregory S Alexander, 'Dilemmas of Group Autonomy', above n 76, 39.

95 See discussion in Chapter 2, 62–5. See also Gregory S Alexander, 'The Publicness of Private Land Use Controls' (1999) 3 *Edinburgh Law Review* 176; Carol M Rose, 'The Moral Subject of Property' (2007) 48 *William & Mary Law Review* 1897; Carol M Rose, 'Property as the Keystone Right Propter Honoris Respectum' (1995) 71 *Notre Dame Law Review* 329.

96 See also Gregory S Alexander, 'The Publicness of Private Land Use Controls', above n 95.

97 Gregory S Alexander, 'Dilemmas of Group Autonomy', above n 76, 50.

98 Group homes or aged care facilities are often run by the private sector in the United States, and HOAs characterise their properties as business, not residential premises. Zoning and covenant litigation in relation to group homes became particularly intense in the late 1970s and early 1980s, after many people had been moved out of large institutions and into the community (as was common in the western world: see, e.g. David T Richmond, New South Wales Department of Health, *Richmond Report*

Recommendations – Inquiry into Health Services for the Psychiatrically Ill and Developmentally Disabled (March 1983)). See Robert D Brussack, 'Group Homes, Families, and Meaning in the Law of Subdivision Covenants' (1981–1982) 16 *Georgia Law Review* 33; Thomas F Guernsey, 'The Mentally Retarded and Private Restrictive Covenants' (1983–1984) 25 *William and Mary Law Review* 430.

99 Roberto Mangabeira Unger, *Knowledge and Politics* (Macmillan, 1975) 266, cited in Gregory S Alexander, 'Dilemmas of Group Autonomy', above n 76, 53.

100 Gregory S Alexander, 'Dilemmas of Group Autonomy', above n 76, 55–6. For a critique of Alexander's communitarian conclusions, see Glen Robinson, 'Communities' (1997) 83 *Virginia Law Review* 269, 295ff.

101 Gregory S Alexander, 'Dilemmas of Group Autonomy', above n 76, 56.

102 Ibid. 59.

103 Ibid. 60.

104 Alexander *et al.*, 'A Statement of Progressive Property', above n 43, 744.

105 Singer, 'Normative Methods for Lawyers', above n 47, 904.

106 Ellickson 'Alternatives to Zoning', above n 54.

107 Richard A Epstein, 'Covenants and Constitutions' (1987) 73 *Cornell Law Review* 906.

108 Stewart E Sterk, 'Foresight and the Law of Servitudes' (1987) 73 *Cornell Law Review* 956.

109 This is the connection between property and morality discussed in Chapter 2, 49–50: Thomas W Merrill and Henry E Smith, 'The Morality of Property' (2007) 48 *William and Mary Law Review* 1849.

110 Ibid. 968.

111 Ibid. 970.

112 Stewart E Sterk, 'Freedom from Freedom of Contract: The Enduring Value of Servitude Restrictions' (1984) 70 *Iowa Law Review* 615.

113 Ibid. 622.

114 Servitudes can be consensually removed if a burdened landowner pays a benefited landowner to release the servitude; however, if there are hundreds of burdened landowners and worse still, hundreds of benefited owners, as there are in large-scale subdivisions, this is an almost impossible task.

115 Sterk, above n 112, 638.

116 O W Holmes, 'Path of the Law' (1896) 10 *Harvard Law Review* 457, cited in Sterk, above n 112, 660.

117 See Chapter 2, n 29.

118 Sterk, 'Freedom from Freedom of Contract', above n 112, 661.

119 Rose, above n 49, 1404.

120 Robert H Nelson, *Private Neighbourhoods and the Transformation of Local Government* (Urban Institute Press, 2005) xiii.

121 Robert Dilger, 'Residential Community Associations: Issues, Impacts, and Relevance for Local Government' (1991) 23(1) *State and Local Government Review* 17, 17.

122 James L Winokur, 'The Mixed Blessing of Promissory Servitudes: Toward Optimizing Economic Utility, Individual Liberty, and Personal Identity' [1989] *Wisconson Law Review* 1.

123 See also Robert Natelson, 'Consent, Coercion, and "Reasonableness" in Private Law: The Special Case of the Property Owners Association' (1990) 51 *Ohio State Law Journal* 41.

124 Winokur, above n 122, 18–23.

125 Houston, the fourth largest city in the United States, has no public zoning, but rather, relies on the market and private land-use arrangements like covenants: Ellickson, 'Alternatives to Zoning' above n 54; Nelson, above n 120, 135–91.

126 One of the most contentious issues in United States constitutional law is the question of when land regulation, such as zoning, goes so far as to amount to a taking of

private property which requires compensation in accordance with the Fifth and Fourteenth Amendments. Zoning was first challenged in *Village of Euclid, Ohio* v. *Ambler Realty Co*, 272 US 365 (1926) and was significantly litigated from the 1980s onwards: A J Casner *et al.*, *Cases and Text on Property* (Aspen Publishers, 5th edn, 2004) 1109–224. Regulation of land by government has not stimulated similar debates in Australia about the need for compensation: Sharon Christensen and W D Duncan, 'Aligning sustainability and the Torrens register: Challenges and recommendations for reform' (2012) 20 *Australian Property Law Journal* 112, 119.

127 Jane Jacobs, *The Death and Life of Great American Cities* (Penguin, 1965), cited in Winokur, above n 122, 22.

128 Despite seeming to be based on empirical research conducted entirely from her doorstep and window in Greenwich Village in the 1950s, Jacob's delightful book has been profoundly influential in urban planning, particularly in the push for denser cities. For example, 60 years on, and 10 000 miles away, the City of Sydney's Laneways Project bears all the hallmarks of Jacob's thinking: City of Sydney, *City Transformation* www.cityofsydney.nsw.gov.au/development/cityimprovements/roadsandstreetscapes/CBDLanewayUpgrade/Default.asp.

129 Winokur, above n 122, 53.

130 Above n 24.

131 Ibid. 57.

132 Ibid. 58.

133 Cf. Buchanan, above n 61; Tiebout, above n 62.

134 McKenzie, *Privatopia*, above n 3, 11–12.

135 Evan McKenzie, *Beyond Privatopia: Rethinking Residential Private Government* (Urban Institute Press, 2011) 75–8. Perkins estimates that 50 per cent of homeowners in Las Vegas now live in an HOA: Casey Perkins, 'Privatopia in Distress: The Impact of the Foreclosure Crisis on Homeowners' Associations' (2010) 10 *Nevada Law Journal* 561, 561–2.

136 Winokur, above n 122, 59.

137 *Sanborn* v. *McLean*, 233 Mich 227 (1925); *Turner* v. *Brocato*, 206 Md 336 (1954), cited by Winokur, above n 122, 61, footnote 254.

138 R Natelson, above n 123, 59.

139 Hyatt and French, above n 4.

140 Winokur, above n 122, 61–2.

141 In the course of the last New South Wales strata review, there were repeated calls for by-laws to be attached to contracts of sale and posted in the foyer of buildings. It seems that many strata 'stakeholders' are actually unaware that by-laws are easily accessible on the Torrens register: see Global Access Partners, 'Strata Laws On-line Consultation Final Report' (2012) 66. It seems that while the majority of strata owners and residents think that their own understanding of their by-law rights and responsibilities is good, their neighbours do not share this view: H Easthope, B Randolph and S Judd, 'Governing the Compact City: the Role and Effectiveness of Strata Management' (City Futures Research Centre, 2012) 46–7.

142 Winokur, above n 122, 63, footnote 263.

143 Kelly Joyce, *Man Accused of Shooting at HOA Board Members* (4 June 2013) myFoxOrlando.com www.myfoxorlando.com/story/22491059/man-accused-of-shooting-at-hoa-board-members; *Trial Date Set for Man Accused of Fatally Shooting 2 Men at HOA Meeting* (3 June 2013) WLKY.com www.wlky.com/news/local-news/louisville-news/trial-date-set-for-man-accused-of-fatally-shooting-2-men-at-hoa-meeting/-/9718340/20401906/-/rb4fhe/-/index.html; Cameron Steele, *Cabarrus County, NC, Standoff Ends with Suspected Shooter and Two Others Dead* (7 April 2013) PoliceOne.com www.policeone.com/rural-law-enforcement/articles/6187650-3-dead-in-NC-standoff/; *State* v. *Glassel*, 211 Ariz 33, 41–2 (2005), in which Richard Glassel was convicted for the murder of two women at an HOA

meeting after ongoing conflict with the Board; Glassel later died of old age on death row; 'Homeowners' Leader Arrested In Injury Of Neighbor', *Orlando Sentinel* (online), 4 July 2004 (HOA president shot a resident in the leg with bird shot after ongoing dispute) http://articles.orlandosentinel.com/2004-07-04/news/0407040302_1_aggravated-assault-presha-woods-homeowners-association.

144 Winokur, above n 122, 63–4; cf. G Korngold 'Reply: Resolving the Flaws of Residential Servitudes and Owners Associations: For Reformation not Termination' (1990) *Wisconsin Law Review* 513, 522, who suggests that the increase in litigation may be the result of an increased number of servitude regimes.

145 Winokur, above n 122, 62, footnote 259; cf. Korngold, above n 144, 521–2, who posits that non-turnout may be a result of resident satisfaction. Non-participation in governance is a chronic problem in New South Wales: see Easthope, Randolph and Judd, above n 141, 54.

146 M Radin, 'Property and Personhood' (1982) 34 *Stanford Law Review* 957, cited by Winokur, above n 122, 71.

147 Ibid. 73.

148 Winokur, above n 122, 74.

149 Ibid.

150 Ibid. 75.

151 Winokur, above n 122, 78–84.

152 Clayton P Gillette, 'Courts, Covenants, and Communities' (1994) 61 *University of Chicago Law Review* 1375.

153 Ibid.

154 Edward J Blakely and Mary Gail Snyder, *Fortress America: Gated Communities in the United States* (Brookings Institution Press, 1997); Setha M Low, *Behind the Gates: Life, Security, and the Pursuit of Happiness in Fortress America* (Routledge, 2003).

155 Above n 79.

156 Ebenezer Howard, *Garden Cities of To-morrow* (Swan Sonnenshein, 1902).

157 McKenzie, *Privatopia*, above n 3, 8.

158 Jacobs described Le Corbusier's city as like 'a wonderful mechanical toy ... [like] a great, visible ego it tells of someone else's achievement. But as to how a city works, it tells, like the Garden City, nothing but lies': Jacobs, above n 127, 31–2. Jacobs derisively labelled the combined ideals of Howard and Le Corbusier the 'Radiant Garden City Beautiful'.

159 Ibid. 32.

160 McKenzie, *Privatopia*, above n 3, 7.

161 Ibid. 8.

162 Graeme Davison, 'The Past and Future of the Australian Suburb' (1993) 31(2) *Australian Planner* 63.

163 Singer notes a mid-nineteenth-century New York case in which a judge struck down a requirement that tenants pay one quarter of the purchaser price to a landlord each time they sold land. The judge said that United States property law did not allow such feudal land arrangements: Joseph William Singer, *The Edges of the Field: Lessons on the Obligations of Ownership* (Beacon Press, 2000) 22–3.

164 See discussion above, 77.

165 278 NY 248 (1938), discussed by McKenzie, above n 3, 51–5.

166 Reichman n 236.

167 McKenzie, *Privatopia*, above n 3, 17–18.

168 Ibid. 19.

169 Singer, *Entitlement*, above n 4, 152–4.

170 *Buchanan* v. *Warley*, 245 US 60 (1917). See also James A Kushner, 'Apartheid in America: An Historical and Legal Analysis of Contemporary Racial Residential Segregation in the United States' (1979) 22 *Howard Law Journal* 547.

171 McKenzie, *Privatopia*, above n 3, 61.
172 Ibid. 65.
173 Ibid. 70–1.
174 Alexander, 'Dilemmas of Group Autonomy', above n 76, 53.
175 334 US 1 (1948).
176 Strangely the Court's reasoning has been restricted to racially restrictive covenants, although in theory it should apply to the judicial enforcement of any covenant that is contrary to the Constitutional guarantee of equal protection of law – for example, single family covenants. United States courts have consistently declined to apply *Shelley* in this way.
177 McKenzie, *Privatopia*, above n 3, 74–8. See Hyatt and French, above n 4, 551–6.
178 See Alexander 'Dilemmas of Group Autonomy' above n 76.
179 McKenzie, *Privatopia*, above n 3, 81.
180 Ibid. 83.
181 Orange County has a very high proportion of HOAs today. Its homogeneity and 'exclusivity' have made it iconic in popular culture. The County featured in the 2002 satirical film *Orange County* and the long-running TV series, *The OC*; tagline: 'It's where all the beautiful people live'. One of Orange County's oldest HOAs, Coto de Caza, is the centre of the Bravo reality TV show, *The Real Housewives of Orange County*. Coto de Caza is set on 5000 acres, with over 4000 homes and 13 000 residents. See City Guide – Coto de Gaza, *The History* www.orangecounty.net/cities/CotoDeCaza.html.
182 McKenzie, *Privatopia*, above n 3, 107.
183 Above n 121.
184 McKenzie, *Beyond Privatopia*, above n 135, 2.
185 Matthew Gordon Lasner, *High Life: Condo Living in the Suburban Century* (Yale University Press, 2012); Notes, 'Cooperative Housing Corporations and the Federal Securities Laws' (1971) 71 *Columbia Law Review* 118; Federal Department of Housing and Urban Development, 'Condominium/Cooperative Study' (1975); Robert M Nelson, 'Examining Cooperative Conversion: An Analysis of Recent New York Legislation' (1982) 11(4) *Fordham Urban Law Journal* 1089, 1093.
186 *National Housing Act 1962* s 234, cited in McKenzie, *Privatopia*, above n 3, 95.
187 McKenzie, *Privatopia*, above n 3, 95.
188 Ibid. 105.
189 Michael C Kim, 'Involuntary Sale: Banishing an Owner from the Condominium Community' (1997–1998) 31 *John Marshall Law Review* 429.
190 Above n 55.
191 W Hyatt and J Rhoads, 'Concepts of Liability in the Development and Administration of Condominium and Home Owner Associations' (1976) 12 *Wake Forest Law Review* 915.
192 Security is one of the most significant services provided by HOAs. It is estimated that there are more private security guards in America than police. Leisure World, a retirement or 'adult' community in California, has more than 300 security guards patrolling the grounds: McKenzie, *Privatopia*, above n 3, 141.
193 The latest edition of Hyatt's textbook states that 'common interest communities are almost never treated as state actors subject to the constitutional constraints placed on public governments': Hyatt and French, above n 4, 113.
194 326 US 501 (1946).
195 See Joseph William Singer, 'How Property Norms Construct the Externalities of Ownership' in Gregory Alexander and Eduardo Peñalver (eds), *Property and Community* (Oxford University Press, 2010) 57.
196 192 NJ 344 (2007).
197 See Duncan Kennedy, 'Residential Associations as State Actors: Regulating the Impact of Gated Communities on Nonmembers' (1995) 105(3) *Yale Law Journal*

761. Kennedy argues that HOAs should be treated as state actors in relation to non-members because the non-members have not consented to their formation or rules.

198 See Gregory S Alexander, 'Dilemmas of Group Autonomy', above n 76.

199 Susan F French, 'Tradition and Innovation in the New Restatement of Servitudes: A Report from Midpoint *Gallivan Conference*' (1994) 27 *Connecticut Law Review* 119.

200 Susan F French, 'Constitution of a Private Residential Government Should Include a Bill of Rights, The Symposium on the Uniform Real Property Acts' (1992) 27 *Wake Forest Law Review* 345.

201 Ibid. 360.

202 *Hidden Harbour Estates, Inc* v. *Norman* 309 So 2d 180 (Fla Dist Ct App, 1975), cited in ibid. 345.

203 Ibid. 349.

204 See *Simovits* v. *Chanticleer Condominium Assoc* 933 F Supp 1394 (ND Ill, 1996), cited in Hyatt and French, above n 4, 536. The Court awarded economic damages to plaintiffs for the reduced sale price of their condominium caused by a covenant banning residents under 18. The plaintiffs were also awarded the value of the additional mortgage repayments they had to make as a result of delays in the sale because they could not sell to interested purchasers with children.

205 See, e.g. Martha C Nussbaum, 'Deliberation and Insight: *Bloch* v. *Frischholz* and the Chicago School of Judicial Behavior Essay' (2010) 77 *University of Chicago Law Review* 1139, discussed in Chapter 2, 59, which examines a case in which an observant Jewish family had been told to remove a mezuzah which had been placed outside their door, in violation of the condominium rule on hallways.

206 Paula Franzese, 'Does It Take a Village? Privatization, Patterns of Restrictiveness and the Demise of Community' (2002) 47 *Villanova Law Review* 553.

207 McKenzie, *Beyond Privatopia*, above n 135, 14.

208 Ibid. 114.

209 Campbell Robertson and John Schwartz, 'Shooting Focuses Attention on a Program That Seeks to Avoid Guns', *New York Times* (online), 22 March 2012 www.nytimes.com/2012/03/23/us/trayvon-martin-death-spotlights-neighborhood-watch-groups.html?_r=1.

210 'Trayvon Martin's parents settle with Fla HOA', *USA Today* (online), 5 April 2013 www.usatoday.com/story/news/nation/2013/04/05/trayvon-martins-parents-settle/2058239/.

211 Susan F French, 'Highlights of the New Restatement (Third) of Property: Servitudes' (2000) 35 *Real Property, Probate and Trust Journal* 225. For a critique of the *Restatement*, written immediately prior to its release, see Tarlock, above n 8.

212 *Restatement (Third) of Property: Servitudes* (2000).

213 French, 'Highlights of the New Restatement', above n 213, 232–3.

214 *Restatement (Third) of Property: Servitudes* (2000) §3.1.

215 Ibid. §3.4–3.7.

216 Ibid. §3.2(a).

217 See discussion of *Casuarina Rec Club Pty Ltd* v. *Owners – Strata Plan 77971* (2011) 80 NSWLR 711 in Chapter 4, 173–80.

218 *Bennett* v. *Behring Corp*, 466 F Supp 689 (SD Fla, 1979) (obligation to pay for recreational facilities owned by developer and leased to community held not to be unconscionable); *Streams Sports Club Ltd* v. *Richmond*, 99 Ill 2d 182 (1983) (obligation to pay dues to a for-profit sports club touched and concerned the land), cited in Casner *et al.*, above n 126, 1004.

219 *Eagle Enterprises, Inc v. Gross*, 39 NY 2d 505 (1976), cited in Casner *et al.*, above n 126, 1001–3.

220 *Restatement (Third) of Property: Servitudes* (2000) §3.7.

221 Ibid.

222 Ibid. emphasis added.
223 Above n 147.
224 *Restatement (Third) of Property: Servitudes* (2000) ch 6 (Introductory Note).
225 Above n 112, 122.
226 A decade before the GFC, this comment was prescient. Many United States HOAs are now struggling to pay for their extensive, private facilities as increasing numbers of owners are unable to afford the levies: Perkins, above n 135.
227 The greater latitude for founding documents indicated acceptance of some of the reasoning of Ellickson and Epstein, that parties should be freer to create communities in accordance with their own values than they might be to create subsequent rules.
228 *Restatement (Third) of Property: Servitudes* (2000) §6.7 (Power to Adopt Rules Governing Use of Property, b. Rationale).
229 Hyatt and French, above n 4.
230 *Hidden Harbour Estates, Inc* v. *Norman* 309 So 2d 180 (Fla Dist Ct App, 1975).
231 Singer, *Entitlement*, above n 46, 215–16.
232 *Keppell* v. *Bailey* (1834) 2 My & K 517, 39 ER 1042, 1048, 1049.
233 Uriel Reichman, 'Toward a Unified Concept of Servitudes' (1981) 55 *Southern California Law Review* 1177, 1233.
234 Alexander *et al.*, 'A Statement of Progressive Property', above n 43, 744, emphasis added.
235 Singer, 'Normative Methods', above n 47, 904.
236 Singer, 'Democratic Estates', above n 25, 1059.

Part II

4 The economic implications of by-laws

A legislative re-cap

In Chapter 1 we saw that orthodox Anglo-Australian property law stymied the development of both high-rise freehold apartments and low-rise freehold estates with private amenities. Positive obligations for the necessary repairs of a commonly owned building or amenities could not be imposed on freehold owners, and so apartment buildings were company title or tenanted, and suburban subdivisions were only subject to restrictive covenants and serviced with publicly owned facilities. However, company title was not an acceptable security for banks and the *Landlord and Tenant (Amendment) Act (1948)* (NSW) made ownership of apartments for lease less profitable. During the post-War housing shortage, pressure mounted on the government to provide a legislative solution to the problem of freehold apartment ownership.

The solution was found in the *Conveyancing (Strata Titles) Act 1961* (NSW) ('*CSTA*'), which was replicated in all Australian states, and gradually extended to horizontal subdivisions. As noted in Chapter 1, the multiple state Acts now regulate close to 2 million strata and community title lots in Australia.[1] The Acts create individual freehold titles, along with tenancies in common in shared property, and overcome the common law prohibition on positive obligations on freehold land by authorising the body corporate to extract annual levies from owners in proportion to their unit entitlements. The legislation also facilitates the creation of rules in by-laws and management statements that regulate land use and behaviour. By-laws and management statements are simply statutorily authorised restrictive and positive covenants on freehold land, or in the terminology of the United States literature we have been considering, C, C & Rs or servitudes.

As noted in Chapter 1, the content of these restrictions and obligations has very little legislative limit in all states, except possibly Victoria. While most states have by-laws provided by the legislation, sometimes called 'model' by-laws, developers do not have to use these, and in any event, all by-laws can subsequently be altered by appropriate vote. All states' legislation prohibits specific by-laws, for example, those that restrict transfer, leasing and mortgaging, or which ban assistance animals, and some states have more general prohibitions on by-laws that are oppressive or discriminatory. Only Queensland and

Tasmania prohibit by-laws that are 'unreasonable'.[2] These legislative restrictions will be considered in the next chapter.

For now, it is sufficient to note that the Acts give private citizens extremely broad powers to regulate freehold land. The phrase used in a number of state Acts is that by-laws must relate to 'the control, management, administration, use or enjoyment' of lots or common property. It is hard to think of any activity that can occur in a strata or community development that does not fit this description. For example, eating meat in an apartment is an act that relates to the use of a lot, as does painting the walls white, keeping a goldfish or inviting friends for dinner. It is important not to dismiss these by-laws as fanciful, assuming 'of course a by-law could not ban someone eating meat in their own home!' The question is, what part of the law would make such a by-law invalid?

A very broad initial power to write by-laws is analogous to the law in the United States, where HOAs at least, depend on general property law, not legislation, for their existence. Associations are created by developers' lawyers and given broad functions and powers in founding documents. Servitudes regulate land use and behaviour, and as a general rule, there is no legislative limit on their content, in the same way that there is no legislative limit on the content of easements or restrictive covenants in Australia. The content of these interests in land is limited by the common law and equity;[3] that is, *judicial* limits have been placed on the ability of private citizens to regulate freehold land. The primary limit on servitudes in the United States was the 'touch and concern' doctrine, which while sometimes abstruse, was used by judges to strike down socially and economically inefficient or unconscionable rights. That function has now been clarified by the *Restatement (Third) of Property: Servitudes* (2000). In response to the rise of HOAs and condominiums, courts created an additional rule, that the subsequent actions of associations, including the writing of rules, must be 'reasonable'.

This raises the question, if the by-laws in Australian strata and community title legislation are the statutory equivalent of servitudes/freehold covenants – property rights that have always been judicially limited (or in the case of positive covenants in Australia, completely prohibited) – what judicial limit has been placed on the by-law-making power by Australian courts? In the context of the debate between Ellickson and Alexander, both of whom accept that at some point judicial review can be used to strike down privately created rules, where do Australian courts stand?

It is difficult to answer this question because in the 50 years of the by-law-making power's existence it has attracted remarkably little superior court consideration. There are a number of possible explanations for this. First, many older strata schemes are relatively small, with minimal common property, the use of which could be a source of dispute. Large schemes with extensive common property are a relatively recent phenomenon. Second, occupation of apartments in Australia is typically transitory, either because residents are tenants and long-term leasing is rare, or because apartments are a stepping-stone into freestanding housing.[4] If by-laws are a source of frustration, simply moving out is a realistic option.

Third and possibly most significantly, in many states, the legislation is tied to a mandatory dispute resolution system. Queensland's is by far the most detailed,[5] specifically excluding superior court jurisdiction on a number of issues.[6] The rationale for mandatory dispute resolution systems is that they are accessible and affordable, putting Australia at the forefront of high-density housing practice.[7] However, they exist at the expense of the development of the law. Mediated and adjudicated decisions are designed to minimise dispute and allow people to move forward and live together harmoniously.[8] Tribunal decisions do not create binding precedent. Resolution of disputes through mediation, adjudication and tribunals also reinforces the erroneous assumption that strata disputes only raise low-level questions of law. In contrast, (and at risk of sounding like a lawyer), with clear winners and losers, court decisions engender appeals, with higher courts empowered to consider the full range of legal and equitable doctrines and to create precedent. The question of whether mediators, adjudicators and tribunals are adequately resourced and qualified to deal with the actual, as opposed to perceived legal complexity of many strata and community title disputes, merits serious consideration.

Fourth, despite the possibility of higher court appeal in some cases, many residential occupants either cannot afford or do not want to spend the necessary money to pursue by-law disputes through higher courts. (This is of course precisely why most states have an alternative dispute resolution system). Although they may have a strong case, raising an important point of law, they do not pursue an appeal.

Finally, as a result of the connection between property and morality, discussed in Chapter 2, most people have a lay concept of property rights that is consistent with liberal democracy. Regardless of whether they can articulate the legal or political theory, owners may hesitate to interfere with their neighbours via by-laws, consequently minimising by-law dispute. For example, in New South Wales, there has been much discussion of smoking in strata schemes and as a result, the new *Strata Schemes Management Act 2015* ('*SSMA 2015*') explicitly states that smoke penetration may constitute a nuisance, contrary to s 153 of the Act. However, it has always been possible to ban people smoking in their own apartments in New South Wales, because such a by-law clearly relates to the 'use' of a lot. There is even a Supreme Court case confirming this,[9] and yet many bodies corporate and owners do not believe they can ban smoking in their scheme. It seems that owners do not realise just how much power by-laws give them over their neighbours, and this has minimised intrusive, contentious by-laws and consequent litigation.

Although not extensive, there has been higher court consideration of by-laws in Australia, concentrated in New South Wales. Readers who are not lawyers should be warned that we are about to descend into the world of detailed case analysis, a complex and somewhat dry process. Unfortunately, in the real world of law, 'your Honour, it's the vibe' will not win a case.[10] Judges must apply legislative provisions created by elected representatives, and thus fine-grained statutory analysis is required. Because of the variation in state statutes, some of the detail of particular cases will not be applicable in other states. However, in analysing judicial decisions and the statutory provisions that empower or contain them, we must not lose sight of the bigger picture: all states' statutes and judicial decisions are ultimately making choices about the legitimacy of conflicting claims to the limited resource that is land. All property rights, including the right to create by-laws by virtue of property ownership, must be judged not simply with reference to the moment of their creation, but with reference to their exercise through time, as well as their effects on third parties. The 'mix of entitlements and obligations we can legitimately claim [as property rights] depends on the kinds of human relationships we can defend, nothing more and nothing less'.[11] Ideally, those rights should promote, not undermine, the norms of liberal democracy.

White v. *Betalli* (2007) 71 NSWLR 381

The leading New South Wales by-law case is *White* v. *Betalli* ('*Betalli*').[12] It raised the unusual question of whether it is possible to create by-laws that grant lot owners rights to use other owners' *lots*. The land was a two-lot residential scheme on a river, with special by-law 20 stating that:

> The Registered Proprietors for the time being of Lot 1 shall have the right to store small watercraft within the area denoted (A) on the sketch annexed to this instrument.

The 'instrument' was the registered strata plan and area 'A' was *inside* the boundaries of Lot 2. The plan had been registered by the developer and thus the special by-law was in place at the time both the appellant and the respondent purchased their respective lots.

Special by-law 20 was expressed to be made pursuant to s 52 of the former *Strata Schemes Management Act 1996* ('*SSMA 1996*') Section 52 permitted the creation of 'exclusive use' by-laws which allowed one or more lot owners exclusive use of part of *common property*, typically with the shifting of maintenance and repair from the body corporate to those individuals. Exclusive-use by-laws are now called 'common property rights by-laws' in New South Wales,[13] but they are the same as their former incarnation, as well as being conceptually identical to similar provisions in all other states' Acts.

Both parties in *Betalli* conceded that as special by-law 20 related to lot property, not common property, it could not in fact be a by-law pursuant to s 52. The strata plan and by-laws had obviously been created by an inexperienced surveyor and lawyer with inadequate understanding of the legislation.

The appellant, the owner of lot 2, argued that the by-law was invalid on a number of grounds. First, she said that the rights created by the by-law were in the nature of an easement and that as the legislature had provided a specific means by which easements could be created in s 88B of the *Conveyancing Act 1919* (NSW),[14] the principle in *Anthony Hordern & Sons Ltd* v. *Amalgamated Clothing and Allied Trades Union of Australia*[15] precluded the operation of general provisions such as s 43 of the *SSMA 1996* to achieve the same effect. This argument was rejected by Santow JA at [32], Campbell JA agreeing at [207], on the grounds that the *SSMA 1996* 'creates an alternative mode for creating what is in the nature of an easement or restrictive covenant, with its own detailed prescription distinct from that under the *Conveyancing Act*'.

The applicant then argued that the by-law was ultra vires, as the *SSMA 1996* did not authorise the making of a by-law that gave rights to one lot owner to use another's lot. Santow JA held at [42] that there is 'a broad capacity under that legislation to create by-laws, subject only to there being no incompatibility or inconsistency with any Act or law'. Campbell JA agreed, saying, in what has become a seminal statement of the law, that,

> There is nothing in the notion of a by-law that, of itself, imposes any kind of limitation on the kind of regulation that might be adopted, beyond that it is for the regulation of the particular community to which it applies. Any limitation on the type of restriction or regulation that can be a by-law must arise from the statute that enables the by-laws to be created, or from the general framework of statute law, common law and equity within which that local community is created and administered.[16]

Santow JA agreed with the trial judge that the by-law was 'appropriate to the type of strata scheme concerned', that is, a by-law relating to the storage of a boat in a residential scheme located on a river.[17] Referring to the decision of the Court of Appeal in *Sydney Diagnostic Services Pty Ltd* v. *Hamlena Pty Ltd*[18] which upheld a by-law prohibiting a proprietor or occupier of a lot from engaging in any enterprise other than the practice of medicine, but excluding pathology, the trial judge had said that as by-laws may be made 'which substantially interfere with the right of an owner of a lot to use the lot, it is hard to see why it should be contrary to the 'scheme' of the *Strata Schemes Management Act* for a by-law to confer on one lot owner the right to use part of another lot'.[19] The trial judge had argued that this did not raise the possibility of injustice because by-laws could not be made for improper purposes, and adjudicators had the power to amend, repeal or declare by-laws invalid under ss 157 and 159 of the *SSMA 1996*. Further, no question of injustice could arise when the by-law was in place *before* purchase of the affected lot.

There are a number of difficulties with the reasoning of the trial judge. First, while adjudicators, (and now only the Tribunal[20]), did have the power to invalidate by-laws under ss 157 and 159 of the Act, because the by-law-making power is so wide, the sections were essentially circular and provided no real limit on the content of by-laws. Second, the wider a by-law-making power, the narrower the area that could be considered 'improper', or fraud on the minority, or fraud on a power. Bar the narrow exceptions in the legislation at the time, any by-law that related to the use or enjoyment of lot or common property was within the legislative power and thus by definition proper. The difficulty of applying the doctrines of fraud on a power or fraud on the minority to strata schemes is evidenced by the fact that they have only been argued successfully in a single case in New South Wales.[21] That this lack of application could be a result of a lack of improper or abusive practice in strata schemes is untenable.[22] The recent inclusion of a prohibition on by-laws that are 'harsh, unconscionable or oppressive' is a significant improvement to the New South Wales Act,[23] but it still sets the bar high. It is unlikely that a by-law for the storage of a boat would meet that description. Finally, that the purchaser had notice of the by-law has some significance, but notice has never been a cure-all in property law. That was the arguable mistake of Lord Cottenham in *Tulk* v. *Moxhay*,[24] and from which subsequent courts consistently retreated. Notice as a justification for binding subsequent purchasers to almost any and all servitudes, was the argument made by Epstein,[25] and as we saw in Chapter 3, with which other United States academics and judges have consistently disagreed.[26] Restrictions and obligations on property can be inefficient and socially undesirable. Giving a purchaser notice of them does not make them any less so.

Santow JA's rejection of the appellant's first argument, that the by-law created an easement and thus should have been effected under the *Conveyancing Act 1919* (NSW), is significant. With all due respect to Santow JA, it is in no way clear that the *SSMA 1996* intended to create 'an alternative mode for creating what is in the nature of an easement [*over an individual lot*] ... with its own detailed prescription distinct from that under the *Conveyancing Act*'. It is clear the *Act* intended to allow people to create rights in the nature of an easement over *common property* because of the enactment of s 52, but it is not clear that s 47 of the *SSMA 1996* or s 8 of the *Strata Schemes Development Act 1973* (NSW) ('*SSDA*') was meant to be used for the same purposes in relation to *private lot* property. This is unclear because of the distinct *lack* of 'detailed prescription' in s 47 of the *SSMA 1996*, and the complete absence of prescription in s 8 of the *SSFDA*. A by-law just has to be for the 'control, management, administration, use or enjoyment' of the lots or common property.

Further, it is not correct to suggest that the *Conveyancing Act 1919* (NSW) provides 'detailed prescription' for easements or covenants. The *Conveyancing Act 1919* (NSW),[27] like the *Real Property Act 1900* (NSW),[28] and Torrens and general property statutes in other states, simply deals with formal requirements of writing and registration, while the bulk of the validity of easements and covenants is determined by the common law and equity respectively. Santow JA's own judgment in *Clos Farming Estates* v. *Easton*,[29] is a perfect case in point.

The case concerned Le Clos Verdun, a failed viticulture development on the Hastings River in northern New South Wales. Purchasers could buy a rural home and part of a vineyard, without responsibility for grape growing and wine production. There were complex contractual arrangements in place between the plaintiff and defendants in relation to the growing, harvesting and sale of grapes, but these had expired, leaving a detailed s 88B registered easement granting extensive rights over servient agricultural land. As noted by the trial judge, Bryson J at [14], the essential distinction between the easement rights and the contractual rights was that the effect of the easement would be 'the same when all concerned now are dead and forgotten and their contracts have turned to dust'.[30]

Bryson J at trial, and on appeal, Santow JA (Mason P and Beazely J A, agreeing), held that the easement was invalid on the grounds that it offended the second and fourth conditions in *Re Ellenborough Park*,[31] which require an easement to 'accommodate' (genuinely benefit) the dominant tenement, and to be capable of forming the subject matter of a grant (inter alia, not be so great as to amount to ownership). As we saw in Chapter 2, judicially created criteria like the second and fourth conditions of *Re Ellenborough Park* are 'boundary rules' that prevent socially and economically damaging restrictions attaching to land. Like the 'touch and concern' doctrine in United States covenant law discussed in Chapter 3, or even the 'touch and concern' doctrine traditionally applied in Anglo-Australian landlord and tenant law,[32] while these judicial formulations can seem abstruse, they actual perform a vital function in maintaining the economic and social utility of land. As Bryon J stated at trial in *Clos Farming Estates*,[33]

> [t]he Common Law and its judges have for centuries favoured land being fully available for its owner to use and sell, and restricted the creation and operation of interests, including new classes of interests, which might have different effects. The enormous complexity of land law from medieval times onwards obscures this policy, but resistance to restrictions on use and alienation of land and *on the wishes of property owners to limit the rights of their successors*, to restrict alienation and to devise new forms by which to do so is of very long standing.... The Common Law and its courts maintain public policy controls over the kinds of easements which may be created.

Bryson J went so far as to say at [50] that the extensive easement in the case was 'a re-invention, and an imposition on freehold title, of the substance of the scheme of manorial and copyhold title which existed in England centuries ago' and that the freeholder owners were 'neutralised and powerless, unable to control or in truth to influence what is to happen on their agricultural land'. By holding the registered easement to be invalid the judges were maintaining the economic viability of the land; put simply, by freeing the land of the easement they were making it saleable again. This was consistent with the values of a modern, democratic capitalist economy, as opposed to a feudal state.

In contrast, being creatures of statute and not the common law or equity, none of this doctrine applies to strata by-laws, and as a result, there is none of the corresponding critical assessment of their burdens and restrictions. Had the development in *Clos Farming Estates* been created under strata or community title legislation with the restrictions in by-laws instead of easements, it would have been no more economically successful, and had the landowners been unable to agree to vote to remove the by-laws, a court would not have been able to assist them.[34] This is because the strata Acts deal with the technical creation of by-laws, but there are no accompanying legislative provisions which empower courts to assess the economic, social or political consequences of by-laws. Despite the fact that by-laws are property regulation, the legislature has opted for a freedom of contract ethos; if parties freely create by-laws or purchase by-law limited land, so be it. Nor have courts followed their practice in relation to easements and freehold covenants and attempted to craft any rules or limitations of their own. Despite centuries of courts protecting the economic and social viability of freehold land, inside a strata or community scheme courts apparently have no role to play. *Betalli* is a perfect case in point: both Santow JA and Campbell JA readily applied the simplistic statutory rule that a by-law need only relate to the use or enjoyment of a lot or common property, to hold a peculiar, misconceived and potentially intrusive by-law valid. The consequences in the case in question were not particularly serious, but the broader consequences are, as we will see in the discussion of cases in the next section.

Before leaving *Betalli*, one final issue should be noted. In addition to arguing that the by-law was ultra vires under the *SSMA 1996*, the appellant argued that the by-law was inconsistent with s 42 of the *Real Property Act 1900* (NSW) and the principle of indefeasibility. Santow JA agreed with the trial judge that this argument could not be dismissed on the grounds that the by-law merely had contractual effect, because the by-law 'clearly represent[ed] a proprietary interest'.[35]

Santow JA held that as the Register was made up of folios and registered dealings,[36] 'dealings' were defined as registrable instruments,[37] and the instrument by which by-law 20 was made was registered, then the interest created by the by-law was recorded on the folio for lot 2.[38] These arguments hold water for the facts in question which involved a pre-existing by-law which was on the Register before the appellant purchased, (remembering that it is not notice that binds new registered proprietors, it is the prior registration of an interest). However, these arguments would not apply to a circumstance where a by-law like this was made *after* a person such as the appellant acquired their lot. Then the question would be, is a by-law creating a proprietary interest in a lot, without the owner's consent,[39] inconsistent with the owner's indefeasible title conferred by s 42 of the *Real Property Act 1900* (NSW) and thus invalid under s 136(2) of the *SSMA 2015*?[40] The answer must surely be yes.[41] At least for by-laws that create interests that courts are prepared to characterise as proprietary,[42] this might be some safeguard against the very broad by-law-making power.

It should be noted that in contrast to other states, the Queensland legislation, somewhat inexplicably stipulates that management statements are not registered instruments under the *Land Titles Act 1994*.[43] It also states that the recording of a management statement does not confer the benefits of indefeasibility.[44] The first is not a necessary precondition of the second. Easements and leases are registered instruments but their registration does not make all of their provisions indefeasible, as demonstrated clearly by the decision in *Clos Farming Estates*.[45] Validity of interests in land does not simply depend on registration, it also depends on compliance with the common law. This is the position of by-laws and management statements in most states. They are registered instruments, as is entirely consistent with their binding nature on subsequent parties, but if they contain impermissible provisions, such as a restriction on transfer, that provision will still be invalid.

Casuarina Rec Club Pty Limited v. *The Owners – Strata Plan 77971* (2011) 80 NSWLR 711

The Court of Appeal decision in *White* v. *Betalli* was approved in *Casuarina Rec Club Pty Ltd* v. *The Owners — Strata Plan 77971*,[46] the most recent superior court consideration of the by-law-making power. This case did not relate to a residential scheme, but rather a tourist strata scheme. While the majority of schemes in Australia are residential, the focus of the book, strata title constitutes a substantial and growing sector of the tourist industry. It is concentrated in Queensland, and coastal areas of New South Wales like Port Stephens and the Northern Rivers region, where the *Casuarina* resort was located, as well as Sydney, Melbourne and Brisbane's Central Business Districts (CBD). Many holiday-makers who book a room in the popular Peppers or Mantra resorts, or business people who book a serviced apartment, do not realise that they are staying in an individually owned apartment in a strata scheme. As tourist schemes are predominantly constructed for investment purposes and substantial sums of money are at stake, they are the ideal context in which to explore the economic consequences of by-laws.

While there are model by-laws for hotel/resort schemes in New South Wales,[47] like all model by-laws, they are optional and frequently not used. Further, the general power to create by-laws is in the legislation and common to all schemes. This is the case in all states, even Queensland, which has a detailed 'accommodation module' for its tourist strata schemes.[48]

The by-law in issue in *Casuarina* stated that:

> The Owners Corporation has the power and function to enter into the Facilities Agreement to provide access to the Facilities for Residents.

Pursuant to this by-law, the owners corporation had entered into a contract with a local gym, which allowed residents of the strata tourist resort to use the gym in return for an annual fee paid by the owners corporation. This contract, along with a caretaker agreement, a letting agreement and a security service agreement and by-laws supporting them, had been put in place by the developer prior to the sale of lots. That is, the developer had written the by-laws purporting to authorise the owners corporation to enter the contracts; then the developer, as the sole owner of all lots, caused the owners corporation to enter the contracts.

The by-laws in *Casuarina* purported to give the owners corporation the power to enter into contracts because the tourist development had been structured around 'management rights', the most litigated area of strata practice. In order to understand properly the implications of the Court of Appeal decision in *Casuarina*, a short excursion into the world of management rights is needed.

Management rights

Management rights could be the subject of an entire book. They dominate parts of the tourist industry and have caused significant community dispute and litigation. However, this is not a book about management rights or even a chapter on the topic. It is a chapter about the economic implications of by-laws. As we have seen, by-laws are simply obligations and restrictions on freehold land. Stripped of all of the fuss and technicality in legislative provisions, management rights contracts, and the by-laws that often support them, are also positive obligations on freehold land. That is why they have caused so much dispute, and it is that aspect of management rights that will be the focus of this section.

As noted in Chapter 1, large, contemporary strata and community title complexes frequently have extensive common property and facilities. These need to be maintained and managed as a matter of practicality and law. In tourist- and/or investor-dominated complexes, it is assumed that absentee owners will need a professional to discharge this obligation on their behalf. In addition, short-term renters or holidaymakers will need assistance checking into their apartments and during the course of their stay. Management rights companies provide these services; in return for annual fees, they arrange for the physical maintenance of common property, and in return for commissions, they run short-term holiday or serviced apartment rental pools. Management rights businesses could be seen as a response to an opportunity presented by common property and short-term leasing. This explanation would hold true for some buildings, for example, large residential strata schemes in Sydney with high numbers of investor owners, but in relation to strata title tourist accommodation, the growth of management rights is more complex.

The driving force for strata tourism is that strata title allows developers to 'spread the risk originally assumed by the developer (and banking institution) among many investors (usually unsophisticated) and many institutions, who provide the capital for the development to proceed'.[49] In other words, rather than having to find all of their own finance for the development, developers have relied on multiple 'Mum and Dad' investors securing individual loans for off-the-plan apartment purchases. People go to the Gold Coast for a holiday and come back a property investor. Like issuing shares in a company, subdividing land into individual strata titles allows developers to transfer fragments of a larger whole to investors in return for funds.[50]

For developers engaged in this process, management rights are the icing on the strata development cake. Prior to the sale of any apartments, developers sell the right to manage the scheme to management rights companies, often for very large sums of money. Initial prices from developers are rarely disclosed, except by subsequent litigation; for example, one of the many cases involving the management rights to Regis Towers in Sydney's Chinatown revealed that Meriton initially sold the rights for $1.75 million.[51] However, on-sale prices are openly advertised by agents specialising in the sale of management rights. The management rights for Q1, the tallest tower on the Gold Coast, were on-sold in 2011 by the Mantra Group for $10 million,[52] despite intractable problems Mantra had had with the body corporate. Sales in excess of $2 million are not uncommon.

Queensland is the birthplace of management rights in Australia. The tourist industry, particularly on the Gold Coast, has been built around management rights and the Queensland legislation is strikingly facilitative of developers' aims. In its style and content, parts of the *Body Corporate and Community Management Act 1997* ('*BCCMA*') read more like a commercial contract drafted by a developer's lawyer than an Act of Parliament. For example, the *BCCMA* statutorily enshrines a developer's right to sell management rights, to the exclusion of the body corporate, an extraordinary legislative windfall for the development industry.[53] The Northern Territory has a similar provision.[54] Presumably, the logic behind this is that having built the physical structure that needs management and has produced a rent roll, it is the developer who has the right to sell the business associated with that management and rent roll. There are two flaws in this theory. The first is that ultimately the developer does not pay for the physical structure, the purchasers do, and then some. The second is that if you sell people property, they quite reasonably expect property rights. As we have seen, the power to control your own property and not be controlled by decisions of a predecessor in title, is a long-standing and legitimate expectation of freehold owners which the sale of management rights flouts. By way of contrast to the Queensland legislation, applying ordinary rules of equity, the New South Wales Supreme Court in *Community Assoc DP No 270180* v. *Arrow Asset Management Pty Ltd*[55] ('*Arrow Asset Management*'), concluded that the proper recipient of the profit for the sale of management rights was the body corporate, not the developer.

Most states do not have legislation that is as facilitative of developers' interests as Queensland. However, many states have provisions that expressly authorise bodies corporate to enter into strata management and caretaker contracts. Prior to the sale of apartments, developers have been more than happy to use this power on bodies' corporate behalf.

Legally, management rights are a bundle of contractual and by-law-created rights which make up a management 'package'. Bugden describes the typical management package thus:[56]

- i A long-term contract for the management company to provide services such as cleaning, security, supervising common property and work on it;
- ii A contractual and/or by-law-based entitlement to run a letting pool (owners cannot be obliged to place their properties in the letting pool but management rights holders will be given an exclusive right to run a letting service from the building);[57] and
- iii Title to a commercial or residential lot; and/or
- iv An exclusive-use by-law over a foyer or reception desk; and
- v Other by-laws which purport to authorise the body corporate forming increasingly diverse service and management contracts (for example, the by-law in contention in *Casuarina*).

Management contracts are rarely short-term. In Queensland, service provider and letting agent contracts in tourist schemes can last up to 25 years,[58] leading many developers to inappropriately register schemes under the accommodation module to take advantage of this generous provision. In New South Wales, 'caretaker' or 'building manager' contracts (as management rights are called under the legislation) were initially unregulated, but now can only last up to 10 years in a strata scheme.[59] Caretaker contract periods remain unlimited in community schemes. In the ACT, strata management contracts cannot exceed three years but service contracts can last for 25 years.[60] A number of states, like Victoria, have no limit on the length of management contracts.

The crucial element of management rights, and what allows developers to sell them, is that the legal apparatus is set up by the developer, as sole owner of all lots, before the sale of any lots is settled.[61] If the ultimate owners are dissatisfied with the management apparatus, they cannot amend the by-laws because to do so would put the body corporate in breach of contract.

The obvious problem with the practice of developers creating contractual rights is that it is not the developer who will discharge the obligations, but the ultimate purchasers of the lots. As a result, not only does the developer have no incentive to make sure that the contract is robustly negotiated by the body corporate, it has a positive incentive to do otherwise. In *The Owners – Strata Plan No 64972* v. *Rinbac Pty Ltd*, McCallum J said:

> the Owners Corporation observed that the factual context was one that is 'familiar in the world of strata developments'. It stated that what often happens, and what has essentially happened here, is that a developer, after registering a strata scheme, often commits the owners corporation to a management agreement with an entity connected to the developer on terms that are disadvantageous to the owners corporation and unduly favourable to the management entity.[62]

The longer the management contract lasts and the higher the fees it entitles the management company to charge, the more a management company will pay a developer for the contract.[63] In addition to overcharging, management contracts typically grant bodies corporate less than robust rights to terminate, effectively locking bodies corporate into contracts even when they are being poorly performed.

In Queensland, developers' ability to conclude poor management contracts is theoretically constrained by a legislative requirement to exercise 'reasonable skill, care and diligence' and to act in the best interests of the purchaser-constituted body corporate when negotiating a management contract.[64] The developer must ensure that the contractual terms achieve 'a fair and reasonable balance' between the interests of the body corporate and the manager; the terms are appropriate for the scheme; and the powers and functions of the manager do not adversely affect the body corporate's ability to carry out its functions. Despite this, management rights disputes and litigation are rife in Queensland. Owners frequently attempt to terminate management contracts, but are unsuccessful, not only because developer-made contracts give them inadequate rights, but because the *BCCMA* is so protective of developers and management rights holders.[65] As McGill DCJ noted in *Corporate for Palm Springs Residences* v. *J Patterson Holdings Pty Ltd*[66] ('*Patterson Holdings*'), the termination of a management contract would be quite straightforward if it were a matter of contract law, but it is much more complicated because the *BCCMA* and accommodation module,

> contain mechanisms designed to make it difficult for a body corporate to terminate an agreement of this nature. This is because property developers who set up these schemes want to be able to sell the management rights for large sums of money, so the rights that are conferred by the body corporate while it is still under the developer's control have to be reasonably secure, otherwise prospective managers will not be willing to pay so much for them.[67]

McGill DCJ was not only concerned by an arguable bias in the legislation. His Honour noted at [17] an assumption on the part of the Adjudicator that the management rights were an 'asset' that should be protected, leading to a predisposition on the part of the Adjudicator to find in favour of the caretaker. In contrast, McGill DCJ approached the dispute by applying fundamental and uncontentious principles of the general law.[68] *Patterson Holdings*, like *Arrow Asset Management*, highlight a concerning phenomenon in strata law and practice; that is, the extent to which strata has been assumed and allowed to operate, quite unjustifiably, as a law unto itself. At the end of the day, disputes in strata schemes are simply disputes about land and contracts and the wider legal system has been successfully adjudicating on these for centuries.

In a management rights session at the Griffith University Strata Title Conference 2015,[69] the overwhelming sentiment in the room was that management rights were not working in Queensland. Apartment owners were angry and frustrated about management contracts forced on them by developers, and good managers (the only managers inevitably present at a policy and education-focused conference) were angry and frustrated about owner resentment. Noticeably absent from the session were developers. Having taken their fees for the sale of management rights, they had little incentive to concern themselves with how the parties to the contract that they created were faring. As an outside observer of the industry, it was hard not to conclude that the only real beneficiaries of management rights were developers, courtesy of considerable legislative assistance. The sale of management rights had left both managers and owners substantially out of pocket, leading them to focus on band-aid solutions, like clarifying manager responsibilities, rather than recognising the fundamentally flawed nature of the industry. That is, a contract that is negotiated by a party who has no responsibility to carry it out is by definition never going to be a good contract. The appropriate entity to be negotiating a management contract is the true owner of the common property, the body corporate *as constituted by purchasers*.[70] These are the people who will pay for the services.

State legislatures are not completely oblivious to the risk of developer exploitation, but they have legislated to protect bodies corporate with limited success. Developers' lawyers have been able to work around prohibitions and limitations on developer-made contracts and their efforts have frequently been approved by courts, in decisions which, with all due respect, do not always seem entirely cognisant of the issues at stake.[71] Legislation and strata lawyers also place great faith in the concept of 'disclosure' or notice, assuming that so long as developer-made agreements are disclosed in contracts of sale or in management statements, purchasers can properly be bound. Australia seems to have had none of the debate that occurred in the United States, and discussed in Chapter 3, over the genuineness of 'consent' to complicated property contracts,[72] despite the fact that the issues are replicated here. Contracts for new developments are typically a lever-arched folder thick, and few purchasers, (or even their lawyers), read or understand the disclosures contained within. Further, because parties never have a crystal ball, no initial disclosure can alert a purchaser to how a management contract will be carried out in future and whether it will prove to be of value. As repeatedly noted, notice has never been a cure-all in property because giving purchasers notice of burdens that are, or transpire to be, inefficient or unfair, does not make them any less inefficient or unfair. There is nothing new about landowners attempting to maximise profit by extracting money from successors in title, (it was the basis of feudal land systems), and property law is well aware of the problems that this causes. As a result, outside the *numerus clausus* principle, modern property law has never allowed predecessors in title to burden land with their contracts.

And this is the salient point about management rights that makes them a perfect illustration of the thesis of this book: management contracts are simply positive obligations on freehold land created by a sleight of hand; that is, the manipulation of the separate legal entity that is the body corporate. By causing the body corporate to enter into contracts, knowing that responsibility for discharging those contracts will fall on the body corporate constituted by purchasers, developers are burdening freehold land, (albeit in a time-limited way). It makes no difference if the obligation to pay money is attached directly to the land with a positive covenant or it is a contractual obligation of a body corporate of which purchasers automatically become a member. The upshot of both is that by virtue of owning land, purchasers have to pay a monetary obligation created by a predecessor in title, a practice that courts and legislatures spent centuries eradicating from our land law. It is crucial to note that bodies corporate were not created for this purpose. They were created to help residents maintain and run their own building/scheme, (which might legitimately include the body corporate as constituted by owners forming management contracts). The *sale* of management rights by developers is arguably an abuse of the entire legislative structure, which is precisely why it has caused so much dispute.

And so back to **Casuarina**

Ineffective as it may be, most states' legislation at least attempts to regulate caretaker and strata management contracts in some way. However, legislation typically does not regulate other contracts that developers have begun to form on behalf of the body corporate, such as the 'Facilities Agreement', which was the subject of the litigation in *Casuarina.*[73] These other contracts must be supported with 'empowering by-laws'.

Empowering by-laws are a problematic practice that originate from obiter comments of judges. In a rare High Court consideration of strata, the Court in *Humphries* v. *Proprietors 'Surfers Palms North' Group Titles Plan 1955*[74] held that the body corporate did not have the power to enter into a management agreement to provide a letting service to lot owners because, at that time, 'there was no statutory power authorising, and there was *no by-law* which might have authorised' the agreement.[75] Strata lawyers have taken this statement, along with small phrases in the legislation, to mean that if a body corporate does not have the power to do something – for example, enter into a contract with a local gym – a by-law can be created authorising it to do so.

Empowering by-laws are arguably deeply flawed, and given an opportunity to consider the issue properly, the High Court may well reconsider its obiter comment. Bodies corporate are not like ordinary companies with the power of a natural person.[76] They typically only have the powers expressly or impliedly given to them by statutes. This is a crucial protection to lot owners. A body corporate has a significant ability to affect the lives, homes and finances of individual owners. If a body corporate can just create its own powers through by-laws, augmenting its operation beyond that expressly or impliedly authorised by statute, it ceases to be a body with statutorily limited operation. Administrative lawyers would call this a 'stream rising above its source'.[77] That empowering by-laws must relate to lots and common property is no real limit on body corporate power; it just means that in relation to the lots and common property (that is, all parts of a strata scheme), a body corporate's power is almost unlimited. The detailed legislative provisions on the funding and management of schemes in all states' Acts indicate that legislatures never intended bodies corporate to have unlimited powers in relation to lots or common property.

In New South Wales, the practice of empowering by-laws was meant to be eradicated by the enactment of s 43(4) of the *SSMA 1996*, which like many states' Acts,[78] expressly states that a by-law has no force or effect if it is inconsistent with the *SSMA 1996* or any other Act or law.[79] If a by-law purports to give a body corporate a power that it does not expressly or impliedly have under the strata legislation, by definition it is inconsistent with the legislation and invalid. However, this argument does not seem to have been raised in litigation and the creation of empowering by-laws, such as the 'facilities agreement' by-law in *Casuarina*, has flourished.[80]

In the earlier, lower court decision, *sub-nom*, *Santai v. Owners – Strata Plan No 77971*, the owners' corporation had successfully argued that the facilities agreement by-law was invalid, not because the by-law was inappropriately 'empowering', but because the by-law did not relate to the lots or common property of the strata scheme. McDougall J held that the by-law-making power:

> does not extend to making a by-law authorising the Owners Corporation to enter into an agreement with an owner of nearby (but not adjacent) facilities that gives lot owners and other residents the use of those facilities, at least where that agreement has nothing to do with the lots or the common property, or the rights of lot owners as between themselves or as between themselves and the Owners Corporation.
>
> In this case, the relevant by-law does not involve the management or control of the common property, or any of the functions of the Owners Corporation under the SSM Act or other relevant legislation.[81]

On appeal, Young JA, with whom McFarlane and Handley JA agreed, cited Campbell J's formulation on by-laws from *Betalli*, that there is 'nothing in the notion of a by-law that, of itself, imposes any kind of limitation on the kind of regulation that might be adopted, beyond that it is for the regulation of the particular community to which it applies'.[82]

Prompted by McDougall J's reference to the gym not being 'adjacent' to the resort, Young JA's judgment includes a discussion of easements, noting the requirement that they benefit the dominant tenement, that they be sufficiently close to do so, and that they not be mere rights of recreation. His Honour acknowledged at [80] that both counsel were reluctant to be drawn into a discussion of easements, and ultimately concluded at [86] that while a consideration of easements was 'illuminating' it was 'a bit of a red herring', and the 'vital question is whether the by-law was within the powers of the Owners Corporation, and a proper exercise of its functions under the SSM Act'.

However, easements were not a red herring. Young JA's instinct about the relevance of easements was entirely justified and astute. What his Honour was drawing attention to, like McDougall J in the court below, was the need for a boundary rule; a rule that would help define the limits of individual power to attach restrictions and obligations to land thus compromising its utility. A rule that gives property law its structure and function, preventing it from sliding into the excessive flexibility of contract law; a flexibility that is acceptable when only two parties are involved, but unacceptable in relation to third parties and the limited, vital resource of land. As Heller says, '[t]he pervasive presence of boundary rules [provides] … overwhelming evidence … that the notion of an open-ended bundle of property rights is wrong'.[83]

Unfortunately, that is exactly what the strata and community title by-law-making power creates: an open-ended bundle of property rights. By granting owners corporations the power to write any by-laws, as long as they just happen to relate to the use or enjoyment of lots or common property, the strata legislation obliterates the *numerus clausus* principle, allowing private citizens to dream up almost any obligation or restriction they please and attach them to freehold land. A by-law empowering the body corporate to enter into a contract to employ an interior decorator to redo *all private lots* is a by-law that relates to the use and enjoyment of lots. Unlike the United States *Restatement*, which maintains boundary rules in the power of courts to strike down servitudes that are illegal, unconstitutional, against public policy,[84] undue restraints on trade, or unconscionable,[85] as well as association rules that are unreasonable, the New South Wales strata legislation, like most states', contains no such provision.

As we saw in Chapter 3, when attempting to simplify the 'touch and concern' doctrine, the United States never tried to eradicate the complicated process of allowing courts to mediate the effect of servitudes. The *Restatement* explicitly states that in replacing the 'touch and concern' doctrine, 'it is not intended to remove courts from their historic role of safeguarding the public interest in maintaining the social utility of land resources'.[86] When assessing a servitude, United States courts have specifically looked at 'the fairness of the arrangement, its impact on alienability and marketability of the property, its impact on competition, and the degree to which it interferes with the fundamental rights and expectations of property owners'.[87] The old 'touch and concern' doctrine looked 'both backwards and forwards', allowing courts to assess not only the validity of a servitude at its point of creation, but also its validity through time.[88] The *Restatement* did not change this. In contrast, in writing the strata legislation, no one seems to have noticed that courts perform a role of maintaining the social utility of land resources. This is no doubt partly because, being banned outright, Anglo-Australian law had little experience of adjudicating some of the most problematic burdens, positive covenants, and thus few people perhaps appreciated their dangers.

That said, Australian courts do have experience adjudicating 'tying arrangements'. If the obligation in *Casuarina* had been attempted using orthodox restrictive covenants, on the authority of *Tooth & Co Ltd* v. *Barker*,[89] the arrangement would have been invalid for failing to benefit the land of the covenantee, as opposed to the business currently conducted on the covenantee's land, (remembering, this rule is not a mindless technicality – it is a well-thought-out prohibition that prevents the problem of uneconomical contracts like the Facilities Agreement in *Casuarina*). This is not to suggest that the old restrictive covenant rule should be applied to strata schemes – the legislation obviously affects a change in the law – it is just to say that there is still an important and difficult conversation that must be had about the desirability of allowing one person to impose obligations on freehold land that oblige other people to continue to purchase a service they do not need or can no longer afford, even when they have been given notice of the obligations. As Reichman said in the United States context:

> There might be nothing objectionable in personal agreements concerning … promises to buy from a certain supplier. When such obligations, however, become permanently enforced against an ever-changing group of owners, the matter acquires different dimensions.[90]

When sidestepping the application of complicated legal and equitable doctrine in relation to freehold covenants, the strata legislation did not miraculously sidestep the difficult questions about legitimate land use that those doctrines were trying to answer. Nothing in the strata and community title legislation resolves the questions about the desirability of tying arrangements.

As Carol Rose warned, when 'devising new strategies in an old war among generations of land owners ... we should not be overly surprised if we wind up in some of the same old trenches'.[91] As she also pointed out, just by changing the label we use (for example, 'servitude', rather than 'real covenant, restrictive covenant and easement'; 'unconscionable' instead of 'touch and concern'; 'by-law' instead of 'covenant'), we do not change the fundamental differences between the various things that servitudes (or by-laws) can demand of people. So, to use Rose's example described in Chapter 3, (1) having to allow someone to come on to land to repair an electricity system (an easement), and (2) not being allowed to put in a wading pool without architectural committee approval (a restrictive covenant), might both be acceptable, but (3) compelling someone to keep a Doberman dog (a positive covenant),[92] or to continue to pay for gym services no one is using (a tying covenant), may not. Lumping all manner of rights and obligations together on the grounds that they all relate to the 'use and enjoyment' of land, does not remove their underlying fundamental differences; it just prevents us from acknowledging those differences and most importantly, assessing them properly.

The *Restatement* acknowledges that there are vast differences in the types of possible servitudes; they do not all impose similar burdens. The *Restatement*'s stipulation that a servitude cannot be unconscionable was specifically included to allow courts to strike down arrangements that are oppressive or unfair, including tying arrangements, even when they have 'little impact on competition'.[93] The comment on the §3.7 prohibition on unconscionability states:

> the developer has the opportunity to create long-term obligations for the property-owners association that will fall on purchasers through the assessments they become obligated to pay. Such obligations can be imposed in the original declaration of covenants governing the project, or they can be created by contracts entered into while the developer retains control of the homeowners association. The servitudes imposing the obligation to pay may be unconscionable with respect to these obligations, particularly if the circumstances include self-dealing by the developer.

The great benefit of the *Restatement* is that 'unconscionability' is a flexible concept that captures a range of conduct, including that which developers and their lawyers are yet to think up. In contrast, single legislative rules, such as the prohibition on creating by-laws in the initial period or the limits on caretaker and strata manager contracts, are static targets that lawyers can draft their way around, as developers in Australia have demonstrated time and again.[94] Legislative provisions have not worked to eradicate undesirable developer practice.[95]

This is the distinction that Alexander makes between 'rules' and 'standards', advocating the latter for judicial review because standards 'create opportunities for those inside and those outside to engage each other in dialogue'.[96]

Using standards, United States courts sometimes uphold developer-made tying arrangements,[97] and sometimes they do not, particularly if the service is no longer needed, or can be obtained elsewhere at a better price.[98] Tarlock argues that the 'touch and concern' doctrine was specifically used by the judiciary 'to protect legitimate financial expectations of purchasers when they are frustrated by the use of funds for purposes with remote benefits to the burdened land'.[99] There seems to be a lack of recent cases on this issue in the United States, quite possibly as a result of the §3–105 of the *Uniform Common Interest Ownership Act*, which when enacted by individual states, allows HOAs to terminate *any* developer-made contract on 90 days' notice.[100] In any event, to reiterate the point made at the end of Chapter 3, what matters is not that individual parties are always successful in challenging servitudes, but that individuals have an opportunity to challenge them because the law permits conversations about servitudes' legitimacy. In contrast, most Australian states' legislation does not allow any opportunity for similar discussion of by-laws, and as a result, the cards are stacked heavily in earlier owners' favour. Of course the earliest owner is invariably the developer.[101]

Faced with an almost completely silent statute, Young JA concluded in *Casuarina* at [89] that the 'power to make by-laws is to be liberally interpreted' and the by-law in question, having some connection to the enjoyment of lots and common property, was a valid by-law for a strata tourist scheme. Interestingly, obviously aware of the potential dangers of a broad by-law-making power, Young JA made an obiter remark that this liberal power is subject to the doctrine of fraud on a power[102] and a proviso that by-laws cannot be 'unreasonable'. As noted above, while the concept of reasonableness is central to United States case law,[103] like most states (with the exception of Tasmania and Queensland), the New South Wales legislation has no provision in this regard. Despite submissions to include a reasonableness provision in the new legislation,[104] the New South Wales legislature declined, opting for the much more limited requirement that by-laws not be oppressive, unconscionable or harsh. In this context, Young JA's obiter comment on reasonableness was both progressive and unusual.[105]

The lack of express provision for broad judicial review of by-laws, and the consequent lack of active mediation of their potentially harmful effects, results in a property regime that fails to do all the things that the theorists discussed in Chapter 2 argue property law should do. Property law must amount to 'justificatory practice', on both a public and private level. We cannot determine justificatory practice solely by reference to individual preference, that is, 'I, the developer (or even we, the community), want this by-law entitlement, therefore I should have it'. Not all preferences are legitimate in property law, because even if legitimate for the original parties, they may not be legitimate for successors in title, or strangers; the 'legitimacy of the exercise of a property right depends on the consequences of exercising the right' through time.[106] Despite the incommensurability of values potentially embodied in property law, 'rational choice remains possible [but only] through reasoned deliberation'.[107]

The simplistic formulation in most states' legislation that by-laws just need to relate to the use or enjoyment of lots or common property at the time of creation means that no 'reasoned deliberation' can occur. Nothing demonstrates this lack of opportunity for proper discussion quite so well as the fact that the brief judgment in *Casuarina* contains no reference to the practice of management rights or the real-world use of the by-laws in question.

Italian Forum Pty Ltd v. *Owners – Strata Plan 60919* [2012] NSWSC 895

The high point of blank slate approval for almost any by-law content is *Italian Forum*, which centred on a stratum subdivision that also contained strata schemes. As explained in Chapter 1, stratum or 'volumetric' subdivision was created in response to the demand for multiple, separate commercial and residential subdivisions in large developments,[108] but like strata and community title, legislation was enacted with arguably insufficient appreciation of its consequences.

Stratum owners (some or all of which may be strata schemes) need to collectively maintain and manage often extensive physical infrastructure and services in complex modern buildings. Building Management Statements (BMSs) and Strata Management Statements (SMSs) were devised for this purpose. BMSs and SMSs are written by the owners of the stratum lots, prior to the registration of any strata plan and sale of any apartments; that is, (in a familiar pattern), they will be written by the developer, not the ultimate members of the owners corporation(s). Statements are registered in order to make them binding on all subsequent owners and lessees. While BMSs and SMSs are a practical necessity, they are another way of attaching registered positive and negative obligations to freehold land, (including non-strata title land).[109] In essence, they are strata by-laws and levies rolled into one largely unregulated, privately created document. While an increasingly popular form of subdivision of valuable urban land, stratum subdivision has been the subject of little academic or judicial attention.[110]

Italian Forum considered the complicated issue of the relationship between a Strata Management Statement (the SMS) and strata by-laws in the Italian Forum at Leichhardt in Sydney's inner west. There were five stratum parcels in the subdivision – a piazza, an Italian Cultural Centre, a car park, a residential strata scheme, and a commercial strata scheme. The plaintiff was the registered proprietor of two stratum lots in the scheme, the piazza and the Italian Forum Cultural Centre. The first defendant was the owners corporation of the residential strata scheme, the second defendant the owners corporation of the commercial strata scheme, and the third defendant the owner of the car park. The SMS required each stratum lot/parcel to pay $60 000 per year in 'promotional levies' to the registered proprietor of the Cultural Centre. This was supported by by-laws in the strata schemes which purportedly empowered the owners corporation to levy owners for promotional levies for 'the administration and promotion of cultural events at Italian Forum' and 'to generally promote the Italian Forum Complex, as a retail and commercial centre'.

The dispute arose initially because it seemed that no one had in fact been paying their levies. The plaintiff sought judgment against the first defendant for $420 000 plus interest, however consent orders were made on 11 September 2009 to amend the SMS to include clauses which required each stratum lot and parcel to pay, but only subject to the other stratum lots and parcels paying. Then on 9 November 2009, in separate litigation, Patten AJ handed down a judgment in favour of two individual lot owners in the commercial strata scheme, holding that the by-laws purportedly empowering the owners corporation to levy money from owners to pay the 'promotional levies' in accordance with the SMS were invalid.[111] The initial Adjudicator had said that if owners and tenants wanted to promote the location as a commercial enterprise, they could, but not via the strata Act.[112] The by-law requiring the payment of money was inconsistent with s 78 of the *SSMA 1996*, which empowered the owners corporation to levy owners, but only in proportion to their unit entitlements, and the money had to be paid into the administrative or sinking fund. Patten AJ confirmed this reasoning. Patten AJ also agreed that the by-law was for a commercial purpose and not for the administration, management or use of a strata scheme as required by s 47; nor was the by-law in the interests of all lot owners. The reasoning of the Adjudicator and Patten AJ used the words 'administration, management or use' and 'in the interests of all lot lots' as a boundary rule, limiting the ambit of by-laws. The decision is important, but unusual in this regard.

As a result of Patten AJ's decision, the owners corporations of the strata schemes argued in *Italian Forum* that the 2009 consent orders should be set aside or that the amendments to the SMS be declared invalid or ultra vires the *SSFDA 1973*. They said that they could not pay the $60 000 levies to the Building Management Committee because they had no power to raise that money from lot owners.

White J dismissed the owners corporations' notice of motion because it was not sufficiently clearly pleaded and did not include any consequential relief, but in the course of his judgment, his Honour made some concerning comments about by-laws and their ambit. Further, the marked disagreement between two superior court judges on a fundamental issue suggests a pressing need for deeper legislative and judicial consideration of the question.

First, in relation to SMSs, White J held at [45] that:

> I doubt that there can be any limit on what can be included in a strata management statement, except that it be reasonably adapted for the purpose for which the statement is to be made. This includes regulating the rights and obligations between different owners of shared facilities on a site. I do not accept that this precludes the raising of funds which may have to be contributed to by lot owners of a strata scheme that forms part of the building or site, to be applied for the purposes of the site as a whole.

In other words, registered SMSs are even more of a blank slate than strata by-laws. Lawyers are free to include whatever seems desirable to their clients, as long as the provisions are 'reasonably adapted for the purpose for which the statement is to be made'. The legislation does not provide particularly clear guidance on the purpose of an SMS beyond mandating certain matters, suggesting certain matters, and in New South Wales, making it clear that their content is limited to neither.[113] There is not even the limitation of strata by-laws that content relate to the 'control, management, administration, use or enjoyment of the lots or the lots and common property'. As a matter of practice, most SMSs probably do relate to the financial and physical management of multi-owned buildings, but as *Italian Forum* demonstrates, this will not always be the case.

White J then went on to note that under s 28W(1)(b) of the *SSFDA 1973*, SMSs bind not only bodies corporate but all lot owners and lessees *personally*. It is likely that s 28W(1)(b) was simply intended to regulate lot owners' and lessees' behaviour within the building, for example, preventing them putting up signs in contravention of the SMS, but in fact, it also seems to make them personally liable for monetary sums imposed by the SMS. White J said at [49] that, '[e]ven if such obligations could not be enforced through the making of by-laws under the *Strata Schemes Management Act* it does not appear to me that they could not be enforced against the lot owners pursuant to s 28W'. In other words, irrespective of any protections in the levying provisions of the strata legislation for lot owners, they can apparently be overridden by an SMS.

However, White J then said that in accordance with s 28W(5), an SMS has no effect to the extent it is inconsistent with 'any other Act or other law' and that a by-law is a 'law'. As a result, an SMS is invalid to the extent that it is inconsistent with strata by-laws. With all due respect to White J, this interpretation is difficult to reconcile with the general structure of the legislation. Even if a by-law is a 'law', it is only law within the physical and administrative boundaries of the strata scheme to which it relates. It cannot regulate anyone outside that scheme. Strata by-laws can no more regulate the owner of a stratum lot within the stratum subdivision than they can regulate the owner of a house next door to a suburban apartment block. If White J were correct, a strata scheme in a stratum subdivision which did not like the SMS could just enact a by-law inconsistent with it, rendering the SMS invalid. This would contradict s 28U(1) of the *SSFDA 1973* (now *SSDA 2015*, s 103), which, not surprisingly, required unanimous agreement of all stratum lot and parcel owners to amend an SMS. White J's reasoning would in effect allow a strata scheme to unilaterally regulate the entire stratum subdivision.

The salient issue was not SMS inconsistency with a *by-law*, but inconsistency with the strata legislation. This is clearly impermissible under s 28W(5) and leads to the perfectly sensible conclusion that an SMS cannot validly require a body corporate to do something that the strata Acts do not authorise a body corporate to do. That is, a document written by private citizens (the developer's lawyers) cannot override an Act of Parliament.

In any event, White J did not seem to think that there was any reason the strata Acts did not allow a body corporate to raise promotional levies from owners by writing a by-law in this regard. It is worth quoting his Honour's analysis of the strata by-law-making power in full:

> 50. These powers are not to be narrowly confined (see *White* v. *Betalli* [2006] NSWSC 537; (2006) 66 NSWLR 690 and on appeal *White* v. *Betalli* [2007] NSWCA 243; (2007) 71 NSWLR 381). The type of strata scheme concerned in the present case is that the commercial strata scheme is part of a wider development involving a car park, a piazza and residential apartments. One would have thought that matters appropriate to the participation of the commercial strata scheme in that form of development are matters for which by-laws could be made. I think there is a serious question as to whether his Honour [Patten AJ in *The Owners Strata Plan Number 60919* v. *Consumer Trader and Tenancy Tribunal* [2009] NSWSC 1158] was correct in his conclusion as to the scope of ss 43 and 47. His Honour was of the view that the *Strata Schemes Management Act* did not permit the levying of promotional levies. If that is so, I do not see how that consideration necessarily limits the matters that can be dealt with in the strata management statement that is enforceable against the lot owners directly under s 28W of the *Strata Schemes (Freehold Development) Act.*
> 51. The *Strata Schemes Management Act* provides for the raising of levies to be applied to the administration fund or to the sinking fund. With respect, it is not obvious that a by-law could not be made under ss 43 or 47 that provided for the levying of other contributions. But in any event, moneys can be raised to be applied to the administrative fund for, among other things, the meeting of 'other recurrent expenses'. As presently advised I do not see why that could not extend to levies properly raised pursuant to a strata management statement.
> 52. His Honour also considered that by-law 11 was invalid because it was not in conformity with s 78(2) of the *Strata Schemes Management Act.* Again, with respect, it is not clear to me why that was said to be so. But assuming that to be so, there is no reason the first defendant could not make a by-law that conformed with s 78.
> 53. For these reasons, I would not accept the argument advanced that the amendments to the strata management statement were inconsistent with the legislative scheme in either the *Strata Schemes Management Act* or the *Strata Schemes (Freehold Development) Act.*[114]

There are very good reasons why the body corporate does not have the power to raise money for promotional levies. However, the vague nature of the legislation and the lack of analysis of it within the framework of general property law, have perhaps made those reasons unclear. Based on black letter, functional and theoretical analyses, the reasons are these.

On a black letter analysis, outside the levying provisions, the general rule that positive obligations to pay money cannot be imposed on freehold land still applies. Strata and community title are not a law unto themselves. Like all property statutes, the strata and community Acts operate over a body of pre-existing common law and equity (see Figure 4.1 below). The Acts will repeal those rules to the extent they are expressly or impliedly inconsistent with them, but no more. Like the Torrens Acts and general property Acts in all states, the strata and community title legislation did not attempt the impossible task of wiping the property slate clean.[115] Contrary to the mantra commonly heard in the strata industry, strata title is freehold,[116] Torrens title, and the ordinary rules that apply to freehold land apply to it unless the Acts expressly or impliedly state otherwise.[117]

The statutory levying provisions override the general rule of law and equity that obligations to pay money cannot be attached to freehold land. These provisions authorise the annual extraction of money from owners to be paid into sinking and administrative funds. Money can only be paid out of those funds if authorised by the legislation. The strata and community title legislation creates this exception to the general rule because it makes financial and social sense to compel owners to pay to maintain the property to a satisfactory standard when freehold land is collectively owned by multiple, unrelated individuals. In the terms of property theorists, this is clearly 'justificatory practice'.[118]

The by-law-making provisions also possibly override the common law and equitable prohibition on positive obligations on land. There is no requirement that by-laws, like restrictive covenants, only be negative. By-laws can require people to wash their windows every two weeks, or in community schemes to mow their lawns on a monthly basis. However, it is not clear that positive obligations in by-laws extend to the requirement to pay money. Patten AJ in *Owners Strata Plan No 60919* v. *Consumer Trader and Tenancy Tribunal* did not think so.[119] The Queensland legislation expressly states that 'a by-law (other than an exclusive-use by-law) must not impose a monetary liability on the owner or occupier of a lot',[120] as does the South Australian community title, (but not strata title) legislation.[121] As noted above, one of the purposes of the United States

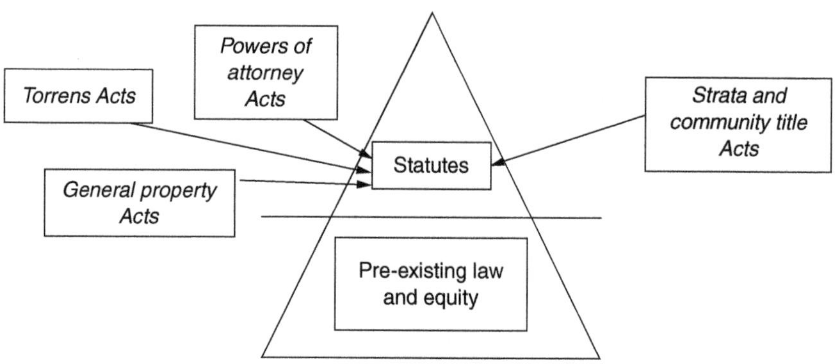

Figure 4.1 The operation of property legislation over pre-existing common law and equity.

touch and concern doctrine was to prevent positive covenants extending to 'funds for purposes with remote benefits to the burdened land'.[122]

On a functional level, if White J's reasoning were correct, an owners corporation could pass by-laws requiring everyone to pay $10 000 a year into a slush fund for entertainment on common property, as it would clearly relate to the use and enjoyment of common property in a residential scheme; the owners corporation could pass a by-law requiring everyone to renovate their kitchens as this would relate to the use and enjoyment of lots; by-laws could be passed requiring everyone to pay for a *feng shui* consultant to re-orient the lobby or to rearrange all furniture *inside* lots, as these would relate to the use of lots or common property. Of course, a special resolution would be needed, but the majoritarian nature of the decision is likely to be small comfort to a hypothetical minority of baby boomer owners who find themselves paying for the very genuine desire of the Gen Y majority to foster community spirit with rock bands playing on common property every Friday night. Further, if the reasoning of the Court of Appeal in *Casuarina* is correct, even if an owners corporation could not do these things directly (as suggested by White J in *Italian Forum*), the owners corporation could do them indirectly by simply writing a by-law authorising the owners corporation to enter into contracts for such services; for example, an empowering by-law purportedly authorising an owners corporation to enter into a *feng shui* contract or an entertainment contract. The by-law would apparently be valid just by relating to the lots or common property, and if the by-law were valid, the contract would be too. Lot owners, as members of the owners corporation, would then have to pay to discharge these contractual obligations.

Finally, on a theoretical level, White J's reasoning fails to take into account property law's long-standing concerns about positive obligations, irrespective of purchaser notice. These concerns remain as valid today as they were 300 years ago, because purchasers and current owners are no more capable of seeing into the future than they were in the past. The original owners of the Italian Forum who genuinely wanted to promote Italian culture, could not know that the development would not be particularly financially successful and that a $60 000 a year promotional levy might be untenable. In addition to lacking foresight, we are also unable to take land with us when we die or move away, thus the presumption that long-gone possessors of land should not determine its current use is not an outmoded idea; in a capitalist economy it is arguably more important than it ever has been before. Without boundary rules, without the ability to limit the obligations that current owners can attach to freehold land (including those created in good faith and disclosed to purchasers), we risk reducing land's use, value, marketability and transferability. As Heller warns, while arrangements might be initially wealth creating (for the developer, the payee, or even the payer), land which is burdened by too many third party rights is prone to underuse and can become an anticommons. Of their very nature all strata and community schemes are at risk of being anticommons because of the presence of multiple owners who may not be able to agree on use, but the addition of further rights holders via long-term contracts or obligations like those in the *Italian*

Forum SMS, intensifies this risk. The economically damaging potential of monetary obligations, extracted by multiple rights holders, is why the United States legal system only allows them with active judicial oversight, typically only permitting covenants with a *genuine* connection to land.

Interpretation of by-laws

Perhaps the most telling illustration of the legislature and judiciary's reluctance to characterise by-laws as positive and negative obligations on freehold land is the confusion over the proper interpretation of by-laws (as opposed to their actual validity, discussed above). While some judges have recognised that at least exclusive-use by-laws create proprietary rights,[123] other judges have assumed that by-laws operate as contracts.

With the exception of the *Unit Titles (Management) Act 2011* (ACT) s 112, which stipulates that the *Legislation Act 2001* (ACT) applies to the rules of an owners corporation as if the rules were an Act and as if each rule were a section of an Act, no state Act determines how by-laws are to be interpreted. This question has been left to courts and tribunals, and their most common conclusion is that ambiguous by-laws should be interpreted as ambiguous contracts. The well-established principles in *Codelfa Construction Pty Ltd* v. *State Land Rail Authority (NSW)* ('Codelfa'),[124] hold that extrinsic evidence and circumstances known to the parties can be used to determine the objective meaning of ambiguous terms.

The most detailed higher court consideration of by-law interpretation is the New South Wales Court of Appeal decision in *Owners of Strata Plan No 3397* v. *Tate* ('Tate').[125] The by-law in question was an exclusive-use by-law under section 52 of the *SSMA*, giving the owner of Lot 1 exclusive use of Lift 4 in a mixed use building. Lot 1 comprised parking spaces in the basement and levels 6–15 of the tower, all of which made up a motel.[126] The owner of Lot 1 had had de facto exclusive use of Lift 4 for many years, but this was formalised in 1989. The by-law provided that the owner of Lot 1 was to pay all sums related to maintenance and repair of Lift 4 and one-quarter of the costs of running the lift system. The respondent thought that he was overpaying because when all owners were levied for the administrative and sinking funds, he paid 19.382 per cent in accordance with his unit entitlement. He then paid an additional 25 per cent of the costs of running the lift system in accordance with the exclusive-use by-law levies, amounting to a total of 44.382 per cent of the lift costs. The owner of Lot 1 asserted that when levied for the administrative and sinking funds, he should not be paying any share of the maintenance costs of Lifts 1–3.

Applying the decision of Rath J in *North Wind Pty Ltd* v. *Proprietors – Strata Plan 3143*[127] that by-laws were to be interpreted as contracts, the trial judge had held that the court had to consider the 'chosen language of and the surrounding circumstances in which By-Law 21 was made in order to determine, objectively, the mutual intention of the parties'.[128] His Honour said that as a matter of common sense, the parties who made the exclusive-use by-law must have intended that 75 per cent of the costs of the lift system would be paid by the remaining lot owners and that such an interpretation gave the by-law a 'commercial businesslike interpretation'.[129]

On appeal, Harrison J (Mason P agreeing) came to the opposite conclusion. He found that the circumstances of the case persuasively suggested that at the time the by-law was formulated, the proprietor of Lot 1 was intent on converting the informal exclusive use of lift 4 into a legally enforceable right to do the same thing.[130] This right was valuable and payment over and above the proportionate cost of maintaining Lift 4 appeared a matter of 'common sense' and to conform to 'a commercial, businesslike' approach.[131]

The actual decision in the case is not important for our purposes, but Harrison J's comments on interpretation of the by-law are. His Honour said that, 'the need for the court to consider the chosen language, and the surrounding circumstances in which By-Law 21 was made, in order to determine, objectively, the mutual intention of the parties' was not a matter of dispute between the parties on appeal. His Honour then went on to cite a number of contract cases, including a significant extract from Tobias JA's judgment in *Ryledar Pty Ltd* v. *Euphoric Pty Ltd*,[132] on the interpretation of commercial contracts.[133] Harrison J then applied Tobias JA's reasoning to the facts at hand. Harrison J clearly thought that by-laws operated as contracts and were to be interpreted as such, an approach that has also been applied in tribunal decisions in other states.[134] His Honour's judgment contains no consideration of the problem of enforcing contracts against people who may not have been party to their creation, or of the difficulty in determining the objective intention of the negotiating parties if none or some of them are not involved in the dispute.[135] In short, there is no reference to the fact that by-laws are not contracts at all; they are statutory property rights.

In contrast, McColl JA held that by-laws should not be interpreted in accordance with contractual principles, identifying two possible acceptable characterisations. The first was that by-laws are delegated legislation, being instruments 'made under an Act'[136] and thus should be interpreted according to principles of statutory interpretation.[137] This is the explicit approach taken by the ACT legislature in the *Unit Titles (Management) Act 2011* (ACT) s 112. It also formed part of the reasoning in the High Court decision in *Dainford Ltd* v. *Smith*.[138] Delegated legislation must not contradict, be repugnant to or inconsistent with the Act under which it is made,[139] a principle that is instructive in determining the meaning of by-laws, as well as their ultimate validity, and which is explicitly included in a number of state Acts.[140] McColl JA cited Lindley LJ in *London Assoc of Shipowners and Brokers* v. *London and India Docks Joint Committee*,[141] who instructively said that, '[a] by-law is not an agreement, but a law binding on all persons to whom it applies, whether they agree to be bound by it or not'. McColl JA at [46] noted that an exclusive-use by-law, such as the one in question, would bind all owners of the common property whether or not they had voted in favour of it, so long as the by-law had been agreed to by the appropriate majority. It would also obviously bind individuals who were not yet members of the strata scheme at the time the vote was taken, that is, successors in title.

The second characterisation of by-laws was as 'a statutory contract, deemed to exist by statute and constituted by the 'bundle of rights and liabilities' created by the 1973 Act, the model by-laws and any special by-laws'.[142] By-laws have been characterised as a statutory contract in Victoria, in a well-reasoned Supreme Court decision of Garde J in *Avranik Pty Ltd* v. *Lloyd*, ('*Avranik*')[143] and also in Western Australia.[144] Like a number of states, the New South Wales legislation stipulates that by-laws bind lot owners and the body corporate as if they were 'mutual covenants' signed and sealed by all parties. In her judgment in *Tate*, McColl JA pointed out at [48] that the deemed covenant provisions were modelled on the deemed covenant provisions of corporations law, which ensured that a company and its members were bound by the memorandum and articles of association (now a constitution). The significant characteristic of a company's constitution is that it has a public, not just private purpose. It informs third parties who deal with the company of its 'permitted range of enterprise',[145] and as third parties cannot be privy to the circumstances in which the constitution was negotiated, a tight rein must be kept on the surrounding circumstances permissible to consider when construing a company constitution.[146]

McColl JA ultimately held that it was not necessary to decide whether by-laws should be interpreted as delegated legislation or a statutory contract. Either interpretation produced the conclusion that the by-law only dealt with the proprietor of Lot 1's liability to contribute to the cost of Lift 4 and did not have any effect on that proprietor's statutory obligation to contribute to the maintenance of common property generally, including Lifts 1–3. Further, while the by-law had a commercial effect for the proprietor of Lot 1, it was not a commercial transaction, but rather:

> an adjustment of statutory obligations which conferred a proprietorial estate upon the proprietor of Lot 1 in respect of Lot 4 [*sic* Lift 4] and, conversely, deprived the remaining proprietors in the strata scheme of the proprietary interest they had hitherto had in that part of the common property. The proprietorial nature of the by-law was indicated by its registration. The public nature of the by-law called for the interpretive exercise to focus on its language and statutory context rather than inferred intentions drawn from the de facto position concerning Lift 4 prior to the making of Special By-Law 21.[147]

McColl JA's discussion is analytically interesting, and impressive in its attempt to move beyond contractual reasoning. However, in relation to the statutory contract or quasi-corporate model her Honour offers, the following should be noted. While the legislation might have been modelled on the corporations law, and while corporations law might have some relevance to strata title, for example, the doctrine of fraud on the minority,[148] corporations law is not always an appropriate model for strata and community title. First, although courts keep a tight rein on the use of extrinsic evidence to aid the interpretation of a company's constitution, the rule is slightly more relaxed than it once was. This is because the ultra vires rule has been abolished in corporate law, and by virtue of s 124 of the *Corporations Act 2001* (Cth), a company now has the powers of natural person.[149] In contrast, a strata body corporate does not; it is still limited by the powers and functions granted to it by legislation.

Second, the rules in relation to extrinsic evidence used to interpret a company constitution relate to surrounding circumstances *known* to, or easily capable of being *ascertained* by third parties.[150] (Similarly, the rules in relation to extrinsic evidence and commercial contracts relate to circumstances *known* to both parties.) In the context of a strata scheme, many owners are unlikely to know anything of the circumstances in which a by-law was created, either because they were not members of the body corporate when the by-law was made or because they did not attend the relevant meeting. Purchasers are required to check the Torrens register, which contains a copy of the by-laws, but they are not required to read the minutes of meetings in which by-laws were made (if indeed the minutes even exist). The fundamental premise in both contract and corporate law is that it is only extrinsic evidence and surrounding circumstances known to the parties or ascertainable by them that can be used to aid interpretation of words, and this presents a very real impediment to applying contract or corporate interpretation case law to strata by-laws.

Third, and more fundamentally, strata and community title relate to land, which endures and has significance for non-owners in a way that a company rarely does. As discussed at the beginning of Chapter 2, there is an entire body of property law that only applies to land, for the simple reason that land is different to all other forms of property. We do not have a choice to use or not use land, (like our choice to own or not own shares); we and future generations need to be physically present on land at all times. If we apply principles that were developed in a commercial context for forms of property that will never have the social and physical significance of land, we will make bad law. And finally, strata and community schemes frequently relate to land that is people's homes, 'the scene of one's history and future, one's life and growth ... one embodies or constitutes oneself there'.[151] If it is misguided to treat land generally just like other property, it is even less defensible to do so in relation to land that is people's homes. While some apartments are investments, large numbers are not, and principles developed to interpret commercial and corporate instruments are not necessarily appropriate to interpret instruments that regulate people's homes. We will explore this concept more in the following chapter.

All of these reservations point towards the logical conclusion that by-law interpretation should be consistent with the principles that guide interpretation of other interests in land. Putting aside the Queensland legislation, which somewhat peculiarly states that management statements are not registered instruments,[152] by-laws are recorded on the Torrens register, and bind people by virtue of ownership of lots.[153] They are positive obligations and negative restrictions on land. This suggests that they should be interpreted as Torrens interests, in accordance with the leading High Court case in the area, *Westfield Management Pty Ltd* v. *Perpetual Trustee Pty Ltd.*[154]

Westfield involved the question of whether extrinsic material could be used to interpret a registered easement. The Court at [37]–[39] said that:

> rules of evidence assisting the construction of contracts inter partes, of the nature explained by authorities such as *Codelfa Construction Pty Ltd* v. *State Rail Authority of NSW*, did not apply to the construction of the Easement.... The third party who inspects the Register cannot be expected, consistently with the scheme of the Torrens system, to look further for extrinsic material which might establish facts or circumstances existing at the time of the creation of the registered dealing and placing the third party (or any court later seized of a dispute) in the situation of the grantee.

This statement rings equally true for strata by-laws. As noted above, a prospective purchaser cannot possibly be expected to know all the circumstances that led to a by-law's creation, and expecting him or her to do so would fly in the face of Torrens principles.[155] The whole point of the Torrens system is to make the purchase of land simpler and more secure through the maintenance of a reliable public register that reveals pre-existing interests in land. The mischief that the Torrens system was designed to cure was expensive, difficult and sometimes ineffective historical searches. The requirement to register by-laws to render them binding is consistent with Torrens schemes, of which strata and community title are a part. The fact that consideration of *Westfield*, the leading High Court decision on the interpretation of interests recorded on the Torrens register, has not to have surfaced in litigation, highlights how disconnected our understanding of strata title and by-laws is from property law.

The principle from *Westfield* would need to be adapted for the specific context of strata and community title. Unlike easements, by-laws are created in the framework of detailed legislation, and consistent with the reasoning of Garde J in *Avranik*,[156] this should rightly be considered when interpreting by-laws. Unlike negotiations of private parties, legislative provisions are objectively ascertainable. For example, in Victoria, the *Owners Corporations Act 2006* s 140 prohibits by-laws that unfairly discriminate against an owner or an occupier of a lot or which are inconsistent with or limit a right or avoid an obligation under the *Owners Corporations Act* or any other Act or regulation. In *Avranik*, Garde J used this provision to aid the interpretation of an ambiguous parking by-law, an approach that was confirmed on appeal. Garde J also reasonably suggested at [29] that by-laws should be read as a whole, not construed narrowly, but as an enduring and flexible document, that constructions that are unworkable should be avoided and that courts should be wary of implying terms. All of these principles are sound, provided they are applied with an understanding that by-laws are not a contract, negotiated by two parties or even akin to a company constitution. They are rules that regulate land use and are binding on people by virtue of them owning or occupying land.

Conclusion

This chapter has set out the expansive nature of the by-law-making power in Australia. Most states' legislation gives bodies corporate a broad power to write by-laws that must simply relate to the use or enjoyment of lots or common property. While the limited legislative prohibitions that have been placed on this power will be considered in the following chapter, this chapter explored the question of whether *judicial* limits have been placed on the by-law-making power. It demonstrated that they have not, and that judges have allowed private citizens, most commonly developers, extensive freedom to craft by-laws as they please.

This is unusual because the judiciary has historically exercised what Bryson J termed 'public policy controls' over permissible interests in land. For centuries, common law judges have favoured 'land being fully available for its owner to use and sell, and restricted the creation and operation of interests, including new classes of interests, which might have different effects'.[157] As noted in Chapter 2, judges have always been mindful of the fundamental distinction between contract and property law, and that while there

> can be no harm in allowing the fullest latitude to men in binding themselves and their representatives ... to answer in damages for breach of their obligations ... great detriment would arise and much confusion of rights if parties were allowed to invent new modes of holding and enjoying real property, and to impress upon their lands and tenements a peculiar character, which should follow them into all hands, however remote.

Restraints and obligations could be as 'as infinite in variety as the imagination can conceive'.[158] While modern recording systems lessen some potential confusion, as we saw in the consideration of material from the United States, recordation and notice do not cure all problems. Recordation does not necessarily lead to purchaser knowledge, and in any event, knowledge does not cure exploitative, inefficient or simply misconceived burdens.

Cases like *Italian Forum* and *Casuarina* demonstrate the clear mischief that the common law worked so hard to prevent – burdens created by predecessors in title that transpire to be economically inefficient. The successors in title in those cases were paying for services that they did not want, need or could not afford. Like all predecessors in title, even acting in good faith, those in *Italian Forum* and *Casuarina* could not have perfect foresight.

More fundamentally, the extensive and intractable litigation that plagues the management rights industry in Australia is a visceral demonstration of the enduring relevance of the common law's dislike of obligations on freehold land. Purchasers are buying freehold Torrens title apartments, only to discover that the control that they would ordinarily exercise over their own property or collectively with other owners, has been stripped from them by obligations created by the former owner. As the former owner was never going to bear the responsibility for discharging those obligations, they are frequently economically exploitative.

What is the answer to this dilemma? It is not to go back to the common law and equitable position and only permit obligations and restrictions on freehold land that conforms with the limited permissible rights in easements and restrictive covenants. Modern high-rise buildings demand more flexible law. The answer is for the judiciary to be mindful of *why* more flexible law is needed. It is only needed in order to maintain and manage large, complex buildings, *fairly and efficiently*. More flexible law is not needed to allow initial owners of freehold land unlimited freedom to burden land with any and all obligations and restrictions. It is certainly not needed to allow developers and their associates to make inequitable and ultimately, deeply economically inefficient profits at the expense of freehold apartment owners. The inevitable result of the latter is dissatisfaction and dispute, as well as purchaser and community resistance to high-rise construction.

Ideally, the judiciary would reassert its traditional role in maintaining the economic viability of freehold land. Judges should assess by-laws for what they are – obligations and restrictions on land – teasing out the full economic consequences of by-laws and contracts that bind bodies corporate, striking down those that are unconscionable or economically inefficient. If this does not occur, Australia will see increasingly large tracts of land and airspace reduced in function and value, for short-term developer profit and at long-term community cost.

Notes

1 Easthope, Randolph and Judd, 'Governing the Compact City', n 22, 11.

2 *BCCMA* (Qld) s 180(7) and *Strata Titles Act 1998* (Tas) s 91(3)(c). Queensland also expressly requires bodies corporate to act reasonably when carrying out their functions: *BCCMA* s 94(2).

3 The judicial limit on the content of easements is found in *Re Ellenborough Park* [1956] Ch 131; [1955] 3 All ER 667, and habitually applied by courts; e.g. *Clos Farming* v. *Easton* [2002] NSWCA 389. The judicial limit on the content of restrictive covenants is found in *Haywood* v. *Brunswick Permanent Benefit Building Society* (1881) 8 QBD 403; *Austerberry* v. *Corporation of Oldham* (1885) 29 Ch D 750 (CA), *Pirie* v. *Registrar General* (1962) 109 CLR 619 and *Clem Smith Nominees Pty Ltd* v. *Farrelly* (1978) 20 SASR 227.

4 Judith Yates, 'The Rhetoric and Reality of Housing Choice: The Role of Urban Consolidation' (2001) 19(4) *Urban Policy and Research* 491.

5 *BCCMA* ss 226–302.

6 E.g. see discussion in *Henderson* v. *The Body Corporate for Merrimac Heights* [2011] QSC 336, [107]–[123].

7 Evan McKenzie, 'Emerging Regulatory Trends, Powers and Competing Interests in US Common Interest Housing Developments' in Ann Dupuis and Jennifer Dixon Sarah Blandy (eds), *Multi-owned Housing: Law, Power and Practice* (Ashgate, 2010) 53.

8 Lisa Toohey and Dan Toohey, 'Achieving Quality Outcomes in Community Title Disputes: A Therapeutic Jurisprudence Approach' (2010) 37 *Monash University Law Review* 298.

9 *Salerno* v. *Proprietors of Strata Plan No 42724* (1997) 8 BPR 15,457 (Windeyer J), in which the Supreme Court upheld a by-law prohibiting smoking in a commercial

strata scheme, irrespective of whether the by-law might reduce the pool of potential tenants.

10 In the Australian comedy, *The Castle* (directed by Rob Sitch, Village Road Show, 1997), when asked by the judge to which section of the Australian Constitution he is referring, lawyer Denis Denuto replies, 'There is no one section, it's just the vibe of the thing'.

11 Singer, Joseph William, *Entitlement: The Paradoxes of Property* (Yale University Press, 2000), 215–16.

12 *White* v. *Betalli* (2007) 71 NSWLR 381.

13 *SSMA 2015* s 142.

14 Section 88B of the *Conveyancing Act 1919* (NSW) allows for the creation of easements by indicating them on a plan of subdivision, including a strata plan of subdivision.

15 *Anthony Hordern & Sons Ltd* v. *Amalgamated Clothing and Allied Trades Union of Australia* (1932) 47 CLR 1. The principle in *Anthony Hordern*, as stated by Gavan Duffy CJ and Dixon J at 7 is that:

> When the Legislature explicitly gives a power by a particular provision which prescribes the mode in which it shall be exercised and the conditions and restrictions which must be observed, it excludes the operation of general expressions in the same instrument which might otherwise have been relied upon for the same power.

16 *Betalli* (2007) 71 NSWLR 381, 419 [205], cited with approval by Young JA, with whom Macfarlan JA and Handley JA agreed, in *Casuarina Rec Club Pty Ltd* v. *Owners – Strata Plan 77971* (2011) 80 NSWLR 711, 716 [30]. See also *Italian Forum Ltd* v. *Owners – Strata Plan 60919* [2012] NSWSC 89 ('*Italian Forum*').

17 *Betalli* (2007) 71 NSWLR 381, 389 [39] (Santow JA).

18 *Sydney Diagnostic Services Pty Ltd* v. *Hamlena Pty Ltd* (1991) 5 BPR 11,432.

19 *Betalli* (2007) 71 NSWLR 381, 391–2 [48] (Santow JA, quoting the trial judge).

20 *SSMA 2015* s 150.

21 *Houghton* v. *Immer (No 155) Pty Ltd* (1997) 44 NSWLR 46. In *Young* v. *Owners S/P 3529* (2001) 54 NSWLR 60, 73 [45], Santow J held that the doctrine of fraud on the minority would apply to the case but his Honour's interpretation of the legislation made it unnecessary. Similarly, Gzell J in *Lin* v. *Owners – Strata Plan No 50276* (2004) 11 BPR 21,463, 21,473 [88] held that fraud on a power would be applicable in the circumstances of the case if his Honour were incorrect in his interpretation of the owners corporation's obligations under *SSMA* s 62. The doctrines were not applied in *Jennifer Elizabeth James* v. *Owners Strata Plan No SP 11478 (No 4)* [2012] NSWSC 590 and *Owners Strata Plan 50276* v. *Thoo* [2013] NSWCA 270. In *Owners Strata Plan No 50411 &* v. *Cameron North Sydney Investments Pty Ltd* [2003] NSWCA 5 (which dealt with the question of owners corporation consent to development applications relating to individual lots), counsel for the plaintiff stated that if the Court did not support the plaintiff's interpretation of the planning legislation, it 'would leave the owner of the lot with the inadequate protections conferred by the doctrines of fraud on a power, and fraud on a minority': at [146].

22 All reviews of strata law and the strata industry have revealed improper practices on the part of developers, executive committees, lot owners, strata managers and caretakers. For example, see the litany of complaints documented in Global Access Partners, 'Strata Laws Online Consultation Report' (2012) and H Easthope, B Randolph and S Judd, 'Governing the Compact City: The Role and Effectiveness of Strata Management' (City Futures Research Centre, 2012).

23 *SSMA 2015* s 139(1).

24 Lord Cottenham said that 'if an equity is attached to the property by the owner, no one purchasing with notice of that equity can stand in a different situation from the

party from whom he purchased': (1848) 2 Ph 774, 776; 41 ER 1143, 1145. While notice looms large in property law, it should not obscure the fact that the *numerus clausus* principle comes first. That is, giving a third party notice of rights will not make them binding in property law: Pamela O'Connor, 'Covenants as Regulation' (2011) 1 *Property Law Review* 145, 147. Those rights must first fall within property's 'closed list' and then, traditionally with notice (or registration in Torrens systems), a third party will be bound. By its unusual reliance on notice, the decision in *Tulk* actually created a new class of property rights. However, it must be conceded that the ability of modern land systems to give purchasers notice of interests more easily can be used as a justification for a less rigid approach to the *numerus clausus* principle: Michael Weir, 'Easements for Storage and Parking: A Time for a Rethink' in Lyria Bennett Moses, Brendan Edgeworth and Cathy Sherry (eds), *Property and Security: Selected Essays* (Lawbook Co, 2010) 249, 256.

25 Richard A Epstein, 'Covenants and Constitutions' (1987) 73 *Cornell Law Review* 906; Richard A Epstein, 'Notice and Freedom of Contract in the Law of Servitudes' (1981) 55 *Southern California Law Review* 1353.

26 Susan F French, 'Toward a Modern Law of Servitudes: Reweaving the Ancient Strands' (1981) 55 *Southern California Law Review* 1261; Uriel Reichman, 'Toward a Unified Concept of Servitudes' (1981) 55 *Southern California Law Review* 1177; Stewart E Sterk, 'Freedom from Freedom of Contract: The Enduring Value of Servitude Restrictions' (1984) 70 *Iowa Law Review* 615; Stewart E Sterk, 'Foresight and the Law of Servitudes' (1987) 73 *Cornell Law Review* 956; *Restatement (Third) of Property: Servitudes* (2000) ('*Restatement*').

27 *Conveyancing Act 1919* (NSW) Pt 6.

28 *Real Property Act 1900* (NSW) ss 46A, 47, 48, 49.

29 *Clos Farming Estates* v. *Easton* [2002] NSWCA 389. Purchasers bought individual lots that were burdened by an easement in favour of Lot 86. The easement gave Lot 86 the right to grow vines on a significant portion of all of the other lots, leaving the servient owners with few rights over the burdened land.

30 *Clos Farming Estates Pty Ltd (Rec and Man Appt)* v. *Easton* [2001] NSWSC 525.

31 *Re Ellenborough Park* [1956] Ch 131; [1955] 3 All ER 667.

32 *Spencer's Case* (1583) 5 Co Rep 16a; 77 ER 72.

33 *Clos Farming Estates Pty Ltd (Rec and Man Appt)* v. *Easton* [2001] NSWSC 525, [15], [30], emphasis added.

34 A good example of the economically crippling nature of by-laws in schemes whose initial 'vision' has failed is *Grant and Owners of Rosneath Farm – Strata Plan 35452* [2006] WASAT 162. Rosneath Farm, a permaculture and pattern language strata development, was the subject of much litigation.

35 *Clos Farming Estates* v. *Easton* [2002] NSWCA 389. 393 [53] (Santow JA).

36 *Real Property Act 1900* (NSW) ss 31B(2)(a)–(b).

37 Ibid. s 3 (definition of 'Dealing').

38 Santow JA made reference to *Bursill Enterprises Pty Ltd* v. *Berger Bros Trading Pty Ltd* (1971) 124 CLR 73, 77–8 in noting that an interest is sufficiently recorded in the folio if the folio states the registration number of the dealing creating it and identifies the interest in generic terms: *Betalli* (2007) 71 NSWLR 381, 395–6 [71].

39 This would have been possible if the scheme were larger and the appellant in the minority.

40 Section 139(2) of the *SSMA 2015* states that a strata by-law has no force or effect to the extent that it is inconsistent with the *SSMA* or any other Act or law.

41 See Cathy Sherry, 'How indefeasible is your strata title? Unresolved Torrens problems in strata and community title' (2009) 21(2) *Bond Law Review* 159.

42 Some judges have been prepared to characterise some by-laws as proprietary interests: see, e.g. *Betalli* (2007) 71 NSWLR 381, 393 [53] (Santow JA); *Chauhan* v. *Jaynrees Services Pty Ltd* [2008] NSWSC 969, [42] (Young CJ in Eq) (in which his

Honour tentatively said an exclusive-use by-law created statutory rights that are 'probably proprietary sui generis'); *Owners of Strata Plan No 3397* v. *Tate* (2007) 70 NSWLR 344, 363 [76] (McColl JA) ('*Tate*'), discussed below. Others have not: see, e.g. *Proprietors Cathedral Village Building Units Plan No 106957* v. *Cathedral Place Community Body Corporate* [2012] QSC 301, [57] (Douglas J), where his Honour noted that, '[t]he right to use its property ostensibly given by the by-law does not amount to a proprietary interest in it and is not registered on the title as such'. This was approved in *Proprietors Cathedral Village Building Units Plan No 106957* v. *Cathedral Place Community Body Corporate* [2013] QCA 264, per curiam at [27]. If we consider the fact that all by-laws affect the use of property, just like freehold covenants, they could all be characterised as proprietary.

43 *Land Titles Act 1994* (Qld) s 115.

44 See *Proprietors Cathedral Village Building Units Plan No 106957* v. *Cathedral Place Community Body Corporate* [2012] QSC 301, above n 42.

45 A lease for uncertain duration is invalid regardless of registration.

46 *Casuarina Rec Club Pty Ltd* v. *Owners – Strata Plan 77971* (2011) 80 NSWLR 711 ('*Casuarina*').

47 *Strata Schemes Management Regulation 2010* (NSW) sch 5.

48 *Body Corporate and Community Management (Accommodation Module) Regulation 2008*. However the by-law-making power for all schemes is in *BCCMA* ss 169, 180, 181.

49 Kelly Cassidy and Chris Guilding, 'Defining an Emerging Tourism Industry Sub-Sector: Who are the Strata Titled Tourism Accommodation Stakeholders?' (2010) 29 *International Journal of Hospitality Management* 421.

50 In fact, the correlation between these strata schemes and share investment is so strong that the Australian Securities and Investment Commission regulates some schemes: ASIC Management Rights Schemes, *CO 02/303, 02/304, 02/305*. See Sophie Riley and Grace Li 'Disclosure Requirements and Investor Protection: The Compatibility of Commonwealth, State and Territory Laws in Serviced Strata Schemes (2009) 16(3) *Australian Property Law Journal* 262.

51 *Meriton Apartments Pty Ltd* v. *Industrial Court (NSW)* (2011) 284 ALR 130, 137–8 [32] (Sackville AJA).

52 Nick Nichols, *Frawleys Buys Q1 Management for $10m* (16 December 2011) goldcoast.com.au www.goldcoast.com.au/article/2011/12/16/374381_gold-coast-business.html.

53 *Body Corporate and Community Management Act 1997* (Qld) ss 112–13.

54 *Unit Title Schemes Act* s 89.

55 *Community Assoc DP No 270180* v. *Arrow Asset Management Pty Ltd* [2007] NSWSC 527, [229] (McDougall J).

56 Gary Bugden, 'Management Rights: Developments in Building Management Agreements for Strata Schemes' (2003) 41(3) *Law Society Journal* 42, 42.

57 See, e.g. *Rinbac Pty Ltd* v. *Owners Corporation SP 64972* (Strata and Community Schemes) [2009] NSWCTTT 302.

58 *Body Corporate and Community Management (Accommodation Module) Regulation 2008* (Qld) ss 117–18.

59 *SSMA 2015* s 68; *SSMA 1996* s 40B(2)(b), as amended by *Strata Schemes Management Amendment Act 2002* (NSW). This section came into force on 10 February 2003, and did not affect contracts concluded prior to this date. Earlier contracts were typically 20–25 years in duration: *West* v. *Owners Strata Plan No 54288* [2009] NSWIRComm 30, [19].

60 *Unit Titles (Management Act) 2011* (ACT) ss 51, 61.

61 For example, in *Rinbac Pty Ltd* v. *Owners Corporation SP 64972* (Strata and Community Schemes) [2009] NSWCTTT 302, a by-law was registered stating:

> The owner of lot 1 [the management rights holder] is the only party entitled to conduct the following activities in the building or on the land on which the building is constructed:
>
> a. Manage the serviced apartments
> b. Operate a hotel
> c. Operate a restaurant
> d. Operate a laundry service
> e. Manage the common property for the owners corporation. The owners corporation will continue to be responsible for the proper maintenance and keeping the common property in a state of good repair.

Note, the legal apparatus is set up before the *settlement* of sales, but not before the signing of *contracts.* Strata and community sales are frequently 'off the plan' which means that purchasers are binding themselves before the strata and community plans have been registered and before any of the by-laws and management contracts have been finalised.

62 *Owners Strata Plan No 64972* v. *Rinbac Pty Ltd* [2009] NSWSC 745, [9]. See also *Owners Strata Plan No 61643* v. *183 On Kent Management Pty Ltd* [2007] NSWSC 281, [17]–[26]; *Meriton Apartments Pty Ltd* v. *Industrial Court of New South Wales* (2011) 284 ALR 130, 134–7 [15]–[31] (Sackville AJA) *Meriton Apartments Pty Ltd* v. *Owners Strata Plan No 72381* [2015] NSWSC 202, [20] (Slattery J) and *Corporate for Palm Springs Residences* v. *J Patterson Holdings Pty Ltd* [2008] QDC 300, [5] (McGill DCJ) for descriptions of similar creations of caretaker contracts by the developer.

63 This phenomenon was explicitly acknowledged by McDougall J at [231] in *Community Assoc DP No 270180* v. *Arrow Asset Management Pty Ltd* [2007] NSWSC 527. Justice McDougall held that a developer owed a fiduciary duty to a body corporate and that having acted in its own interest in the sale of management rights, the developer had breached that duty. It was ordered to disgorge the profit made. See Cathy Sherry, 'Long-Term Management Contracts as Developer Abuse' in S Blandy, J Dixon and A Dupuis (eds), *Multi-Owned Housing: Law, Power and Practice* (Ashgate Publishing, 2010); Michael Kleinschmidt, 'Falling Short of the Target: Some Implications of Fiduciary Duties for Developer Practice in Queensland and New South Wales' (2011) 19 *Australian Property Law Journal* 263.

64 *BCCMA* s 112. The Northern Territory has a similar provision in the *Unit Title Schemes Act* s 88.

65 E.g. *Henderson* v. *Body Corporate for Merrimac Heights* [2011] QSC 336.

66 *Corporate for Palm Springs Residences* v. *J Patterson Holdings Pty Ltd* [2008] QDC 300.

67 *Corporate for Palm Springs Residences* v. *J Patterson Holdings Pty Ltd* [2008] QDC 300, [5].

68 *Owners – Strata Plan No. 44999* v. *Premier Holdings Corp Pty Ltd and Gregg Ritchie and Partners Pty Ltd* [2012] NSWSC 171 (Black J) is another good example of a judgment that applies the general law to the creation and assignment of management rights, finding that developer and management rights holder practice did not comply with the requirements of the general law in relation to the assignment of contracts.

69 2–4 September 2015, Gold Coast, Queensland.

70 A presentation by the Owners Corporation Network (OCN) described this successful process in relation to a large scheme in Chatswood on Sydney's North Shore: Gerry Chia, OCN, 'Strata building performance is a lottery: If I hadn't bought into my new building 10 years ago, what would be its standing today?', Strata and Community Title for Australia in the 21st Century, Griffith University, 2–4 September 2015.

71 *Bondlake Pty Ltd* v. *Owners – Strata Plan No 60285* (2005) 62 NSWLR 158 was a failed attempt to invalidate a caretaker contract on the grounds that its creation incurred a debt in the initial period in contravention of the *SSMA 1996* s 113(1)(b). Giles JA held (Handley JA and McColl JA agreeing), that it did not follow that just because the debt was improperly incurred by entering the contract, the contract itself was invalid. The Court of Appeal overturned the decision of Pearlman AJ in *Owners – Strata Plan 60285* v. *Bondlake Pty Ltd* [2004] NSWSC 1037. Her Honour had held at [70]:

> In my opinion, the obvious public purpose of s 113 points to the intention of the legislature that a contract contravening its provisions will be void and unenforceable. The mischief that s 113 intends to avoid is the casting of onerous obligations upon future lot owners. That mischief can be effectively avoided only if no such burden can be cast upon them. One way to achieve that, in a case of contravention of s 113(b)(1), is the recovery of the amount as a debt. But that remedy may not be sufficient. The original owner may not be in existence, or it may not have the means to meet the debt, or it may have a defence under s 113(4). Another way to achieve the avoidance of the mischief is to render the contravening contract void and unenforceable. In my opinion, s 113 contemplates such a legal consequence.

Pearlman AJ's approach was arguably more attentive to the reality of developer practice. See also *Owners – Strata Plan No. 44999* v. *Premier Holdings Corp Pty Ltd and Gregg Ritchie and Partners Pty Ltd* [2012] NSWSC 171, [43], where Black J held that a deed appointing an unlicensed strata manager was contrary to the strata legislation and that the proper result was to hold that the deed was unenforceable.

72 Gregory S Alexander, 'Freedom Coercion and the Law of Servitudes' (1987) 73 *Cornell Law Review* 883; Robert Natelson, 'Consent, Coercion, and "Reasonableness" in Private Law: The Special Case of the Property Owners Association' (1990) 51 *Ohio State Law Journal* 41; James L Winokur, 'The Mixed Blessing of Promissory Servitudes: Toward Optimizing Economic Utility, Individual Liberty, and Personal Identity' (1989) 1989 *Wisconsin Law Review* 1; Sterk, 'Freedom from Freedom of Contract', above n 26; Sterk, 'Foresight and the Law of Servitudes', above n 26; cf. Epstein, 'Covenants and Constitutions', above n 25. See discussion in Chapter 3, 121–8. For analysis of the role of disclosure in real estate transactions generally, see Sharon A Christensen, William D Duncan and Amanda P Stickley, 'Evaluating Information Disclosure to Buyers of Real Estate – Useful or Merely Adding to the Confusion and Expense?' (2007) 7(2) *Queensland University of Technology Law and Justice Journal* 148. Also for an analysis of the misplaced faith in real estate disclosure regimes generally, particularly in relation to information overload, see Sharon A Christensen, William D Duncan and Amanda P Stickley, 'Behavioural Biases and Information Disclosure Laws Relating to Residential Property Sales: Narrowing the Gap Between Existing Laws and Calls for Future Reforms' (2009) 9(2) *Queensland University of Technology Law and Justice Journal* 251.

73 In fairness to developers it should be noted that some service contracts are being required by local or state government as a condition of development consent; for example, the developer is required to provide a bus service in a new residential area that has no existing public transport. The only way to do this is to bind the body corporate to a service provider via contract, prior to the sale of lots. For example, the Management Statement for Macquarie Links in south west Sydney, DP270152, by-law 48, requires the Community Association to provide a bus service to the local railway station and to local schools. By-law 48.4 states that the Community Association 'may enter into agreements with third parties to provide the transport facilities under this By-Law'. The 'may' is misleading; the community association members

are hardly going to drive people themselves; they *must* enter a contract for transport services. The same process occurs in relation to 'green utilities'; for example, Central Park, the redevelopment of the old Carlton United Brewery site in inner Sydney, has a membrane bioreactor (MBR) recycled water facility in the base of the residential building, owned by the Water Factory Company: Pitt Town Water, *Central Park* http://pitttownwater.com.au/community/central-park/. Residents cannot run a blackwater treatment plant themselves, and thus it is inevitable that the body corporate will be contractually bound to purchase water and sewage services from the Company.

74 *Humphries* v. *Proprietors 'Surfers Palms North' Group Titles Plan 1955* (1994) 179 CLR 597.

75 Ibid. 604 [14] (Brennan and Toohey JJ) (emphasis added). In *Travis* v. *Proprietors of Strata Plan 374* (1969) 90 WN (Pt 1) (NSW) 711, Street J held that a body corporate did not have the power to build a swimming pool on common property, and ordered that the body corporate be restrained from doing so unless authorised by a change in the Act or *in the by-laws.*

76 *Corporations Act 2001* (Cth) s 124.

77 One characterisation of by-laws is that they are a form of delegated legislation and the principles of administrative law apply to them: *Lynch, Glenn and Jenny Owners of Lot 75a* v. *Owners Strata Plan No 36458* [1999] NSWSSB 55 (17 August 1999); *Tate* (2007) 70 NSWLR 344, 352–62 [29]–[72] (McColl JA). See discussion below.

78 *BCCMA* (Qld) s 180(1), *Owners Corporations Act 2006* (Vic) s 140(b), *Unit Titles Act 2011* (ACT), s 108(3)(a), *Unit Title Schemes Act* (NT) s 95(3)(a), and *Community Titles Act 1996* (SA) s 41(1)(a), all provide that by-laws must not be inconsistent with the parent Act or any other Act or law. The New South Wales provision is now *SSMA 2015* (NSW) s 136(2). cf. *Strata Titles Act 1988* (SA) and *Strata Titles Act 1998* (Tas), which contain no provision on by-laws with reference to the parent Act or any other Act, and *Strata Titles Act 1985* (WA) s 42(1) and *Community Land Management Act 1989* (NSW) s 14, which contain provisions prohibiting by-laws that are inconsistent with the parent Acts but not other Acts.

79 In her Second Reading Speech for the *Strata Schemes Management Amendment Bill 2003*, Ms Meagher said:

> Many owners' corporations use by-laws to deal with matters specific to their own complexes. By-laws are intended to enhance and utilise the laws that are already in place so that the circumstances of a particular scheme can be accommodated. They are not able to change fundamentally what the general law already provides. It is recognised that some owners' corporations may attempt to stretch by-laws further than their intended limit, and the bill contains a provision that will stress that by-laws cannot be used in an endeavour to go beyond the provisions of the Strata Schemes Management Act or any other relevant law. It is also made clear that by-laws that conflict with any existing law are invalid.

New South Wales, *Parliamentary Debates*, Legislative Assembly, 4 December 2003, 6093 (Reba Meagher).

80 From discussions with strata lawyers it seems that many lawyers, including those who work for bodies corporate, not developers, use empowering by-laws as a precautionary measure; if they are not sure that a body corporate has the power to do something (that is, the body corporate arguably has the implied, rather than express power), they will enact an empowering by-law. This practice is no doubt a product of modern lawyers becoming accustomed to being guided by specific legislative provisions rather than more general principles of the common law and equity. This is a phenomenon in law generally but is particularly marked in a statute-dependent area like strata and community title.

81 *Santai* v. *Owners – Strata Plan No 77971* [2010] NSWSC 628, [116]–[117].

82 *Betalli* (2007) 71 NSWLR 381, 419 [205] (Campbell JA), cited in *Casuarina* (2011) 80 NSWLR 711, 716 [30] (Young JA).
83 Michael A Heller, 'The Boundaries of Private Property' (1999) 108 *Yale Law Journal* 1163, 1192.
84 *Restatement* §3.1.
85 Ibid.
86 Ibid.
87 Ibid.
88 Dan Tarlock, 'Touch and Concern is Dead, Long Live the Doctrine' (1998) 77 *Nebraska Law Review* 804, 818.
89 *Tooth & Co Ltd* v. *Barker* (1960) 77 WN (NSW) 231 ('*Tooth*'). In this case, the Court struck down a covenant requiring the exclusive purchase of beer from a specific brewer on the grounds that it did not benefit the land of the covenantee. In *Casuarina* the positive nature of the covenant would also have been a problem. In *Tooth*, the land of the covenantor was a pub and so not buying beer at all was not an option, thus a restriction not to buy from any other supplier was a meaningful restrictive covenant; ('brick and tile covenants operate the same way; the owner of burdened land could not build at all, but if they build, it must be in brick and tile; of course, not building on residential land is not an option'). In *Casuarina*, realistically, the body corporate could not have provided the guests with gym facilities; while perhaps desirable, gym facilities are not essential in a holiday resort in a way that beer is in a pub.
90 Uriel Reichman, 'Toward a Unified Concept of Servitudes' (1981) 55 *Southern California Law Review* 1177, 1233.
91 Carol M Rose, 'Servitudes, Security, and Assent: Some Comments on Professors French and Reichman' (1981) 55 *Southern California Law Review* 1403, 1404.
92 Ibid. 1408.
93 *Restatement* §3.7.
94 Above nn 51 and 71.
95 Cathy Sherry, 'Long-Term Management Contracts as Developer Abuse', above n 63.
96 Gregory S Alexander, 'Dilemmas of Group Autonomy: Residential Associations and Community' (1989) 75(1) *Cornell Law Review* 1, 56.
97 *Bennett* v. *Behring Corp*, 466 F Supp 689 (SD Fla, 1979); *Streams Sports Club Ltd* v. *Richmond*, 99 Ill 2d 182 (1983), cited in A J Casner *et al.*, *Cases and Text on Property* (Aspen Publishers, 5th edn, 2004), 1004.
98 *Eagle Enterprises, Inc* v. *Gross*, 39 NY 2d 505 (1976) (requirement to pay for water in summer not enforceable as a covenant after homeowner sunk own well), cited in Casner *et al.*, above n 97, 1001–3.
99 Tarlock, above n 88, 813.
100 This section was the end product of consistent litigation in the 1970s and 1980s over management and other contracts that favoured developers or their associates. There had been piecemeal legislative enactments attempting to tackle the problem, but the United States ultimately solved the problem by allowing communities to terminate all developer-made contracts. Having no security of tenure through developer-made contracts, management companies offered their services directly to associations, a healthier market solution: Sherry, 'Long-Term Management Contracts as Developer Abuse', above n 63, 169–74.
101 The initial owner might be a landowner whose land the developer has agreed to develop. The developer typically controls the process, but the land is not transferred to the developer in order to avoid unnecessary stamp duty.
102 See above n 21.
103 *Hidden Harbour Estates* v. *Norman* 309 So 2d 180 (Fla Dist Ct App, 1975). See also *Body Corporate and Community Management Act 1997* (Qld) s 180(7).

104 Cathy Sherry, Submission, Strata and Community Scheme Laws, November 2012, www.fairtrading.nsw.gov.au/biz_res/ftweb/pdfs/About_us/Have_your_say/Review_of_strata_community_scheme_laws/Individual_submissions/Sherry_cathy.pdf.
105 Cf. *Williams* v. *City of Melbourne* [1933] HCA 56; (1933) 49 CLR 142, in which the High Court held that courts could not assess the validity of non-strata by-laws on the basis of reasonableness.
106 Singer, above n 11.
107 Gregory S Alexander *et al.*, 'A Statement of Progressive Property' (2008) 94 *Cornell Law Review* 743.
108 Gary Bugden, 'All about Strata and Building Management Statements and Building Management Committees: A Complete Guide to Working With Building Management Committees' (Institute of Strata Management, 2006).
109 Cathy Sherry, 'Building Management Statements and Strata Management Statements: Unholy Mixing of Contract and Property' (2013) 87 *Australian Law Journal* 393. In some ways, BMSs and SMSs are analogous to new public authority covenants, e.g. *Conveyancing Act 1919* (NSW) ss 88D–F.
110 *Owners Corporation Strata Plan 70672* v. *The Trustees of the Roman Catholic Church for the Archdiocese of Sydney* [2011] NSWSC 973; *Owners – Strata Plan 78102* v. *Owners – Strata Plan 78101* [2010] NSWSC 973. There are a number of adjudicator decisions in Queensland about volumetric subdivisions, and a small number of higher court decisions (e.g. the multiple disputes in relation to the Cathedral Place development in Brisbane) but none deal with the fundamental nature of building management statements.
111 *Owners Strata Plan No 60919* v. *Consumer Trader and Tenancy Tribunal* [2009] NSWSC 1158.
112 Ibid. [7].
113 *SSFDA 1973* sch 1C, now *SSDA 2015* sch 4.
114 *Italian Forum* [2012] NSWSC 895, [50]–[53] (White J).
115 For an explanation of the *Real Property Act 1900* (NSW) in this context, see Lyria Bennett Moses and Brendan Edgeworth, 'Taking it Personally: Ebb and Flow in the Torrens System's in Personam Exception to Indefeasibility' (2013) 35(1) *Sydney Law Review* 1, 3–6.
116 With the exception of limited numbers of leasehold strata developments.
117 For example, courts have had to have reference to the general law on tenants in common to determine disputes about common property: *Lin* v. *Owners – Strata Plan No 50276* (2004) 11 BPR 21,463; *Houghton* v. *Immer (No 155) Pty Ltd* (1997) 44 NSWLR 46.
118 Hanoch Dagan, 'The Limited Autonomy of Private Law' (2008) 56 *American Journal of Comparative Law* 809, 814–15.
119 *Owners Strata Plan No 60919* v. *Consumer Trader and Tenancy Tribunal* [2009] NSWSC 1158.
120 *Body Corporate and Community Management Act 1997* (Qld) s 180(6).
121 *Community Titles Act 1996* s 37(1)(b).
122 Tarlock, above n 88.
123 See above n 42.
124 *Codelfa Construction Pty Ltd* v. *State Land Rail Authority* [1982] HCA 24; (1982) 149 CLR 337.
125 *Owners of Strata Plan No 3397* v. *Tate* (2007) 70 NSWLR 344.
126 Strata Plan No 3397 is the Park Regis building in the Sydney CBD. The strata plan was registered in 1966, so it predates the stratum subdivision provisions in *Conveyancing Act 1919* (NSW) Pt 23 Div 3B. As a result, it is a single strata scheme, rather than separate stratum lots and parcels.
127 *North Wind Pty Ltd* v. *Proprietors – Strata Plan 3143* [1981] 2 NSWLR 809.
128 *Tate* (2007) 70 NSWLR 344, 365 [90] (Harrison J).

129 Ibid. 366–7 [95] (Harrison J).
130 Ibid. 372 [116] (Harrison J).
131 Ibid.
132 *Ryledar Pty Ltd* v. *Euphoric Pty Ltd* (2007) 69 NSWLR 603.
133 *Tate* (2007) 70 NSWLR 344, 372 [115] (Harrison J).
134 E.g. *Taunton Enterprises Pty Ltd Appeal Against Order No C579394 Strata Title Act* [2005] TASRMPAT 76.
135 While assignees of leases or reversions will have contractual terms they did not negotiate enforced against them as a matter of property law, leases are not an appropriate analogy with by-laws for two reasons. First, traditionally not all original terms of leases are enforceable against assignees, but only those that satisfy the property boundary rule of 'touching and concerning' the land: *Spencer's Case* (1583) 5 Co Rep 16a; 77 ER 72; *Gumland Property Holdings Pty Ltd* v. *Duffy Bros Fruit Market (Campbelltown) Pty Ltd* (2008) 234 CLR 237. Second, even if assignees are bound by all terms of a lease in accordance with *Real Property Act 1900* (NSW) s 51 and *Karacominakis* v. *Big Country Developments* (2000) 10 BPR 18,235, it is likely that an assignee has read and genuinely consented to all of the terms of the lease by buying the lease or reversion. The lease is the complete delineation of the interest acquired by an assignee of the lease and a delineation of a substantial part of the interest acquired by the assignee of the reversion. In contrast, even if a purchaser reads the by-laws registered at the time of purchase, unlike covenants in leases, those by-laws can subsequently change without the purchaser's consent.
136 *Interpretation Act 1987* (NSW) s 3 (definition of 'Instrument'), cited in *Tate* (2007) 70 NSWLR 344, 353 [35] (McColl JA).
137 *Collector of Customs* v. *Agfa-Gevaert Ltd* (1986) 186 CLR 389, 398.
138 *Dainford Ltd v. Smith* (1985) 155 CLR 342, per Wilson J. Also see posters on *Owners Corperation PS 503919* v. *Balcome* [2016] VSC 384 in Chapter 1, 37.
139 *Re Taylor* [1995] 2 Qd R 564.
140 *SSMA 2015* (NSW) s 136(2), *BCCMA* (Qld) s 180(1), *Owners Corporations Act 2006* (Vic) s 140(b), *Unit Titles Act 2011* (ACT) s 108(3)(a), *Unit Title Schemes Act* (NT) s 95(3)(a), and *Community Titles Act* 1996 (SA) s 41(1)(a), provide that by-laws must not be inconsistent with the parent Act or any other Act or law. cf. *Strata Titles Act* 1988 (SA) and *Strata Titles Act* 1998 (Tas), which contain no provision on by-laws with reference to the parent Act or any other Act, and *Strata Titles Act* 1985 (WA) s 42(1) and *Community Land Management Act* 1989 (NSW) s 14, which contain provisions prohibiting by-laws that are inconsistent with the parent Acts but not other Acts. In *Owners Corporation SP 68751* v. *Community Assoc DP* 270281 [2015] NSWCATCD 99, the Tribunal had to consider the validity of a provision of a community management statement that required all subsidiary strata schemes to employ the same managing agent as the community scheme, a provision put in place by the developer. If the community title legislation contained a provision that prohibited management statements from containing any provision which was inconsistent with any other Act or law, such a provision would presumably have been inconsistent with the *SSMA 1996* s 13(1), which stated that 'An owners corporation may employ such persons *as it thinks fit* to assist it in the exercise of any of its functions' (emphasis added).
141 *London Assoc of Shipowners and Brokers* v. *London and India Docks Joint Committee* [1892] 3 Ch 242, 252.
142 *Tate* (2007) 70 NSWLR 344, 355–6 [47] (McColl JA).
143 *Avranik Pty Ltd* v. *Lloyd* [2012] VSC 306, confirmed on appeal in *Avranik Pty Ltd v. Lloyd* [2013] VSCA 244.
144 *Re Carey; Ex parte Exclude Holdings Pty Ltd* [2006] WASCA 219, per Martin CJ, [14].

145 *Egyptian Salt and Soda Co Ltd* v. *Port Said Salt Assoc Ltd* [1931] AC 677, 682 (Lord Macmillan), quoted in *Tate* (2007) 70 NSWLR 344, 358 [61] (McColl JA).
146 *Lion Nathan Australia Pty Ltd* v. *Coopers Brewery Ltd* (2005) 56 ACSR 263, 276 [79] (Finn J), quoted in *Tate* (2007) 70 NSWLR 344, 359–60 [65] (McColl JA).
147 Ibid. 363 [76] (McColl J).
148 *Houghton* v. *Immer (No 155) Pty Ltd* (1997) 44 NSWLR 46.
149 *Lion Nathan Australia Pty Ltd* v. *Coopers Brewery Ltd* [2006] FCAFC 144, Kenny J, [114].
150 Ibid. per Weinberg J at [66].
151 Margaret Jane Radin, 'Property and Personhood' (1981–1982) 34(5) *Stanford Law Review* 957, 992; discussed in Chapter 3, 125–6.
152 See above n 43.
153 In *Owners of Pearl Beach Survey Strata Plan 49019 and Heyns* [2011] WASAT 66, the Tribunal member came tantalisingly close to recognising the significance of by-laws being interests recorded on the Torrens register at [43(b)(d)], when he said, 'the registered by-laws, including the strata plan and s 6 ST Act use restriction thereon, are lawful encumbrances registered against the title and any proprietor is therefore deemed to have knowledge thereof'. The reference to notice is unfortunate, as notice is irrelevant in a Torrens system, as was his subsequent reliance on contractual principles to justify relying on a range of extrinsic evidence such as contracts of sale and the evidence of the professional who drafted the by-laws, none of which could ever be accessible to purchasers. His reliance on the use restriction in the strata plan was entirely acceptable, the strata plan being part of the Torrens register.
154 *Westfield Management Ltd* v. *Perpetual Trustee Co Ltd* (2007) 233 CLR 528.
155 The one exception to this general rule might be that if the by-law is an exclusive-use by-law, negotiated by a particular owner who is also involved in a dispute about its interpretation, it would be possible and logical to consider the circumstances surrounding the by-law's creation. This would do no violence to Torrens principles as it would be consistent with the *in personam* exception: *Frazer v. Walker* [1967] 1 AC 569, 585.
156 *Avranik*, [29].
157 *Clos Farming Estates Pty Ltd (Rec and Man Appt)* v. *Easton* [2001] NSWSC 525, [15], [30].
158 *Keppell* v. *Bailey* (1834) 2 My & K 517, 39 ER 1042, 1048, 1049. See discussion in Chapter 2, 51–7.

5 Privacy and personal autonomy

The social and political implications of by-laws

Introduction

This chapter will move beyond a consideration of the economic effect of by-laws to explore their social effects; in particular, it will examine whether current Australian law is able to respond adequately to the social consequences of by-law-making power. This is consistent with the property theory that was outlined in Chapter 2 which argued that the legitimacy of property entitlements cannot be determined by a single metric, but must account for the full range of values about which we care, and that we cannot judge property rights with sole reference to the moment of their creation, but must consider their exercise through time. A broader consideration of by-laws is also consistent with traditional property doctrine, which is instrumental in the creation of more than economic freedom; it helps to safeguard political and personal freedom, particularly in the home. While strata and community title were designed to overcome specific limitations in orthodox property law, this does not mean that they operate entirely outside the ordinary principles that govern property on either a doctrinal or theoretical level.

As we have seen, strata and community title allow private citizens to write by-laws that regulate the use of both common property and lot property. In relation to the use of common property by current owners, the by-law-making power is a statutory formalisation of the existing rights co-owners have to regulate their own land. Co-owners can make contractual agreements about the use of co-owned land (for example, joint venture agreements, agreements to pay occupation rent or to divide rents and profits) and courts will enforce them. However, the by-law-making power differs from these agreements because, as we have seen in previous chapters, by-laws are not contracts. They bind people who have not agreed to them, including minority owners and all subsequent owners (who may or may not have agreed to them, depending on how much weight we give to the notion of consent to by-laws by notice and purchase). In this sense, the by-law-making power in relation to common property is novel.

In relation to individually owned lots, the by-law-making power is completely novel. By granting bodies corporate the power to make by-laws, the legislature has authorised private citizens to make rules regulating other people's homes, investment properties and work places. This is extremely unusual. In liberal democracies, the power to regulate land is invariably a public function performed by elected representatives, at municipal, state and federal level. Private organisations like companies, universities and voluntary organisations have the power to write by-laws,[1] but they rarely relate to land, at least on any great scale. As we saw in Chapter 2, land has a significance that other forms of property lack. First, we all need to be present on land at all times; that is, everything we do has to be done somewhere;[2] and second, when we die or move on, we cannot take land with us. When we write rules that affect land, we are affecting people who have no choice but to be on some land, not just now but in the future.[3] While everyone has a choice to be a member of a university or to buy shares in a company, nobody has the choice not to have a home or even a workplace (homeless and unemployed people are forced into that position). As increasing numbers of properties are located in strata and community title schemes, people's ability to avoid strata and community title will be compromised, particularly when entire suburbs are created within the legislative framework. Thus, the ability of private citizens to write rules through a strata or community title body corporate has a significance which is absent in corporations and other associations.

There are risks in granting private citizens the power to make rules for each other. The most obvious one is self-interest. While the doctrines of fraud on a power and fraud on the minority both apply to strata and community title schemes theoretically preventing blatantly exploitative behaviour, in most states, lot owners are entitled to act in their own interests. For example, in *Houghton* v. *Immer (No 155) Pty Ltd*, the New South Wales Court of Appeal held that 'the power of a proprietor to vote at general meetings of the body corporate is not fiduciary, and within limits it may be exercised by the proprietor for his or her own benefit'.[4] As a result, when voting on a body corporate, owners are not performing a public duty – they are exercising a private right. They are not required to have reference to the well-being of others even inside, let alone outside the scheme.

Queensland has attempted to confront the potential for self-serving behaviour on bodies corporate, although some provisions are more successful than others. The regulations for the standard, accommodation and commercial modules prohibit owners voting if they have a direct or indirect interest in an issue. As noted by the Adjudicator in *Paloma* [2012] QBCCMCmr 37, that would technically disqualify all members from voting on many issues.[5] Further, it is nonsensical to prevent people from voting on issues that affect them personally; that is precisely when their vote matters most. More successfully, the Queensland legislation includes a code of conduct for members of the body corporate committee, which requires them to act in the body corporate's best interests and to act fairly and honestly,[6] and a legislative stipulation that when carrying out its general function, a body corporate must act 'reasonably'.[7]

However, irrespective of the standard imposed on people when they vote as ordinary members of a body corporate or members of a committee or executive, they remain lay people with limited expertise in relation to land regulation, and with no obligation to the wider community. For example, when a privately run community makes a decision about a playground or the use of open space by children, no one on the executive committee or body corporate may know anything about the developmental significance of open space for young people. In contrast, when a local council makes a decision about open space or playground facilities, in accordance with its statutory obligation to consider multiple community interests, it will use the skills of a professional urban planner who is trained to consider a wide range of issues for individual children and the community at large when determining optimal use of space. The Victorian Civil and Administrative Tribunal acknowledged this lack of expertise in deciding whether a body corporate had the power to write a by-law banning short-term letting. In *Owners Corporation PS501391P* v. *Balcombe* (Owners Corporations)[8] Member Rowland said that,

> The enforcement of the planning use of private lots is an onerous and fraught task and would place an unreasonable burden on volunteer lay committees, the members of which have a vested interest in the outcome. It would do little to enhance the harmony of lot owners. That task is best left to local Councils.

Unfortunately, legislatures have not been as ready to acknowledge the inherent problems in lay governance.

In addition to professional expertise, when public authorities make rules and decisions that affect private citizens, citizens are protected by the statutory obligations of public authorities, as well as the armoury of public law. The latter does not generally apply to the private sphere of strata and community title schemes. For example, in *Jennifer Elizabeth James* v. *Owners Strata Plan No SP 11478 (No 4)*, Ball J held that a compulsorily appointed strata manager, carrying out the functions of a body corporate, did not have a duty to afford procedural fairness to lot owners. Justice Ball found that:

> The duty to afford procedural fairness is an obligation that is generally imposed as a matter of public law on statutory or administrative bodies. In some cases, the duty has also been applied to private bodies which are sufficiently public in nature. So, for example, in *Forbes* v. *New South Wales Trotting Club Ltd* [1979] HCA 27; (1979) 143 CLR 242 it was accepted that the New South Wales Trotting Club was under an obligation to afford procedural fairness before excluding someone from its race course in accordance with its rules. On the other hand, it has been held that a casino operator was not under an obligation to exercise procedural fairness in deciding whether to exclude a patron from the casino, even though a right of exclusion is given by s 79 of the *Casino Control Act 1992*: *Hinkley* v. *Star City Pty Ltd* [2010] NSWSC 138. That was because the casino operator was not exercising a 'public power' but merely a common law right: at [169] per Ward J.

> In the present case, Mr Anderson [the compulsorily appointed strata manager] is not exercising a public power or something analogous to it. Although he is exercising a statutory power, that power is concerned with the administration of private property in which a number of individuals have an interest. His position is analogous to that of a receiver or an administrator of a company. When an owners corporation makes a decision that affects other owners, it is not exercising a 'public power' and does not need to afford procedural fairness. In particular, the making of a by-law under ss 52 or 65A only affects the private rights, interests and expectations of other lot owners. The SSM Act sets out procedures to ensure that those rights, interests and expectations are not unfairly defeated. So long as the decision is made in accordance with those procedures, it is valid. The appointment of a strata scheme manager under s 162 does not convert what was otherwise a private power to make decisions for the management of the strata scheme into a 'public power'.[9]

While the absence of an obligation to afford procedural fairness to a lot owner may not seem a cause for concern in the six-lot strata scheme in question, it might be in a 1000-lot community scheme comprising an entire new suburb or even a 350-lot high-rise strata scheme housing over 500 residents. The larger the scale of the development, the more 'public' it becomes, and the more closely the body corporate resembles a miniature local government. It seems feasible that French's comments, noted in Chapter 3, in relation to unrestrained HOA power in the United States, will be equally applicable to strata and community schemes:

> As the resemblance [of HOAs] to government grows stronger, the more people will fear the lack of controls that limit the powers of public governments. People will feel comfortable with this most local form of government only if it is subjected to the same kinds of restrictions we, as a people, have imposed on our other levels of government.[10]

As discussed in Chapter 3, the United States has confronted, but not solved, the problems of private citizens exercising quasi-governmental functions without the constraints that apply to public power. McKenzie identified the way in which 'cluster subdivisions' were seen as a solution to the challenge of providing traditional suburban amenities in the face of rising land prices,[11] and how it was optimistically assumed that privately owned common space could be managed as well, if not better, by private residential associations. McKenzie argued that governments and the real estate industry failed to anticipate the danger of private citizens wielding significant power over their neighbours. Twenty years on, in his most recent book *Beyond Privatopia*, McKenzie said that the volunteer nature of HOA governance cannot be overemphasised: 'the entire institution of common interest housing rests on the volunteer directors, yet they are unpaid, untrained, often unqualified, and almost entirely unsupported by the governments whose work they are often doing'.[12] We saw that Winokur, in an article that grappled with both the doctrinal rationale for HOAs, as well as their lived reality, had argued that the implementation of community rules by inexperienced association boards, in combination with a lack of initial understanding of servitude regimes on the part of purchasers, led to significant increases in HOA litigation. He said that on the ground, HOAs were not the small participatory democracies they were promoted to be and that many residents simply did not accept the legitimacy of their neighbours regulating their homes.[13]

In Australia, research conducted to date, (as well as substantial litigation, mediation and adjudication), has demonstrated that all of these problems of lay governance have emerged in our strata and community schemes.[14] In New South Wales, a survey of over 1000 strata owners found that executive committee performance was highly variable, with dissatisfaction centring on perceptions that committee members acted in their self-interest.[15] The 2012 on-line community consultation that preceded the review of the New South Wales legislation, revealed concerns with a lack of skills, training and knowledge among executive committee members; a lack of knowledge of by-laws on the part of all owners and tenants; unwillingness to serve on executive committees or attend general meetings; proxy farming; conflicts of interest and improper dealings; a lack of transparency in decision-making and a lack of accountability of executive committee members.[16] Ever expanding state legislation, particularly in Queensland, with vast provisions on voting, meeting procedure, secret ballots, notice, financial, administrative and property management matters, is clear evidence of a gradual legislative realisation that strata and community title vest significant powers in the hands of private citizens that are open to abuse. In short, all of the problems that private governance by lay people could reasonably be anticipated to produce have been realised.

In the absence of the application of public law to HOAs, McKenzie identifies civil litigation between associations and individual owners as the 'main check' on association power in the United States.[17] Many cases have challenged new rules, in particular where there are allegations of unreasonableness or infringements of existing property rights (for example, by closing the golf course or restricting the leasing of lots), as well as challenging the validity of initial C, C & Rs (for example, those that exclude all but single family occupation of lots). Some of the United States issues may not arise in New South Wales as a result of our history and national psyche,[18] and by continually expanding legislation, Australian state legislatures have been more proactive in attempting to regulate body corporate powers and functions than their counterparts in the United States. However, the by-law-making power remains a source of confusion for Australian legislatures and courts. Some elements of the legislation and case law reveal a recognition of the potentially harmful nature of the by-law-making power and a consequent need to limit that power, while others display a repeated default to contractual norms and the erroneous assumption that, so long as by-laws are voluntarily agreed, they are legitimate.

The current law

As described in Chapter 1 and illustrated in the preceding chapter, most states use a single basic rule to determine the validity of by-laws. In New South Wales, Tasmania, Western Australia and Queensland, it is that by-laws must relate to the administration, management, use and enjoyment of lots and common property. The Northern Territory, South Australia and the Australian Capital Territory simply empower bodies corporate to create rules. As noted in Chapter 1, Victoria is the only state that limits by-law-making power to a predefined list of matters, set out in sch 1 of the *Owners Corporations Act 2006*.

All states then limit the general by-law-making power with express prohibitions. These prohibitions are important because they constitute a legislative acknowledgement that the by-law-making power can be harmful and that private citizens should not be given an unlimited quasi-legislative power. However, their protection is limited because most are narrow and specific, relating only to activities that the legislature has already identified as unacceptable. They do not cover as yet unimagined by-laws that might also be harmful. To prohibit these, a more flexible prohibition is needed.

Prohibitions on by-laws restricting transfer and leasing

Most states have a prohibition on strata by-laws restricting transfer and leasing to prevent from re-emerging the very problem that the strata legislation was designed to fix– limitations on transfer of company title shares had made banks reluctant to lend money with a company title unit as security.[19] However, by banning restrictions on transfer and leasing, Australia has also avoided the social segregation problems associated with such restrictions that have plagued the United States.[20]

United States HOAs and condominiums frequently had outright bans on leasing, including bans that were subsequently introduced by communities after people had purchased.[21] Associations were also given a right of first refusal in relation to transfers. Sometimes these rights of veto were simply used to ensure that new members were able to meet the financial obligations of living in a private community, but sometimes they were used to control social characteristics of new members and to promote homogeneity. For example, it was (and still is) often assumed that tenants, as temporary members of a community, are less likely to care for their environment, less likely to participate in community activities and more likely to engage in disruptive behaviour.

Restrictions on transfer and leasing raise questions about the legitimacy of 'small group autonomy', the subject of Alexander and Ellickson's articles discussed in Chapter 3.[22] While some argue that there are benefits in a community being able to choose its members, there are obvious risks in people being able to exclude others, retreating into homogenous enclaves regulated by values disconnected from the wider community.

There are some exceptions to the prohibition on by-laws that restrict transfer and leasing. In New South Wales, the community title legislation not only does not contain an express prohibition on restrictions on leasing and transfer, it expressly authorises by-laws 'limiting occupancy under the scheme to persons of a particular description'.[23] The ACT and Tasmania have a prohibition on restrictions on dealing but expressly permit them if they are necessary to preserve the character of the scheme.[24] The second reading speech for the New South Wales Act made reference to equestrian or golfing 'themed' developments,[25] a concept presumably copied from United States' developments where themes are common. To my knowledge, other than age limitations for retirement communities, communities restricted by theme have not emerged in any state, presumably because it is uneconomical to reduce the potential pool of purchasers.

This even seems to be the case inside communes or 'intentional communities', some of which have utilised community title, and whose very existence depends on membership of like-minded people.[26] Fisher argues that while section 17(1)(a) of the *Community Land Management Act 1989* (NSW) ('*CLMA*') provides intentional communities with the tools to limit membership, the enthusiasm for community title in 'Rainbow Region' in the Byron Shire[27] 'appears less to do with an interest in community living than with the acquisition of a legal interest in property located in an attractive rural region'.[28] While many established communes traditionally controlled membership,[29] this does not seem to have translated into section 17(1)(a) by-laws when the community title legislation is used. Rather than limiting membership, communities encourage self-selection. For example, in what has to be one of the more poetic parts of the Torrens register, the management statement for Jindibah in Byron Shire begins with the statement 'this community is being created with ecological sensitivity, and hopefully, the wisdom of the owls for a few people who share the same vision and objectives'.[30]

It is possible that controlled membership has also been limited by cl 5(1)(c) of sch 3 of the *Community Land Development Act 1989* (NSW) ('*CLDA*'), which prohibits provisions in community title management statements based on 'race, creed, ethnic or socio-economic grouping'.[31] The Macquarie Dictionary defines 'creed' as 'any system of belief or of opinion', and this is arguably so broad that it would encompass the shared values of an intentional community. Clause 5(1)(c) of sch 3 of the *CLDA* has (possibly wisely) avoided many interesting questions raised by 'small group autonomy'.[32]

Restrictions on transfer and leasing, like other limitations on by-laws discussed below, raise questions about the application of discrimination legislation to strata and community schemes. Would a restriction on transfer or leasing be contrary to state or Federal discrimination legislation, rendering the express prohibition on these by-laws otiose?

The answer is no, the express prohibition in strata legislation is not otiose. First, discrimination legislation only covers specific issues – race, disability, gender, sexual orientation, age, breastfeeding, etc. Some characteristics are not included in any legislation. For example, if a community wanted to restrict transfers or leases to people with a golf handicap under 9, there is no legislation prohibiting discrimination on the basis of sporting ability. Discrimination legislation would not forbid this, but the prohibition on restrictions on transfer and leasing in the strata Acts would. Further, some issues are grounds for unlawful discrimination in some states but not others. For example, religion is covered in Queensland,[33] but not in New South Wales. Without the express prohibition on transfers in the strata legislation, a strata scheme in New South Wales, (but not a community scheme) could create a by-law prohibiting sales to Muslims or Christians.

A by-law prohibiting transfers on the basis of race, gender, age or disability and other characteristics covered by state or Federal discrimination legislation, would generally fall foul of both strata Acts and discrimination Acts. While discrimination legislation usually operates in the public sphere, the disposition of land is one of the few 'private' activities that is covered.[34] The blurred public/private nature of strata schemes and the difficult question of when or if bodies corporate are captured by discrimination legislation will be discussed further below.

It is essential to acknowledge that, while the express prohibition on by-laws restricting transfer and leasing or a provision like cl 5(1)(c) of sch 3 of the *CLDA* may prevent overtly discriminatory by-laws, they will not prevent indirectly discriminatory by-laws. For example, in the United States, while expressly racially restrictive covenants prohibiting sale or leasing to non-Caucasians were declared unconstitutional,[35] subtle racially restrictive covenants still abound. It is arguable that the fundamental economic theory of HOAs is inherently racially discriminatory, in effect at least. C, C & Rs aim to maintain property values, which if effective, will limit occupancy to wealthier citizens and, in the United States, low-income families are more likely to be African American and Hispanic than white. Prohibitions on subdivisions prevent housing lots from being divided into smaller, more affordable units such as duplexes or apartments; fees for upkeep of extensive facilities, in particular 'country club' membership fees, automatically exclude low-income purchasers.

In Australia, the obligation to maintain privately owned facilities imposed through a complex web of by-laws, legislation and body corporate contracts, will have the same effect as in the United States, excluding lower income families, and may in turn indirectly discriminate against Indigenous or recent migrant families. Research on Macquarie Links, a community title development in south-western Sydney with tennis courts, pools, a golf course, 24-hour security and a private bus service, concluded that:

> Living in Macquarie Links would not be affordable for those on average, or below-average incomes and income data from the 2006 Census shows that average (gross) weekly household earnings in the suburb of Macquarie Links are $2299, which is twice the average for Sydney as a whole at $1154.[36]

Irrespective of the express prohibition on 'socio-economic grouping' by-laws in New South Wales, community title developments with private facilities have the potential to create enclaves of privilege, dividing cities and compromising the equality of services that has traditionally existed between Australian suburbs.[37]

The other form of C, C & Rs that have had an indirectly racially discriminatory effect in the United States are single family covenants, restricting ownership and occupation to a married couple and their children. These are very common. While discriminatory to all unmarried citizens, this restriction has a particularly negative impact on African American women who are more likely than other women to be living in a de facto relationship, as well as other non-white families who live in extended, rather than nuclear families.[38] While it would be very unusual to see an attempt to regulate family composition via by-laws in Australia, one difficult occupancy issue with racial or cultural implications has emerged recently in Sydney.

Many strata schemes are experiencing overcrowding, specifically, large numbers of overseas students from south east Asia sharing two bedroom apartments, either through hot-bedding or the illegal division of rooms.[39] The problem is a consequence of governments and universities exploiting the earning capacity of our education system, without addressing the fact that Australian universities do not traditionally provide housing for all students, pushing students into the private rental market. Overcrowding in strata schemes is a serious fire risk, and results in overuse of common property; both are a source of legitimate complaint from other owners and tenants. Some strata lawyers think that by-laws limiting occupancy of apartments would be invalid, although it is unclear why as the number of occupants clearly relates to the use and enjoyment of lot property. However, as a result of these concerns, the government enacted a provision expressly authorising the creation of by-laws that limit occupancy to two adults per bedroom.[40] Occupancy limiting by-laws will have a disproportionate effect on the Asian student community, and potentially other migrant communities, members of whom might not consider their living conditions overcrowded, or even if they do, have no economic alternative. As property theorists continually remind us, the exercise of property rights produces externalities; a right in one person, (to limit occupancy of lots), can be to the 'derogation, exclusion, and often injury of others',[41] (by rendering them homeless). There is no easy answer to this dilemma, except perhaps to suggest that this kind of regulation, like minimum allotment size or land use, is arguably better done by government via planning or fire regulations, than self-interested private citizens via private by-laws.

Prohibitions on by-laws banning children

New South Wales is the only state to have an express prohibition on by-laws banning children.[42] It was introduced during the major 1996 reforms, although the Second Reading Speech does not make it clear why.[43] Sydney may have begun to experience some of the problems of the United States, where widespread bans on children were a source of great family hardship. Some United States commentators had argued that so long as bans on children were clearly stated in the C, C & Rs, people who bought with knowledge of them had no grounds to object to their terms,[44] even if that meant selling a home when a woman fell pregnant. Some United States courts supported bans, while others disagreed, arguing that a covenant banning children was either unconstitutional or an unreasonable intrusion on the private life of a family.[45] Ultimately, as a result of the difficulties families were experiencing finding housing (another example of an externality produced by a private property right), the Federal government passed legislation outlawing age restrictions except for retirement communities.[46]

Attempts have been made by some Queensland schemes to retrospectively restrict ownership and occupation to people over the age of 50 or 55, motivated by an apparent aversion to living near children or teenagers. Cases seem to have arisen in circumstances in which the ordinary government-controlled process of designating a scheme as a retirement village has gone awry, or developers have simply misled purchasers. This raises the question of whether private citizens (as opposed to government), should have the power to exclude children from housing. It is arguable that the Queensland BCCMA provides a clear negative answer in its express prohibition on by-laws that 'restrict the type of residential use' or which 'discriminate between types of occupiers',[47] although QCAT has declined to interpret the BCCMA in that way.[48] Schemes have asked for an exemption from the *Anti-Discrimination Act* 1991 (Qld) s 7(f), which prohibits age discrimination in relation to the supply of goods and services, accommodation and the transfer of land. While some schemes have succeeded,[49] others have failed.[50] In *Body Corporate for Village Green (Caloundra)*[51] the Tribunal member viewed the majority of the evidence presented by the body corporate in favour of the exemption as unsubstantiated, discriminatory assertions about people under the age of 55. The body corporate's 'evidence' arguably highlighted the danger of allowing people to retreat into homogenous enclaves in which they are not required to be either informed or tolerant of different social groups. There was also evidence that people under the age of 55 either wanted or needed to live in these developments, again illustrating that body corporate by-laws, like all property law, affect parties both inside *and outside* the transaction. Their legitimacy cannot be solely determined with reference to the desires of owners.

Disability, guide, hearing and assistance animals

As noted above, while there are some private activities like disposing of interests in land that are captured by discrimination legislation, as a general rule discrimination legislation operates in the public sphere. For example, while I may not be admired for refusing to let someone into my home because they are gay, transgender, sight-impaired, etc. I am not doing anything illegal. This is because discrimination legislation is not only limited to particular characteristics (e.g. disability, not sporting ability), it is limited to particular spheres such as employment, the supply of goods and services, public education, public spaces and registered clubs. Outside those spheres it is legally, (if not morally) permissible to discriminate.

Strata schemes are private property and prima facie, in the same way that I can refuse to allow someone to bring a guide dog into my house, so too can the collective owners of a strata scheme refuse entrance to their home, that is, the common property. That is why all states have provisions in their strata legislation prohibiting by-laws that restrict access to guide dogs or assistance animals.

The inapplicability of some discrimination law to privately owned common property was highlighted by the Federal Court in *Hu* v. *Stansure Strata Pty Ltd* [2014] FCCA 905. The Court had to consider whether s 18C of the *Racial Discrimination Act* 1975 (Cth) (made notorious by the successful action against *Daily Telegraph* columnist, Andrew Bolt[52]), could apply to actions that occurred inside a strata scheme. Section 18C provided that it is an offence to offend, insult, humiliate, or intimidate someone on the basis of race, 'otherwise than in private'. While the Court was adamant that the alleged acts had not occurred, concluding that the applicant suffered from delusions, (e.g. that her octogenarian neighbour had been walking on her roof), it was equally adamant that the strata scheme was not a public place.[53] Even if the statements had been made, s 18C had no application to strata common property.

The potential non-applicability of discrimination legislation to strata schemes was also illustrated by the New South Wales strata legislation, which until recently only prohibited by-laws that restricted guide or hearing dogs, not by-laws that restricted assistance animals. In *Thornton & Farnham* v. *The Owners SP 30653 (Strata & Community Schemes)*,[54] Member O'Keefe held that the body corporate had not engaged in unlawful disability discrimination by excluding the applicants' assistance dog because a strata scheme was a private place.

Member O'Keefe declined to apply the reasoning of *Hulena* v. *Owner's Corporation Strata Plan 13672*,[55] in which the Administrative Decisions Tribunal held that a body corporate was guilty of indirect discrimination by failing to provide the applicant with disability access to the common property. The Tribunal reasoned that the body corporate *was* captured by the *Anti-Discrimination Act 1977* (NSW) because by maintaining access to the common property, the body corporate provided a 'service'. Like all other discrimination statutes, the *Anti-Discrimination Act 1977* (NSW) s 49M, defines the provision of goods and services as an activity in which discrimination is prohibited.

Arguments that bodies corporate are captured by discrimination legislation through their provision of services have been more readily accepted in Queensland,[56] possibly because bodies corporate in resort-style schemes with extensive facilities more clearly provide a service than bodies corporate in four- or six-lot schemes. However, the law remains unclear. I would argue that the existence of express prohibitions on by-laws restricting guide dogs demonstrates a legislative assumption that strata schemes are not generally covered by discrimination legislation. However, this position will need to be addressed by legislatures and courts as it is not tenable to define all strata and community schemes, particularly large schemes, as exclusively private space. A person refusing entry to an assistance animal to their own home is not the same as a group of owners refusing entry to collectively owned space or to *someone else's home*. The fact that all of that space is technically private property does not justify the non-application of discrimination legislation. Characterising bodies corporate as providing a 'service' is arguably strained and the issue would be much better addressed directly with amendments to either discrimination or strata legislation.

The most pressing issue for schemes, (both small and large), will be failure to retrofit or allow retrofitting of common and lot property for disability access. This may become a common source of dispute as our population ages, particularly if aging in place is the option preferred by the elderly and governments. Discrimination legislation typically includes an 'unjustifiable hardship' exception,[57] which would protect schemes from bearing the costs of extensive and expensive retrofitting.

Clarification will also be needed to determine the validity of by-laws. For example, if a body corporate banned children playing on *all* common property (examples of which will be discussed in the next chapter), although this could be characterised as age-discrimination, if a body corporate is not providing a 'service', then it is not captured by state Acts. If the scheme is a large community scheme, it may be captured by section 27 of the *Age Discrimination Act 2004* (Cth) in relation to space that is open to the public, but it would not if the community were gated. This is an unpredictable and unsatisfactory result.

Prohibition on environmentally harmful by-laws

In what may become a more common by-law prohibition, the ACT *Unit Titles (Management) Act 2011* s 108(d), forbids by-laws that prohibit or restrict 'the installation, operation or maintenance of sustainability or utility infrastructure'. Sustainability or utility infrastructure is defined as infrastructure that 'improves the environmental sustainability of the units' or 'reduces the environmental impact of the owners corporation and the unit owners'. Queensland has a similar restriction in its *Building Act 1975* Pt 8 Div 2, which renders by-laws or management statements ineffective if they prevent particularly sustainable building practices.[58] The Division also expressly prohibits bodies corporate from withholding consent to the installation of solar hot water systems or panels in order to preserve the aesthetics of the building.

These provisions are clear legislative recognition that by-laws do not simply affect those inside the scheme, but like all property entitlements, they potentially produce externalities. The provisions are consistent with the insistence of progressive property theorists that we must be attentive to 'the effects of claiming and exercising property rights on others, including future generations, and on the natural environment and the non-human world'.[59] Satisfaction of individual preference (for example, owners' aesthetic preference) is not the sole goal of good property law. The provisions demonstrate a flash of legislative understanding that it is both legally feasible and desirable to limit a body corporate's by-law-making power.

Prohibition on by-laws that are inconsistent with the strata legislation, any other act or law

Most states have an express prohibition on by-laws that are inconsistent with the parent Act and any other law. These are designed to ensure the subordinate place of by-laws, and to maintain limits on the body corporate's powers and functions. However, as noted in the preceding chapter, in New South Wales at least, the prohibition has not prevented the creation of 'empowering by-laws' which give a body corporate powers that it does not expressly or impliedly have. This is despite the fact that such by-laws must, by definition, be inconsistent with the strata Acts and threaten to undermine the necessarily limited nature of a body corporate.

Prohibition on by-laws that are discriminatory or unreasonable

While the express prohibitions discussed above are necessary, they are not sufficient. Being specific, many potentially harmful by-laws will fall outside their ambit. For example, a prohibition on by-laws that ban children will have no effect on by-laws that prohibit children from playing on any and all common property.

The solution is not more specific prohibitions, but broader, conceptual prohibitions such as prohibitions on by-laws that are 'unreasonable', 'unfairly discriminatory' or 'harsh, unconscionable or oppressive'. These capture a range of by-laws, including those we cannot yet imagine schemes creating. Some states already have these provisions, with Queensland, Tasmania, South Australia and Victoria prohibiting by-laws that unfairly discriminate,[60] while Queensland and Tasmania also prohibit by-laws that are unreasonable.[61] New South Wales has recently enacted a prohibition on by-laws that are 'harsh, unconscionable or oppressive',[62] and in what is one of the best provisions of all states, Tasmania prohibits by-laws that adversely affect 'the health, welfare or safety of *any* person'.[63]

However, as we will see in the case of pet litigation in Queensland, the provisions are relative terms that operate within an existing legal, social and political context in strata schemes, courts and tribunals. The question of what is 'reasonable' or 'unfair' is affected by the dominance of freedom of contract, notice and consent in strata law. If we place great faith in the notion of consent, by-laws are much more likely to be considered reasonable so long as they are voluntarily agreed. However, if we acknowledge that land is a limited resource, that people's consent by purchase or leasing may be compromised by economic and social factors, and that land law has never placed absolute faith in consent, we may assess the question of 'reasonableness' objectively, with reference to a wider range of factors than that the majority has agreed to the by-law.

In order to examine this question further, we will look at the kinds of by-laws that should concern us in a liberal democracy and which may need to be limited. Those by-laws have been divided into two categories: the most problematic, those that regulate activities that do not affect others, (what property law would call 'self-regarding acts'), and those that affect others in ways that property law has traditionally not concerned itself; what I have termed 'borderline self-regarding acts'.

Problem 1: regulation of self-regarding acts

In Chapter 2, we saw that the progress of land law, gradually eradicating doctrines that limited possessors' control and use of land, promoted more than economic freedom; it promoted one of the most cherished principles of liberal democracy, negative liberty. The classic exposition of negative liberty is found in John Stuart Mill's *On Liberty*, in which he said:

> the sole end for which mankind are warranted, individually or collectively, in interfering with the liberty of action of any of their number, is self-protection. That the only purpose for which power can be rightfully exercised over any member of a civilised community, against his will, is to prevent harm to others. His own good, either physical or moral, is not a sufficient warrant. He cannot rightfully be compelled to do or forbear because it will be better for him to do so, because it will make him happier, because, in the opinions of others, to do so would be wise, or even right.[64]

We noted that while many have argued that negative liberty is insufficient for human flourishing, no one actually argues we should be *needlessly* interfered with by government or others.[65] The concept of negative liberty is fundamental to modern land law. Today, it is axiomatic that our land use will be regulated, but only so far as our activities have the potential to harm others, for example, immediate neighbours or the wider community. As noted at the beginning of the chapter, this regulation is usually public, enacted by local, state, and more rarely, Federal government.

Consequently, if we are not harming or affecting others in any meaningful way, there is a sphere of private property that remains free from the interference of government, neighbours, landlords and family. This is why ownership of land in a democratic community enables us 'to exercise autonomy, to enjoy our liberties, to shape our destiny, to form relationships with others'.[66] Property secures the freedom to 'live one's life on one's own terms'.[67] While these principles are important for the land on which people work, in particular, run their own businesses, it is most obviously significant for the properties that are our homes.

As others have noted, Margaret Jane Radin's personhood theory of property is relevant in this context.[68] Radin argues that as people can be bound up with 'things' in a constitutive and not necessarily negative or 'fetishist' sense, those things that are closest to personhood should be accorded the greatest legal protection.[69] Along with progressive property theorist Laura Underkuffler,[70] Radin argues that as a matter of morality, different rules must apply to different kinds of property. Not surprisingly, Radin identifies the home as a prime example of property which is inextricably bound up with a person's sense of themselves. Radin describes the home as the 'moral nexus between liberty, privacy, and freedom of association'.[71] She argues that:

> It would be an insult for the state to invade one's home, because it is the scene of one's history and future, one's life and growth ... one embodies or constitutes oneself there. The home is affirmatively part of oneself – property for personhood – and not just the agreed-on locale for protection from outside interference.[72]

Radin argues that a personhood theory demonstrates why protecting people's expectations of continuing control over objects is so important. She says that if

> an object you now control is bound up in your future plans or in your anticipation of your future self, and it is partly these plans for your own continuity that make you a person, then your own personhood depends on the realization of these expectations.[73]

If it is 'an insult for the state to invade one's home', presumably, it is would be an equal insult for the 'mini-state' of a strata scheme to do the same.

Consistent with a theory like Radin's, some strata and community title legislation makes an effort to differentiate between residential, commercial, industrial, tourist or retirement property, but those efforts are relatively ineffectual. In New South Wales, this is because the differentiation only exists in the model by-laws, which do not apply to community schemes at all, and are optional in strata schemes. The provisions that are compulsory, namely those in the Acts, apply equally to all schemes. This includes the broad by-law-making power, which is no different for a home than an industrial site. The paradox of placing 'model' provisions in optional by-laws that can then be excluded, demonstrates a tendency by the legislature, from which courts have taken their lead, to fall back on the concept of freedom of contract, despite the ultimate inappropriateness of freedom of contract in relation to land.

Queensland uses regulation modules to differentiate between residential, tourist, commercial and small schemes, however the system is open to abuse, for example, when developers register residential schemes under the accommodation module in order to sell longer-term management rights. Further, the BCCMA, s 169 by-law-making power is exactly the same for all schemes. The power of the developer and collective owners to regulate a private lot is identical, whether the lot is a serviced apartment, a commercial office or a home.

The core problem with the universally applicable by-law-making power is that it violates the principle of negative liberty and the sanctity of the home by allowing the regulation of behaviour *inside* people's homes which does not affect others *at all*, or does not affect them *in any meaningful way*. As noted at the beginning of the previous chapter, a by-law banning the eating of meat in all apartments is prima facie valid in New South Wales.[74] All states, bar Victoria, give the body corporate the power to make rules in relation to lot property without any requirement that the behaviour affects others; the by-law must simply relate to the use or enjoyment of a lot. As Justice Edelman observed at [22] in the leading Western Australian case, *Mackie* v. *Henderson*,[75]

> ... [by-laws] interfere with a myriad of personal rights in relation to the lot. The range of possible by-laws can be extremely broad.

Victoria is the sole exception. Under s 138 of the *Owners Corporations Act 2006* (Vic), the body corporate is only given the power to make rules 'for or with respect to any matter set out in sch 1' and those rules must be for 'the control, management, administration, use or enjoyment' of lots or common property. Schedule 1 contains a number of matters relating to common property, committees, dispute resolution, but in relation to lots, it only stipulates change of use, external appearance, notice of renovations and times in which work can be carried out. Schedule 1 allows rules in relation to behaviour, but only on common property, although there is a general power to make rules in relation to noise on both lot and common property.

Unfortunately, the case law in Victoria is inconsistent in relation to the body corporate's power to regulate individual lots. In *Owners Corporation PS331362S* v. *Rhodes (Owners Corporations)*,[76] VCAT held that an existing rule prohibiting commercial use of a lot was valid, however in *Owners Corporation PS501391P* v. *Balcombe (Owners Corporations)*[77] and in *Lawandi* v. *Owners Corporation 21842D (Owners Corporations)*[78] VCAT held that an existing rule and a changed rule, respectively, prohibiting commercial use, were beyond the power of the body corporate. These cases are not an ideal illustration of the importance of limiting a body corporate's power to regulate self-regarding acts inside lots because short-term letting, the subject of the latter two cases, is not a self-regarding act as it inevitably affects other residents. However, the Victorian legislation and case law is important as both evince an understanding that there should be limits on the power that private citizens have to regulate other people's property. Vice President Davis said in *Lawandi* at [38] that

> Schedule 1 shows that the content of the OC's powers to make rules is not identical as between common property and lots. Rather, the OC has a wider power to make rules in relation to common property than it does in regulating or proscribing individually owned lots that do not form part of common property. This is consistent with the very nature and purpose of the creation of OCs.

The idea that the power to regulate lot property should be narrower than the power to regulate common property might seem obvious, but it is in fact absent from all other states' legislation.

Perhaps the best illustration of by-laws that inappropriately regulate self-regarding behaviour are pet by-laws. These include blanket bans, bans based on size or weight, and the requirement of pre-approval by a body corporate or committee to keep a pet.[79] If readers will excuse the pun, pets are not a petty issue. The freedom to keep a companion animal is central to many people's lives, and the inability to keep a pet is a source of significant distress and subsequent litigation in states where pet bans are allowed. In the leading United States case on a condominium rule banning pets, *Nahrstedt* v. *Lakeside Village Condominium Assoc Inc*,[80] Arabian J (in dissent) said that:

> the value of pets in daily life is a matter of common knowledge and understanding as well as extensive documentation. People of all ages, but particularly the elderly and the young, enjoy their companionship. Those who suffer from serious disease or injury and are confined to their home or bed experience a therapeutic, even spiritual, benefit from their presence. Animals provide comfort at the death of a family member or dear friend, and for the lonely can offer a reason for living when life seems to have lost its meaning.... Not only have children and animals always been natural companions, children learn responsibility and discipline from pet ownership while developing an important sense of kindness and protection for animals. Single adults may find certain pets can afford a feeling of security. Families benefit from the experience of sharing that having a pet encourages. While pet ownership may not be a fundamental right as such, unquestionably it is an integral aspect of our daily existence, which cannot be lightly dismissed and should not suffer unwarranted intrusion into its circle of privacy.[81]

Even if we do not accept Justice Arabian's reasoning about the value of pets, if we are committed to the values of liberal democracy we must concede that our own view of pets is irrelevant to the question of whether someone else is allowed to keep one. What others do in their own home is their business. The only way that it will become our business is if what they do disturbs us. So, if their dog barks continually, this is an other-regarding act that we can legitimately regulate, but if they have a goldfish, a cat that never leaves the house, or even a Labrador who lies comatose on the sofa all day, it is no concern of ours. If their pet crosses our (common) property without damaging it, we are not being affected in any meaningful way. As a tenant, keeping a pet would be a use of property that affects the landlord's reversion and could rightly be regulated in a lease, but it is not automatically a use of property that affects the apartment next door, or the common property in any significant way.

To illustrate the problem with the current power to regulate animals, let's imagine that a local council banned all pets on the grounds that as some pets are a nuisance, it is easier and more certain to ban all, (reasoning that has been accepted in strata litigation.[82]) Alternatively, the local council requires pre-approval to keep any pet, although there is no evidence that a particular pet is going to be a nuisance to the neighbours, (the 'model' by-law in many states). In a public context, neither rule would be tolerated. In keeping with time-honoured liberal principles, people would argue that regulation of land use is only justifiable to prevent harm to others. Not surprisingly, current public regulation is consistent with liberal principles. While a pet may need to be registered, no one is required to ask the council's approval to acquire a pet. Animals only become the subject of regulation if and when they prove to be harmful to others. So, while I can exclude my neighbour's cat from my house, I cannot legally exclude him or her from my yard unless the cat meets the description of a 'nuisance' cat; one that makes

> a noise that persistently occurs or continues to such a degree or extent that it unreasonably interferes with the peace, comfort or convenience of any person in any other premises, or repeatedly damages anything outside the property on which it is ordinarily kept.[83]

As a member of a community, I am expected to tolerate other people's cats on my property. I do not have to tolerate my neighbour's dog repeatedly barking at 3 am,[84] but his cat sunning itself on my lawn, and even less picturesque activities,[85] I am required to accept. The ludicrous nature of strata by-laws that permit blanket bans on pets is illustrated by the fact that a body corporate can ban an owner from keeping a cat which never leaves the apartment, but can do nothing about neighbourhood felines using the common property garden to their heart's content.

Does it really matter that strata schemes are out of step with the wider community? Yes, very much so. First, if as Smith and Merrill assert, property cannot survive without a connection to basic morality, or what is widely believed to be 'right',[86] strata schemes will be subject to ongoing dispute about pets, as owners rail against prohibitions they consider unjustifiable.[87] The striking contrast between extensive pet litigation in New South Wales and Queensland, and its almost complete absence in Victoria, supports this view. Property entitlements are meant to be compatible with our way of life. In a democracy, most people do not expect to be able to control what their neighbours do in the privacy of their own home, and so a property right that allows people to do that, *without adequate justification*, must be called into question.

In Queensland, where pet bans have been the subject of vast litigation, adjudicators and tribunal members have been turning themselves in knots to find ways around blanket bans. Unlike New South Wales, Queensland has a provision expressly prohibiting by-laws that are 'unreasonable'.[88] Relying on freedom of contract and consent by notice, earlier Queensland cases had upheld blanket pet bans,[89] but gradually decision-makers began to find that they were 'unreasonable' and thus contrary to section 180(7) of the *Body Corporate and Community Management Act 1997* (Qld).[90] The most recent decisions have held that as a body corporate only has the power to 'regulate' lots, a blanket ban is beyond power because 'prohibition' is not 'regulation'.[91] Reading the cases, it is obvious that the essence of all disputes is negative liberty and the discomfort both residents and decision-makers have with the regulation of self-regarding behaviour. Numerous cases quote Member Dorney in *Tutton* v. *Body Corporate for Pivotal Point Residential CTS 33550*, who said that:

> Since there is clearly no rational basis upon which it can be said that the keeping of a goldfish in a safe and healthy environment could be a matter which could cause any difficulty to any other lot owner, yet is the subject of an 'absolute' ban, the conclusion is fairly open that such a by-law is 'unreasonable'.[92]

Second, regulation of self-regarding acts affects not just those who are being regulated, but those who are regulating. As noted above, by *not* regulating animals who do not affect others in any significant way, state legislation requires citizens to be tolerant. We must be tolerant of dogs who occasionally bark, who might occasionally defecate on our property or who might jump up excitedly on us on the footpath. We must be tolerant of cats who yowl at night, roam across our yards and sleep or defecate in our gardens.[93] To be sure, some cranky citizens struggle with such tolerance, but they are the exception rather than the rule. The state legislation dovetails with a community norm of tolerance and strengthens that norm in turn.

In contrast, strata and community title runs the very real risk of fostering intolerance. In *Ephraim Island – Subsidiary 105*,[94] lot owners who supported the body corporate's rejection of the applicant's request to keep their dog said 'we most definitely would not have purchased if we had known there was the possibility of sharing the elevator with a large dog'. The dog was a golden retriever. When those lot owners leave their Island scheme and venture out into the wider community, they will be required to share footpaths, parks and other public spaces with multiple dogs. Within their own home they can maintain a dog-free zone, but on their shared property, is it really too much to ask that they occasionally share a lift with a golden retriever?[95] As Arabian J said in *Nahrstedt*:

> Our true task in this turmoil is to strike a balance between the governing rights accorded a condominium association and the individual freedom of its members. To fulfill that function, a reviewing court must view with a skeptic's eye restrictions driven by fear, anxiety, or intolerance. In any community, we do not exist in vacuo. There are many annoyances which we tolerate because not to do so would be repressive and place the freedom of others at risk.[96]

Remembering our discussion of property theory in Chapter 2, this is not simply a matter of one intolerant couple. Property entitlements, multiplied thousands of times across a physical landscape and through time create social and political systems. If thousands of bodies corporate have the power to prevent fellow citizens keeping pets within their own home in circumstances in which the pet has minimal or no effect on others, we are creating a community of control and intolerance. As Singer says:

> [a] change in legal rules will bring about changes in the character of social relations over time, regardless of whether any immediate effects are visible.

> This in turn will influence public perceptions about what kinds of conduct are allowable and appropriate and what kinds are not. It may even affect beliefs about right and wrong.[97]

McKenzie suggests in the United States context, '[i]t is conceivable that children raised in CID housing may be undergoing a form of differential political socialization', in which they are learning that it is legitimate to regulate the traditionally autonomous acts of their neighbours.[98] The failure to limit by-law-making power in relation to individual lots to activities that have some meaningful effect on others is a serious flaw in the legislation of all states, bar Victoria, and as increasing numbers of Australians live in strata schemes, it may produce long-term, detrimental effects on our community.

Problem 2: borderline self-regarding acts

A more complicated question than that pertaining to by-laws that regulate self-regarding acts is the one in respect of by-laws that regulate behaviour, which is not readily classifiable as 'self' or 'other-regarding'. These include by-laws that regulate blind colour, balcony furniture, building materials, plant types, paint colour, window washing, gardening methods and lawn mowing. They are most common in community title developments, many of which are modelled on United States HOAs. In the United States, by-laws and C, C & Rs operate as a 'sort of privatized zoning smorgasbord',[99] from which developers can pick and mix restrictions and conditions to mould the character of the development.

The rationale for this regulation is generally overtly liberal and economic, although paradoxically, it can also be communitarian. As we saw in the discussion of United States literature, theorists like Epstein[100] and Ellickson[101] argue that individuals should be free to create or join the kind of communities that they, not we, believe are desirable, and the law should support them in this regard. Public and rational choice theorists argue that communities with private associations and regulations allow developers to offer purchasers a range of housing 'products', so that residents can 'shop' between neighbourhoods choosing the housing and community style appropriate for them.[102] Further, unlike local government controls, private regulations are extremely detailed and ensure residents can maintain the environment they purchased, leading to increased resident satisfaction and property values.[103] Some argue that this ability to control the physical environment and run one's own community also increases social capital,[104] an asset that has been at the forefront of sociologists' and urban planners' minds in recent years.[105]

Theories of privatisation dovetail with developments in urban theory, like New Urbanism, a neo-traditional approach to architecture and planning which promotes human-scale, resident-driven communities in which people can walk or ride between home, work and public spaces.[106] New Urbanism aims to create ecologically and socially sustainable communities, by sacrificing individual space for collective space (for example, small individual yards and large, accessible 'public' facilities). With its focus on community, New Urbanism, in turn, dovetails with communitarianism. Communitarianism is sceptical of liberals' focus on the individual and insists that as individuals can never function outside society, their needs and desires must sometimes be subservient to the greater good. Paradoxically, some communitarians, like liberals, have adopted the private community form to further their ends. United States proponents of intentional communities and co-operatives see private legal structures as a 'form of socialism by contract',[107] which allows them to restrict membership of communities to like-minded individuals, to create collective living spaces and to regulate activities that the group might consider ecologically or socially undesirable.

In contrast to these positive perceptions of the ability to privately regulate an unlimited range of activities on land, others do not have such a benign view. Sociologists have argued that privately regulated communities, in particular gated communities, promote intolerance and fear, dividing society into the privileged few inside the gates and the undesirable hordes outside.[108] Legal critics discussed in Chapter 3, like McKenzie and Winokur, suggest that the ability to control other residents' behaviour does not create harmonious communities of shared values, but rather fosters resentment, frustration and litigation among residents who neither understand nor accept the legitimacy of regulation by their neighbours. The limited research that has been conducted in Australia to date reveals similar findings. In their case study of a Victorian community title master planned estate (MPE), Goodman and Douglas concluded that:

> the lived experience of communal ownership is one of frustration and disappointment for a section of residents ... [and] that due to the long-term ramifications such legal entities in MPEs [bodies corporate] should be avoided where possible.[109]

Winokur and McKenzie also argue that C, C & Rs controlling aesthetics are responsible for the spread of dull, homogenous, low-rise sprawl, which is architecturally valueless. More fundamentally, McKenzie challenges rational choice theory arguing that it has little empirical evidence to support its claims that people genuinely want private regulation of their homes.[110] Finally, others like Alexander, have focused on the darker side of private regulation, with its potential to inappropriately control, oppress and exclude.[111]

The first group of thinkers, (liberals and communitarians with a positive view of private regulation), might argue that a wide range of behaviour is other-regarding. They might say that the non-conforming colour of the neighbour's blinds reduces property values or their lack of permaculture undermines the community's agreed ecological aims. As such, both are the subject of legitimate regulation. They would argue that purchasers have voluntarily joined the community, knowing the existing restrictions and obligations, as well as the potential for new restrictions and obligations, and that people who flout them are harming those who bought on the promise that others would comply.

The second group of thinkers, (those with a more critical view of private regulation), might argue that in reality, a choice of blind colour or permaculture garden are self-regarding acts that do not affect others at all, or at least in any way that should be legally meaningful. As we saw in Chapter 2, Singer argues that property law does not prevent us from *ever affecting* others in the exercise of our rights; it just prevents us from harming them.[112] Complaints that we are affected by another's use of property can be divided into the legitimate and illegitimate, the latter being traditionally ignored by property law. Deciding what is or is not a self-regarding act comes back to a considered judgement about the legitimate and reasonable expectations of the parties in democratic society. Thus, those sceptical of private regulation would argue that, consistent with the principles of democratic society, in particular negative liberty, the choice of blind colour or the non-use of permaculture, is legally harmless behaviour that should not be the subject of any regulation.

The first group might then counter that the principle of negative liberty in fact gives everyone the freedom to form or join communities with rules; that is, we are all free to voluntarily agree to be less free.[113] As Mill said, a person 'cannot rightfully be compelled to do or forbear because it will be better for him to do so, because it will make him happier, because, in the opinions of others, to do so would be wise, or even right'.[114] If some people want to live in communities where they are required to mow their lawn every two weeks and can compel others to do the same, who are we to decide they should not?

As has become clear, the validity of private regulation of borderline other-regarding behaviour can cause us to turn endlessly in circles. As the discussion of Alexander and Ellickson in Chapter 3 demonstrated, private regulation produces strange paradoxes such as liberals arguing for less individual freedom and communitarians warning of the dangers of community. This book will not attempt to solve these dilemmas or decide definitely in favour of either camp; there is some validity in both perspectives. I accept the importance of people having the freedom to form their own associations and live by their own values, and I certainly do not want to be guilty of hypocritically supporting some private regulation (for example, solar panels and recycled water), while dismissing others (for example, lawn mowing and white blinds). The satisfaction of individual preference and the facilitation of the 'freedom to live one's life on one's own terms' are the proper purpose of property law.[115]

However, consistent with Alexander, Underkuffler, Peñalver and Singer, I would argue that they are not the *only* purpose of property law. Both in theory and in practice, property law is much more complex than the satisfaction of individual preference. As we saw in Chapter 2, and in the express prohibitions on specific by-laws discussed above, property entitlements affect not just parties to a transaction, but those outside it as well. As Singer says:

> Every legal right should be understood not merely by reference to the powers and rights it gives the owner but by reference to the impacts of the

exercise of those powers on others and the shape and character of the social relationships engendered by those rights and powers.[116]

We must attend to the systemic effect of those rights,[117] and judge them not solely with reference to the 'the magical moments of acquisition and transfer' (that is, the voluntary creation or joining of a community), but by 'the continuing moments in which property rights are exercised'.[118] We can only do this through reasoned deliberation with reference to both the real world and the plural values of our democratic community. This process is, of necessity, messy and will not yield a neat, universally applicable rule; however, it is the only way that we can determine if property institutions and their regulations are 'compatible with our form of life'.[119]

By way of example, we might consider the validity of the following commonly made by-laws:

- A by-law that restricts external paint colours: this might be acceptable because the impact on the rest of the community is significant and there are comparable provisions in some public regulation (e.g. heritage areas); alternatively, we may decide that we do not want to create a community like the United States, in which we end up with 'a generation ... that doesn't know you should be able to paint your house any color you want'.[120]
- A by-law limiting blind colours: this might be considered a considerable intrusion into the private sphere that produces limited benefit for the community. If '[i]t would be an insult for the state to invade one's home, because it is the scene of one's history and future, one's life and growth',[121] then it must be as great, or even greater an insult for other citizens to do the same. Professor French's Bill of Rights for HOAs included the provision that activities within the individually owned property should not be restricted except to the extent they interfered with other people.[122] While the colour of blinds has some impact on others, it is hard to say that it genuinely interferes with them; it is not difficult to conclude that it interferes with the people regulated.
- A by-law prohibiting trucks or commercial vehicles being parked on or near lots:[123] we might decide that this by-law displays class prejudice, something that most democratic societies aspire not to be. Property law 'can render relationships within communities either exploitative and humiliating or liberating and ennobling. Property law should establish the framework for a kind of social life appropriate to a free and democratic society.'[124] A by-law that prevents a tradesperson parking their truck outside their house because it supposedly lowers the neighbours' property values does not create 'ennobling and liberating' relationships between neighbours, but rather tends towards the 'exploitative and humiliating'.
- A by-law requiring people to be dressed 'appropriately' on common property: regulating the way someone dresses is a very significant intrusion into the personal sphere. The democratic state rarely attempts to regulate dress

other than prohibiting indecent exposure. Private clubs and restaurants have dress codes and so by analogy, a dress code in a community title country club might be acceptable, but any dress code for the park or streets may not be so. We may say that no by-law should exist that creates the power for an estate manager to order a teenage girl from a common property park for wearing a short skirt. While it is not necessarily likely that such a power would be abused, the very existence of such a power is incompatible with liberal democracy.

- A by-law banning all signs: while this by-law is generally aimed at minimising visual clutter on streetscapes, if it bans political signs, it has free speech implications, particularly if the private community is the size of an entire suburb. We may decide that a private rule that limits political signs is unacceptable because it compromises the democratic process.

The trigger to engage in this process of deliberation is a statutory requirement of 'reasonableness' and judicial review. Only two states have this requirement, Tasmania and Queensland. New South Wales has set the bar too high, only stipulating that by-laws not be 'harsh, unconscionable or oppressive'; unreasonable by-laws are apparently OK. Other states stipulate that by-laws must not be discriminatory, however this does not prevent the operation of harmful by-laws if they apply to all owners and residents equally.

The case law on by-law 'reasonableness' is limited in Australia, (although extensive in the United States), but it demonstrates that the concept works. For example, in *Taunton Enterprises Pty Ltd Appeal Against Order No. C579394 Strata Title Act*[125] the Tasmanian Tribunal was faced with a by-law that required owners to use the same type of materials for additions, and in similar proportion, as were used in the original construction. The limitation created a problem familiar to restrictive covenants, that is, enduring limitations on construction can become not just aesthetically, but technically out of date. 'Brick and tile' covenants limit construction in COLORBOND, which unlike the corrugated iron that covenants were designed to prohibit, does not rust. Here, the by-law would have required construction in cladding and fascia that had already proved inadequate and would have prevented the owner from accommodating his growing family. Although there was a voting mechanism, which theoretically allowed the body corporate to change the by-law, the body corporate had declined to do so. (This is an example of the 'one-way ratchet' effect of fragmentation identified by Heller and discussed in Chapter 2;[126] it is much easier to create by-laws than agree to their removal.) The Tribunal stated at [32] that 'the issue of reasonableness is whether the by-law is fair and sensible' and the statutory provision allowed the Tribunal to conclude that a by-law requiring building in substandard material, and effectively prohibiting any additions, was unreasonable and invalid.

In Queensland, a preparedness to declare a by-law unreasonable has been affected by the concepts of freedom of contract, notice and consent. For example, in *Beattie Place South* [2009] QBCCMCmr 27, the Adjudicator relied on the fact that the applicants knew or should have known that a registered by-law banned all pets, and that the vast majority of owners supported the by-law, in deciding that a blanket pet ban was not unreasonable. While notice and consent are relevant, by relying on them exclusively the Adjudicator simply sidestepped the central question: is it reasonable or legitimate for private citizens to decide that one of their neighbours cannot have a pet that does not affect anyone else in any meaningful way?

As noted above, QCAT subsequently decided in *McGarvey*, that blanket pet bans go beyond a body corporate's permitted power to 'regulate' lot property and are invalid.[127] However, QCAT considered the question of reasonableness in *McGarvey*, and rightly concluded at [61]–[62] that the concept can only be determined with reference to the specific circumstances of the case. In other words, decision-makers just have to engage with the substance of the issue. The Queensland Court of Appeal recently considered a body corporate's obligation to act reasonably when carrying out its functions in *Albrecht* v. *Ainsworth*;[128] although not a decision on by-laws, the case is an important higher court clarification of the concept of reasonableness and judicial review in the context of bodies corporate. First, the Court accepted at [84] that 'views as to what was reasonable or unreasonable involved value judgments on which there was room for reasonable differences of opinion, with no opinion being uniquely right'. In other words, '[c]hoices about property entitlements are unavoidable'[129] and that is OK. Second, the Court held that the Adjudicator was entitled to substitute her own judgment and did not have to defer to the private views of property owners as expressed through voting. That is, objective, public oversight of the activities of private communities through judicial review, the position argued by Alexander, is entirely appropriate.

What about the states that do not have a legislative or judicial requirement of reasonableness for by-laws? Have these states avoided a whole range of messy questions about the reasonableness of by-laws that ban pets, determine blind colour or personal dress? No, they have not. Those questions are still there, but *they have been given a blanket affirmative answer* ('yes, you can determine how your neighbour dresses'), *without any engagement with the content of the questions, or the consequences of the answers*. This point cannot be overemphasised and is one of the primary conclusions of this book. The current legislative rule of allowing almost any by-law, so long as it relates to lots or common property, has not and *cannot* eradicate the difficult and inevitable questions about what constitutes acceptable private regulation of land; it just fails to address them. Most state legislatures and courts have not appreciated that when 'devising new strategies in an old war among generations of land owners … we [will inevitably] wind up in some of the same old trenches',[130] or that if we do not limit the kinds of by-laws that attach to land, we will create 'modern variations of feudal serfdom'.[131] Property rules unavoidably contain values and thus choices about which values they should embody are inevitable.

All states need express legislative provisions allowing tribunals and courts to invalidate socially and politically undesirable by-laws. Such provisions should require by-laws to be 'reasonable', but decision-makers must judge reasonableness with reference to the multiple values of a pluralist democracy, not the single contraction value of consent. A by-law should not be considered 'reasonable' simply because a purchaser/tenant was aware (or should have been aware) of it and voluntarily chose to purchase/rent. This is because 'some relationships are out of bounds … some contract terms are off the table … [and] there are some things you should not ask of others; there are some demands that cannot justly be made'[132] via property law in a pluralist democracy. If a by-law causes harm to others, either inside or outside the community, voluntary consent to its terms (particularly as that consent is invariably only majoritarian) cannot justify its existence.

A provision that required by-laws to be reasonable would clarify that it is the proper role of courts to adjudicate on privately created rules. It would establish that it is never appropriate for private citizens to have an almost unfettered quasi-legislative power in relation to land, but that the by-law-making power (just like the power to create easements and covenants) is always subject to public oversight.

States should also have a default rule that by-laws cannot regulate activity inside lots unless the activity has a meaningful and harmful effect on others. This default rule would be consistent with the most basic principles of liberal democracy.

Notes

1 For example, under *University of New South Wales Act 1989* (NSW) s 27, the University Council has the power to make by-laws in relation to listed matters.
2 Jeremy Waldron, 'Homelessness and the Issue of Freedom' (1991) 39 *UCLA Law Review* 295.
3 It is important to note here that by-laws that affect strata title only exist as long as the building stands. If a building is destroyed, the strata scheme must be terminated and the by-laws will disappear. This is because the strata plan of subdivision is drawn by reference to a building. Community title is different. Like ordinary subdivisions, community title plans of subdivision are drawn by reference to land, not buildings. If the houses in a community title scheme are all destroyed by fire, there is no need to terminate the scheme as the land to which the plan of subdivision and management statements relate still exists.
4 *Houghton* v. *Immer (No 155) Pty Ltd* (1997) 44 NSWLR 46, 52–3.
5 *Paloma* [2012] QBCCMCmr 37 [29]–[30].
6 *BCCMA* sch 1A.
7 *BCCMA* s 94(2).
8 *Owners Corporation PS501391P* v. *Balcombe* [2015] VCAT 956.
9 *Jennifer Elizabeth James* v. *The Owners Strata Plan No. SP 11478 (No 4)* [2012] NSWSC 590, [52]–[53].
10 Susan F French, 'Constitution of a Private Residential Government Should Include a Bill of Rights, The Symposium on the Uniform Real Property Acts' (1992) 27 *Wake Forest Law Review* 345.
11 Evan McKenzie, *Privatopia: Homeowner Associations and the Rise of Residential Private Government* (Yale University Press, 1994) 83.

12 Evan McKenzie, *Beyond Privatopia: Rethinking Residential Private Government* (Urban Institute Press, 2011) 14.

13 James L Winokur, 'The Mixed Blessing of Promissory Servitudes: Toward Optimizing Economic Utility, Individual Liberty, and Personal Identity' [1989] *Wisconsin Law Review* 1, 63–4.

14 H Easthope, B Randolph and S Judd, 'Governing the Compact City: The Role and Effectiveness of Strata Management' (City Futures Research Centre, 2012); City Futures Research Centre, 'Common Ground: The Costs and Provision of Community Infrastructure in Community Title Schemes in NSW' (UNSW, 2008) 31; Therese Kenna and Deborah Stevenson, 'Negotiating Community Title: Residents' Lived Experiences of Private Governance Arrangements in a Master Planned Estate' (2010) 28(4) *Urban Policy and Research* 435.

15 Easthope, Randolph and Judd, above n 14, 48.

16 NSW Fair Trading, 'Making NSW No 1 Again: Shaping Future Communities: Strata and Community Title Law Reform Discussion Paper' (2012) 8–21; Global Access Partners, 'Strata Laws On-line Consultation Final Report' (2012).

17 McKenzie, *Beyond Privatopia*, above n 11, 14. Kennedy argues that civil litigation is an inadequate check on HOA power, because it cannot be utilised by non-owners who are still affected by HOAs. He argues that HOAs must be treated as state actors vis-à-vis non-owners: Duncan Kennedy, 'Residential Associations as State Actors: Regulating the Impact of Gated Communities on Nonmembers' (1995) 105(3) *Yale Law Journal* 761.

18 There is no doubt that US housing development is more fixated on exclusionary practices like gating and single family covenants than is the case in Australia. This is the legacy of slavery, Jim Crow laws and racially restrictive covenants, as well as the legacy of the Founding Fathers and pioneering forebears. The idea that groups of people should be free to strike out and live as they please, separately from mainstream or dominant powers, seems more entrenched in the American than Australia psyche. For example, it is safe to say that the issues discussed in John-Patrick Fritz, 'Check Your Rights and Your Guns at the Door: Questioning the Validity of Restrictive Covenants against the Right to Bear Arms' (2005–2007) 35 *Southwestern University Law Review* 551 are unlikely to arise in Australia. That said, it is not uncommon to see by-laws relating to firearms in rural body corporate communities.

19 Alex Kondos, 'The Hidden Faces of Power: A Sociological Analysis of Housing Legislation in Australia' in Roman Tomasic (ed.), *Legislation and Society in Australia* (Unwin Hyman, 1980); see discussion in Chapter 1, 19–22.

20 Wayne S Hyatt, *Condominium and Homeowner Association Practice: Community Association Law* (American Law Institute, 2000) 253–60. Interestingly, 25 years later, there is little mention of restraints on transfer and leasing in the most recent edition of Hyatt's textbook: Wayne S Hyatt and Susan F French, *Community Association Law: Cases and Materials on Common Interest Communities* (Carolina Academic Press, 2nd edn, 2008) 249, possibly because it became uneconomical for communities to have these restrictions.

21 David E Grassmick, 'Minding the Neighbor's Business: Just How Far Can Condominium Owners' Associations Go in Deciding Who Can Move into the Building?' [2002] *University of Illinois Law Review* 185.

22 Gregory S Alexander, 'Dilemmas of Group Autonomy: Residential Associations and Community' (1989) 75 *Cornell Law Review* 1; Robert C Ellickson, 'Cities and Homeowners Associations' (1981) 130 *University of Pennsylvania Law Review* 1519. See also Nomi Maya Stolzenberg, 'Return of the Repressed: Illiberal Groups in a Liberal State, The Liberalism and Illiberalism' (2001) 12 *Journal of Contemporary Legal Issues* 897.

23 *CLMA* s 17(1)(a).

24 *Community Title Act 2001* (ACT) s 49(1)(b)(i); *Strata Titles Act 1998* (Tas) s 91(1) (b)(i).

25 New South Wales, *Parliamentary Debates*, Legislative Assembly, 21 November 1989, 12920 (Ian Causley).

26 John Page, 'Common Property and the Age of Aquarius' (2010) 19(2) *Griffith Law Review* 172.

27 Many existing Byron Shire communes have recently converted to community title: Byron Shire Council, *Multiple Occupancies to Community Title* (February 2009) www.byron.nsw.gov.au/multiple-occupancies-to-community-title.

28 Warwick Fisher, 'The Future of Rural Land Sharing Communities in Far North Coast New South Wales' (2004) 8 *Southern Cross University Law Review* 51, 66.

29 ABC Radio National, 'Sanctuary', *360documentaries*, 30 September 2012 (Kirsti Melville) www.abc.net.au/radionational/programs/360/sanctuary/4268164; ibid.; Page, above n 26.

30 DP286220, Management Statement for Jindibah. 'Jindibah' means both wisdom and tawny frogmouth (an owl-like bird) in a local Aboriginal language.

31 This was expressly included to eliminate discriminatory by-laws: New South Wales, *Parliamentary Debates*, Legislative Assembly, 21 November 1989, 12920 (Ian Causley).

32 There are intentional communities in Australia which exclude people on the basis of belief but do not use the community title legislation: ABC Radio National, above n 29. Religious communities, such as the Twelve Tribes, a Christian community, with properties at Picton and Katoomba, do not offer members title to land.

33 *Anti-Discrimination Act 1991* (Qld) s 7(i).

34 For example, the most fulsome coverage is in *Racial Discrimination Act 1975* (Cth) s 12.

35 *Shelley v. Kraemer*, 334 US 1 (1948).

36 Kenna and Stevenson, above n 14, 441; Robin Goodman and Kathy Douglas, 'Life in a Master Planned Estate – Community and Lifestyle or Conflict and Liability?' (2010) 28(4) *Urban Policy and Research* 451.

37 Davison argues that as services like schools, fire services and roads were always funded by state and not local governments in Australia, there was little difference between suburbs in this regard. In contrast, in the United States, where many services, including schools, were funded by local property taxes, services were connected to resident wealth and hence inequality between suburbs was much greater: Graeme Davison, 'The Past and Future of the Australian Suburb' (1993) 31(2) *Australian Planner* 63. See also B Gleeson, *Australian Heartlands: Making Space for Hope in the Suburbs* (Allen and Unwin, 2006).

38 Gerald Korngold, 'Single Family Use Covenants: For Achieving a Balance between Traditional Family Life and Individual Autonomy Symposium: The Changing Role of the Family in the Law' (1988) 22 *UC Davis Law Review* 951; Hyatt and French, above n 20, 544–56.

39 NSW Fair Trading, above n 16, 32–3; Esther Han and Natalie O'Brien, 'Living on the Edge in Danger High-Rises', *Sydney Morning Herald* (online), 30 September 2012 http://news.domain.com.au/domain/real-estate-news/living-on-the-edge-in-danger-highrises-20120929–26s6i.html. The problem is not restricted to strata properties, but includes many freestanding houses around Macquarie University as well.

40 *SSMA* 2015 s 137.

41 Laura S Underkuffler, 'Property: A Special Right' (1996) 71 *Notre Dame Law Review* 1033, 1043; Jeremy Waldron, 'Homelessness and the Issue of Freedom' (1991) 39 *UCLA Law Review* 295. See discussion in Chapter 2, 62–5.

42 A by-law for a residential strata scheme has no force or effect to the extent to which it purports to prohibit or restrict persons under 18 years of age occupying

a lot. This subsection does not apply to a by-law for a strata scheme for a retirement village or housing exclusively for aged persons.

SSMA 2015 s 139(4)

43 Parliament of New South Wales, *Parliamentary Debates*, Legislative Assembly, 13 November 1996, 5922 (Lo Po).
44 Hyatt, above n 20, 265, 268–9.
45 *Franklin* v. *White Egret Condominium, Inc*, 358 So 2d 1084 (1977), cited in Hyatt, above n 20, 264.
46 Nicole Napolitano, 'Fair Housing Act Amendments and Age Restrictive Covenants in Condominiums and Cooperatives' (1999) 73 *St John's Law Review* 273.
47 *BCCMA* s 180(4), (5).
48 *Caloundra Gardens Village Body Corporate Committee* [2012] QCAT 98, [27].
49 Ibid.
50 *Body Corporate for Village Green (Caloundra)* [2015] QCAT 101.
51 Ibid.
52 *Eatock v. Bolt* [2011] FCA 1103.
53 *Hu* v. *Stansure Strata Pty Ltd* [2014] FCCA 905, [36]–[50].
54 *Thornton & Farnham* v. *The Owners SP 30653 (Strata & Community Schemes)* [2010] NSWCTTT 511.
55 *Hulena* v. *Owners Corporation Strata Plan 13672* [2010] NSWADTAP 27.
56 *C* v. *A* [2005] QADT 14 and *Ondrich* v. *Kookaburra Park Eco-Village* [2009] FMCA 260. See also, *Sutherland* v. *Tallong Park Assoc Inc* [2006] NSWADT 163, in which the Tribunal held that a residential association was akin to a body corporate and that it provided a 'service' within the meaning to the *Anti-Discrimination Act 1977* (NSW).
57 E.g. *Disability Discrimination Act 1992* (Cth) s 29A.
58 Michael Weir, 'Wither Covenants in Queensland in the Name of Sustainability' in Penny Carruthers, Sharon Mascher and Natalie Skead (eds), *Property and Sustainability: Selected Essays* (Lawbook Co, 2011) 145.
59 Gregory S Alexander *et al.*, 'A Statement of Progressive Property' (2008) 94 *Cornell Law Review* 743, 744.
60 *BCCMA* (Qld) s 180(5); *Strata Titles Act 1998* (Tas) s 91(3)(b); *Strata Titles Act 1988* (SA) s 19A(1)(b); *Owners Corporations Act 2006* (Vic) s 140(a). The Northern Territory *Unit Titles Schemes Act* s 95(3)(b) prohibits by-laws that 'unlawfully discriminate' but the use of the word 'unlawful' renders this section ineffective. If it is unlawful for a body corporate to discriminate in some way, then it is already prohibited from doing that act. As we saw in the discussion of discrimination law above, the question is not simply 'is this an attribute covered by discrimination legislation?' (e.g. disability, gender, etc.), but additionally, 'is this an area in which discrimination is prohibited?' (e.g. education, goods and services, etc.). It is only unlawful for a body corporate to discriminate if it is operating in an area in which discrimination is illegal (e.g. provision of services). By only prohibiting 'unlawful' discrimination, the Act does not go beyond existing discrimination law.
61 *BCCMA* (Qld) s 180(7); *Strata Titles Act 1998* (Tas) s 91(3)(a).
62 *SSMA 2015* s 39(1).
63 *Strata Titles Act 1998* (Tas) s 91(3)(c), emphasis added.
64 John Stuart Mill, *On Liberty* (Longman, Roberts & Green, 4th edn, 1869); Isaiah Berlin, 'Two Concepts of Liberty' in Henry Hardy (ed.), *Liberty* (Oxford University Press, 2002) 167, see Chapter 2, n 77. Rudden explores Hegel's philosophy, concluding that 'for Hegel, we are fully free only when our property is (relatively) free': Bernard Rudden, 'Economic Theory v. Property Law: The Numerus Clausus Problem' in John Eekelaar and John Bell (eds), Oxford Essays on Jurisprudence (Clarendon Press, 1987) 250.

65 See discussion of Martha C Nussbaum, 'Deliberation and Insight: Bloch v. Frischholz and the Chicago School of Judicial Behavior Essay' (2010) 77 *University of Chicago Law Review* 1139 in Chapter 2, 59.

66 Joseph William Singer, *The Edges of the Field: Lessons on the Obligations of Ownership* (Beacon Press, 2000), 27.

67 Gregory S Alexander *et al.*, 'A Statement of Progressive Property' (2008) 94 *Cornell Law Review* 743, 743.

68 Winokur, above n 13, discussed in Chapter 3, 95–8.

69 Margaret Jane Radin, 'Property and Personhood' (1981–1982) 34(5) *Stanford Law Review* 957, 986.

70 Laura Underkuffler, 'A Theoretical Approach: The Lens of Progressive Property' in Susan Bright and Sarah Blandy (eds), *Researching Property Law*, Palgrave, 2016, 11, 15–21.

71 Ibid. 991.

72 Ibid. 992.

73 Ibid. 968.

74 Meat-eating bans are a live issue in India where developers create rules prohibiting sales to non-vegetarians. The bans effectively operate as caste and religious discrimination: Vishwas Waghmode, 'Refused a flat in Mumbai because you eat meat? BMC will rescue you', *First Post*, 28 November 2014, www.firstpost.com/india/refused-a-flat-in-mumbai-because-you-eat-meat-bmc-will-rescue-you-1825497.html.

75 *Mackie* v. *Henderson* [2011] WASC 197.

76 *Owners Corporation PS331362S* v. *Rhodes (Owners Corporations)* [2011] VCAT 642.

77 *Owners Corporation PS501391P* v. *Balcombe (Owners Corporations)* [2015] VCAT 956. See postscript on the appeal, Chapter 1, 37–8.

78 *Lawandi* v. *Owners Corporation 21842D (Owners Corporations)* [2015] VCAT 1810.

79 An example of the lengths to which schemes' lawyers have gone to regulate pets can be seen in the by-law from *Pivotal Point Residential* [2008] QBCCMCmr 55:

> 16.1 Subject to this By-Law and the provisions of the Act, a Lot Owner or occupier is not to bring or keep an animal on the lot or on the Common Property without the Body Corporate's written approval.
> 16.2 A Lot Owner of a Lot as at three (3) months following the date of establishment of the Scheme, or a Lot Owner, having purchased the Lot after the date of establishment of the Scheme, such Lot being owned by the original owner and not previously registered or owned by another owner may keep one (1) dog in the lot, so long as:
>
>> 16.2.1 Not less than seven (7) days prior to bringing the dog into the Lot, the Lot Owner must:
>> 16.2.1.1 Provide to the satisfaction of the Body Corporate, evidence of past ownership of the dog, exceeding one (1) year prior to the earlier of the date of establishment of the Scheme or occupancy of their Lot;
>> 16.2.1.2 Register with the Body Corporate, particulars of the dog, including name, sex, weight, description, breed and two coloured photographs, one showing the face, and one the side of the animal;
>> 16.2.1.3 The dog be registered with the Gold Coast City Council and provide a copy of that registration to the Body Corporate, each year on renewal.
>> 16.2.2 Is dry and free of sand, dirt or garden material;
>> 16.2.3 Weighs less than ten (10) kilograms;
>> 16.2.4 Does not disturb others;

16.2.5 Is a domesticated pet;
16.2.6 Is toilet trained;
16.2.7 Only passes over or through the Common Property for the purposes of ingress and egress to a lot and at all times kept on a lead which is no more than 1 meter in length;
16.2.8 Is kept healthy and free of parasites;
16.2.9 Where a lot is owned by one or more owners, only one animal per Lot is permitted;
16.2.10 No owner or joint owner of more than one Lot in the Scheme is permitted to keep a second or subsequent animal in their second or subsequent Lot;
16.2.11 This By-Law is for the personal benefit of an owner as defined in By-Law 16.2 above and lapses immediately upon the death of the dog or upon the sale of the Lot.

16.3 The Body Corporate may order an animal to be removed from the Scheme if the animal does not comply with all of the criteria set out in this By-Law.
16.4 However, a deaf or blind person, shall be permitted to keep or bring into a lot or onto the Common Property, a guide dog, as referred to in the Guide Dogs Act 1971.
16.5 The owner indemnifies and shall keep indemnified the Body Corporate against any loss, damage, injury or claim whatsoever made against the Body Corporate regarding any act on the part of the dog.

80 8 Cal 4th 361, (1994) ('*Nahrstedt*'). Although Justice Arabian's judgment was the dissent and the majority upheld a blanket pet ban in the condominium, the Californian legislature subsequently enacted legislation protecting people's right to keep at least one pet in common interest developments: Hyatt and French, above n 20, 168.
81 *Nahrstedt*, 8 Cal 4th 361, 392–4 (1994).
82 This was the argument accepted by the Adjudicator in *Beattie Place South* [2009] QBCCMCmr 27 to justify a blanket ban on pets. See also *Pivotal Point Residential* [2008] QBCCMCmr 55.
83 *Companion Animals Act 1998* (NSW) s 31.
84 Ibid. s 32A(1)(b).
85 Interestingly, a dog is defined as a 'nuisance' if it repeatedly defecates on property outside that which it is ordinary kept: s 32A(1)(c), but a cat is not.
86 Thomas W Merrill and Henry E Smith, 'The Morality of Property' (2007) 48 William & Mary Law Review 1849.
87 The Cat Protection Society of New South Wales states that pet bans in strata schemes have 'a terrible impact on animal welfare' with many owners being forced into conflict with their body corporate, often resulting in the surrender of wanted and cared-for cats: Email from Kristina Vesk (Chief Executive Officer, Cat Protection Society NSW Inc) to Cathy Sherry, 5 November 2013.
88 *Body Corporate and Community Management Act 1997* (Qld) s 180(7). Queensland adjudicators also had a power that predates this section to invalidate by-laws that were oppressive or unreasonable.
89 *Beattie Place South* [2009] QBCCMCmr 27.
90 *Tutton* v. *Body Corporate for Pivotal Point Residential CTS 33550* [2008] QCCTB-CCM 12; *212 on Margaret* [2011] QBCCMCmr 400; *Central Park Apartments Varsity Central* [2011] QBCCMCmr 107; *Scholars Cover* [2011] QBCCMCmr 310.
91 *Body Corporate for River City Apartments CTS 31622* v. *McGarvey* [2012] QCATA 47, [49] (Kenneth Barlow, Member). Realistically, this characterisation of a 'prohibition' as not being 'regulation' is untenable. Logically and as a matter of legal reality, regulation frequently prohibits certain activities; e.g. the fisheries legislation which was the subject of the High Court decision in *Akiba on behalf of the Torres*

Strait Regional Seas Claim Group v. *Commonwealth of Australia* [2013] HCA 33, was clearly 'regulation' and it implemented a blanket prohibition on commercial fishing without a licence.

92 [2008] QCCTBCCM 12, [35]. In *Mornington Quays* [2012] QBCCMCmr 576, the Adjudicator said at [17] that 'an absolute ban on the keeping of pets within this scheme is unreasonable *because there are circumstances in which some animals could be kept in the scheme without causing inconvenience to other residents'* (emphasis added). The same finding was made in *Hillcrest Apartments* [2013] QBCCMCmr 211 at [22]. It should be noted that in the strange world of strata title, keeping fish has been a cause for legitimate communal concern. A family in New South Wales had an above ground pool in their apartment which they were using to fatten trout for their restaurant: Alexandra Smith, 'Sydney's Strata Surprises: Saga of Horse Next Door, the Trout Farm and 'Niagara Falls', *Sydney Morning Herald* (online), 21 March 2012 www.smh.com.au/business/property/sydneys-strata-surprises-saga-of--horse-next-door-the-trout-farm-and-niagara-falls-20120320-1vi2l.html.

93 E.g. *Companion Animals Act 1998* (NSW) s 32A.

94 *Ephraim Island – Subsidiary 105* [2007] QBCCMCmr 205.

95 It is perhaps necessary to stress that this argument is not motivated by a personal love of dogs. I am not a dog owner and do not wish to be, but I am aware that whether I want a dog is irrelevant to whether other people keep them, so long as they do not unreasonably disturb me.

96 *Nahrstedt*, 8 Cal 4th 361, 396 (1994).

97 Joseph W Singer, *Entitlement: The Paradoxes of Property* (Yale University Press, 2000) 136–7.

98 McKenzie, *Privatopia*, above n 11, 143–4.

99 Paula Franzese, 'Does It Take a Village? Privatization, Patterns of Restrictiveness and the Demise of Community' (2002) 47 *Villanova Law Review* 553, 555.

100 Richard A Epstein, 'Covenants and Constitutions' (1987) 73 *Cornell Law Review* 906; Richard A Epstein, 'Notice and Freedom of Contract in the Law of Servitudes Comments' (1981) 55 *Southern California Law Review* 1353.

101 Robert C Ellickson, 'Cities and Homeowners Associations' (1981) 130 *University of Pennsylvania Law Review* 1519.

102 James M Buchanan, 'An Economic Theory of Clubs' (1965) 32(125) *Economica* 1; Charles M Tiebout, 'A Pure Theory of Local Expenditures' (1956) 64(5) *Journal of Political Economy* 416. Tiebout was not writing about HOAs but his theory has been taken up by their proponents. See Fred Foldvary, 'Proprietary Communities and Homeowner Associations' in David T Beito, Peter Gordon and Alexander Tabarrok (eds), *The Voluntary City: Choice, Community, and Civil Society* (University of Michigan Press, 2002) 258. Noted in Chapter 3, n 61, 62, 63.

103 Robert H Nelson, *Private Neighbourhoods and the Transformation of Local Government* (Urban Institute Press, 2005).

104 Foldvary, above n 102; ibid.

105 Robert Putnam, *Bowling Alone: The Collapse and Revival of American Community* (Simon & Schuster, 2000).

106 Andres Duany, Elizabeth Plater-Zyberk and Jeff Speck, *Suburban Nation: The Rise of Sprawl and the Decline of the American Dream* (Farrar, Straus, and Giroux, 2000). Duany and Plater-Zyberk's iconic New Urbanist community is Seaside, Florida: Seaside: A Simple Beautiful Life www.seasidefl.com/.

107 McKenzie, *Beyond Privatopia*, above n 11, 35.

108 Edward J Blakely and Mary Gail Snyder, *Fortress America: Gated Communities in the United States* (Brookings Institution Press, 1997); Setha M Low, *Behind the Gates: Life, Security, and the Pursuit of Happiness in Fortress America* (Routledge, 2003).

109 Goodman and Douglas, above n 36, 467.
110 McKenzie, *Beyond Privatopia*, above n 11, 52.
111 Alexander, 'Group Autonomy' above n 22. See also Stephanie M Stern, 'The Dark Side of Town: The Social Capital Revolution in Residential Property' (Working Paper, Chicago-Kent College of Law, April 2013) http://ssrn.com/abstract=2243134 or http://dx.doi.org/10.2139/ssrn.2243134.
112 Singer, Joseph William, 'How Property Norms Construct the Externalities of Ownership' in Gregory Alexander and Eduardo Peñalver (eds), *Property and Community* (Oxford University Press, 2010), 57.
113 Larry Alexander, 'Illiberalism All the Way Down: Illiberal Groups and Two Conceptions of Liberalism' (2001–2002) 12(2) *Journal of Contemporary Legal Issues* 625, 627.
114 Mill, above n 64.
115 Alexander *et al.*, 'A Statement of Progressive Property', above n 67, 743.
116 Joseph William Singer, 'Democratic Estates: Property Law in a Free and Democratic Society Essay' (2008) 94 *Cornell Law Review* 1009, 1047.
117 Ibid. 1049–50.
118 Singer, *Entitlement*, above n 97, 174.
119 Singer, 'Democratic Estates', above n 116, 1050.
120 Richard Louv, *Childhood's Future: Listening to the American Family* (Houghton Mifflin, 1990) 317, cited McKenzie, *Privatopia*, above n 11, 144.
121 Radin, above n 69, 992.
122 Susan F French, 'Constitution of a Private Residential Government Should Include a Bill of Rights, The Symposium on the Uniform Real Property Acts' (1992) 27 *Wake Forest Law Review* 345.
123 The Architectural Standards and Landscape Design Guidelines, Item 17, Macquarie Links Management Statement, DP270152, states that 'no trucks or commercial vehicles exceeding 3 tonnes are to be kept within a lot, on a roadway or any communal area'. No trailers and caravans are to be kept unless garaged. This is to 'ensure the visual character and amenity of the neighbourhood is not prejudiced by the presence of large vehicles'.
124 Alexander *et al.*, 'A Statement of Progressive Property', above n 67, 744.
125 *Taunton Enterprises Pty Ltd Appeal Against Order No C579394 Strata Title Act* [2005] TASRMPAT 76.
126 Michael A Heller, 'The Boundaries of Private Property' (1999) 108 *Yale Law Journal* 1163, 1165–6. See Chapter 2, n 61.
127 *Body Corporate for River City Apartments CTS 31622* v. *McGarvey* [2012] QCATA 47, applying the authority of *Mineralogy Pty Ltd* v. *Body Corporate for the 'The Lakes Coolum'* [2002] QCA 550.
128 *Albrecht* v. *Ainsworth* [2015] QCA 220; *BCCMA* s 94(2) requires a body corporate to act reasonably when exercising its functions.
129 Gregory S Alexander *et al.*, 'A Statement of Progressive Property' (2008) 94 *Cornell Law Review* 743, (emphasis added).
130 Carol M Rose, 'Servitudes, Security, and Assent: Some Comments on Professors French and Reichman Comments' (1981) 55 *Southern California Law Review* 1403, 1404.
131 Uriel Reichman, 'Toward a Unified Concept of Servitudes' (1981) 55 *Southern California Law Review* 1177, 1233.
132 Singer, 'Democratic Estates', above n 116, 1048.

6 By-laws' effect on well-being

A case study in children's play

Introduction

In this final chapter, we will expand our analysis again, beyond economics, privacy and autonomy, to the more amorphous concept of well-being. As noted in the preceding chapter, this is consistent with a theoretical approach to property that accounts for 'the full range of human values we care about'.[1] In a modern, pluralist democracy, in addition to the traditional liberal values of the free market, privacy and autonomy, we care about human relationships, 'health, friendship, [and] human dignity';[2] we care about our communities, both large and small; we care about the environment and future generations; and we care about opportunities for citizens, particularly children, to live an optimal existence. In short, we care about 'human flourishing'.[3] Bearing a close connection to common morality,[4] so too does property law, and as a result, it typically requires us to be attentive to 'the effects of claiming and exercising property rights on others, including future generations'.[5] The exercise of property rights should never seriously compromise the well-being of others.

Through a case study on children, this chapter will explore the potential of by-laws to do exactly that. Children have been chosen because their presence in strata and community schemes is increasing, and their well-being is fundamental not just to themselves and their families, but to the entire community, which needs them to become functional adults of the future. Further, a case study on children will allow us to expand our consideration to activities that are other-regarding, but despite this, should not be the subject of by-law regulation.

The chapter includes substantial background information so that the legal issues can be assessed from an informed position. It begins with a description of the presence of children in strata and community schemes; how many children are potentially affected by by-laws? Then, because we need to be clear about the precise values at stake, it includes an explanation of the benefits of play for children, as established by research. Play is not simply an activity that children enjoy, it is essential for their intellectual, physical and emotional development. Finally, the chapter examines two areas in which by-laws have affected children's well-being, outdoor play on common property and falls from windows and balconies.

Children in strata and community title schemes: physical structures and demographics

As described in Chapter 1, Australian governments had a long-standing antipathy to apartment living both in relation to public and private housing.[6] This antipathy was particularly strong in relation to families and the well-being of children was at the forefront of policymakers' minds. In 1908/1909 the *Royal Commission for the Improvement of the City of Sydney and Its Suburbs* heard evidence that 'the flat system tended to destroy family life, and was not conducive to morality'.[7] The flats in the Strickland Buildings, the first large public housing apartments in Australia, built in 1914 and still standing in Chippendale in Sydney, were criticised as an option for working-class families. J J C Bradfield, Chief Engineer for the Sydney Harbour Bridge and City Railway, said that while the flats were at least sanitary, 'there are no gardens, no yards, and, as will be seen, the family washing is often dried by hanging on the balconies, or from the windows'.[8] He advocated for railways to the suburbs where 'children [could] enjoy fresh air and sunlight in healthy surroundings', rather than having their playground confined to the gutter.[9]

As is obvious from the current form of Sydney, as well as all other Australian cities, Bradfield and like-minded thinkers were successful in their promotion of the suburban ideal.[10] The vast majority of residential housing in Australia in the twentieth century was built on the suburban model, with freestanding, freehold Torrens title homes, in generous private gardens, set on quiet, curvilinear streets. Above all else, they were designed for families with children. As described in Chapter 1, most suburbs fulfilled the Garden Suburb ideal of open space with the provision of publicly owned playing fields or bush reserves. While Australian suburbia may not have always been cosmopolitan or culturally stimulating, it did afford children active, unstructured childhoods. Many Australians who grew up in the post-war period remember fondly that the only restriction on their freedom was that they 'be home in time for tea'.

As we saw in Chapter 1, despite strong government preference for suburban development, apartment buildings were constructed in Australian cities in the twentieth century, particularly in the inter-war period in Sydney, and after the passage of strata titles legislation, in other capital cities as well as Queensland's Gold Coast. However, these apartments were never constructed with children in mind, and as a result, they were rarely built with outdoor space, let alone children's play space. Apartment blocks were frequently built on single housing lots, with the new building filling almost the entire parcel. The only outdoor space was typically for parking or a washing line. They were 'barrack-type buildings, their long dimensions filling the depth of the narrow allotment. What used to be yards at the back and on the sides … are denuded of vegetation and paved for cars'.[11] In Sydney in particular, most apartment blocks were built by small builders with no licence or formal training, and combined with minimal council regulation,[12] the end product was usually suboptimal in terms of design and amenities.

In the late twentieth century, apartment construction altered with the development of much larger schemes, including multiple apartment or townhouse

Figure 6.1 Strata schemes in Eastlakes, Sydney.[13]

schemes inside community title subdivisions. In Queensland, 'resort-style' schemes mushroomed on the Gold Coast and Sunshine Coast, with extensive grounds and private facilities. Many were built on hitherto undeveloped land such as the northern end of the Gold Coast. In Melbourne, extensive apartment construction occurred on disused industrial land in the Docklands area, adjacent to the CBD, similar to construction on brownfields sites in inner Sydney and up the Parramatta River.[14] In Sydney, large, disused industrial sites have presented the opportunity to build community title subdivisions such as Jacksons Landing, a community scheme comprised of 20 strata schemes with 3000 residents,[15] on the old CSR Sugar Refinery site at Pyrmont; Balmain Shores, a community scheme made up of 11 strata schemes on the old Balmain power station;[16] and the strategically renamed Breakfast Point, a low- and high-rise community scheme, on the old AGL Mortlake Gas Works.[17]

Unlike individual housing lots, large, disused industrial sites or hitherto underdeveloped areas allow site planning, sensible positioning of high and medium density housing, and the provision of recreational facilities and open space. In particular, large sites can theoretically account for the needs of children. However, in apartment-dominated developments the provision of play and other space for children has been compromised by a continuation of the assumption that families with children will not, and should not, live in apartments.

Policies of urban consolidation have gone hand in hand with assumptions that family housing is oversupplied in Australian cities and that apartments should be constructed for rising numbers of smaller households of singles, couples and the elderly. As a result, new apartments are overwhelmingly one or two bedroom, and rarely include facilities for children. Even outside individual developments, urban regeneration has been 'child-blind'.[18] However, while these demographic assumptions have proved partially true, they have not proved as consistent as government assumed. There is evidence that smaller households in Australia do not necessarily want, or choose to live in smaller dwellings,[19] and that older couples do not consider their homes 'under-utilised'. They have not 'downsized', making their homes available for younger families,[20] thus limiting the supply of detached homes in established suburbs. In combination with sharply rising house prices, many families, particularly in Sydney, could not buy or rent a detached house or townhouse, even if they wanted to, pushing them into the apartment market.[21]

As a result, there are Australian families living in apartments, both old and new. For example, in greater Sydney, a quarter of all apartment households have children compared to just over half of all households living in a separate house.[22] Recent migrant families with young children occupy some of the aging apartment stock at the lower end of the private rental market,[23] and significantly, growing numbers of families now live in both Sydney and Melbourne's new apartment stock, built pursuant to policies of consolidation.[24] In some areas of Sydney like Liberty Grove, on the Rhodes peninsular, and in Ultimo, high proportions of families have a parent born in south east Asia, where family apartment living may have been the norm.

The clearest illustration that children are living in new apartments, irrespective of the assumption of government and planners, is the crisis in inner-city schooling in Sydney and Melbourne.[25] Ultimo Public School, the school serving the exclusively high-density suburb of Jackson's Landing in Sydney, has struggled for over a decade as a result of inadequate facilities and growing student numbers. In the years following the mid-1990s urban regeneration of the Pyrmont-Ultimo peninsular, the school tripled in size.[26] The government has announced plans to build a new school for 1000 students, transforming Ultimo into one of the biggest primary schools in the state. A new 600-student primary school was finished in 2015 to service the growing and apparently unanticipated child population on the redeveloped, high-density Rhodes peninsular in Sydney's inner west, and another school is planned for nearby Wentworth Point. In Melbourne, parents at the Docklands have become increasingly frustrated with the absence of a primary school and the apparent failure to plan for one.[27]

In contrast, governments have anticipated and planned for children in new, low-rise community title developments in New South Wales and Queensland. For example, the New Rouse Hill High School was built to service the large community title development in the north west growth sector of Sydney, and Wilton Primary School was built in Bingara Gorge to service the south west. Government assumptions have proved correct about the attraction of these developments for families. At the time of writing, children made up almost 50 per cent of the population at Bingara Gorge.[28] Similarly, the new residential developments on the inland and northern end of the Gold Coast, many of which are community title, have seen such high increases in the number of young children that they have been dubbed Nappy Valley.[29]

It may never be possible to ascertain accurate numbers of children in strata and community title schemes, but the following generalisations are supported by existing housing patterns:

- apartment buildings are unlikely to be dominated by families, but the assumption that they will house no or negligible numbers of children has proved erroneous;
- families may live in apartments because they are prepared to trade space for proximity to work, services and the cultural life of the city;
- aversion to apartment living with children may be culturally specific, with parents born in south east Asia less likely to view apartments as incompatible with family life;
- some families will live in apartments as a result of economic necessity;
- low-rise community title developments will continue to be dominated by families with children.

We can confidently conclude that a significant proportion of Australian children will be subject to privately written strata and community title by-laws, for part or all of their childhoods.

Children, play and space

If there are children present in strata and community schemes and we want to consider what affect that title might have on their lives and well-being (for example, what externalities the exercise of property entitlements specific to strata and community title may produce), we need to think about things that are essential to children's well-being. Aside from food, care, shelter and formal education, the most important element of a healthy, happy childhood is play.

Research has repeatedly demonstrated that children's play is not a frivolous matter.[30] It is fundamental to children's social, intellectual and physical development,[31] and the quality of their play opportunities will determine their functional abilities as adults.

Not surprisingly, there is a direct relationship between physically active play and children's health, as 'early childhood physical exercise helps build strong bones, muscle strength and lung capacity.... It may also increase cognitive function, improve academic achievement and accelerate neurocognitive processing'.[32] Physically active play is an important factor in avoiding childhood obesity. In 2010, 22.8 per cent of New South Wales children were overweight or obese, with only half of primary school-aged children meeting recommended guidelines for physical activity. Urban children were less active than their rural counterparts, and children from Asian or Middle Eastern families and/or lower socio-economic backgrounds (those also most likely to live in apartments)[33] were significantly less active than children from Anglo-Australian families or higher socio-economic backgrounds.[34] Childhood obesity leads to significantly increased risk of adult obesity, a concern not just for individual children and their families, but also the wider community which will bear the costs of care for obesity-related health problems.[35]

Play is essential for cognitive development, contributing to children's 'developing vocabulary, their understanding of different concepts, their ability to solve problems, their self-confidence and motivation, and an awareness of the needs of others'.[36] Imaginative play, and (perhaps surprisingly) rough play, are both crucial for the development of empathy, a fundamental attribute for functional interaction with other people throughout life. Imaginative play is also particularly important for developing lateral and creative thinking. If a child can pretend to be a fairy, simultaneously knowing she is actually a little girl, she is developing her capacity to think not just about what is, but what could be. Risk- taking play minimises anxiety in children by allowing them to test boundaries and experience their own survival.[37]

Play, particularly with other children, is also essential for social development:

> Playing with other children affects the ways in which children relate to each other, form groups and feel part of a group or part of their local community. When children play they use their own language, rules and values and play helps them to develop their own identities.... Children who are able to play freely with their peers develop skills for seeing things through another person's point-of-view, for cooperating, helping, sharing, and solving problems.[38]

Play can overcome cultural and other boundaries, not only between children, but between their parents, 'who meet and talk to other parents when accompanying their children to play spaces, which helps to foster community relations and friendships'.[39] However, play spaces away from parents' eyes – private cubby houses, secret trees, even alleys – are important for children's enjoyment and for a development of mastery over their own lives; this is especially so for teenagers.[40]

There is ample research on specially constructed children's play spaces, with most research suggesting:

> a move away from uniform, fenced, static and sterile places. The new places recognise the nature of play and use design principles that accommodate a variety of space and landscapes, natural features, different heights, a range of flexible and variable materials, access to the elements. They contain a collection of loose parts with which children can *experiment freely*, creating an environment that gives children the feeling that the world is full of things to explore and *where space and resources can be adapted* to what is needed at the time. Children can have *a significant influence* on the play space and engage all five senses.[41]

Natural playgrounds that allow this kind of experimentation and agency are rarely seen in Australia in either public or private space, with most councils opting for ubiquitous, primary-coloured, plastic modular play equipment.

Research suggests that children do not only need specially constructed play spaces, but that play in public spaces or 'community play' greatly aids the formation of friendships and networks, as well as helping children learn the rules of social life.[42] Play in public spaces helps children to navigate unfamiliar environments, and most importantly, to develop public trust.[43] There is widespread concern among play researchers about the decline in community play with parents' increasing fear of traffic and 'stranger danger'.[44] This in turn has led to a loss of acceptance for children's use of public space, and researchers argue that 'hanging out' unsupervised in public needs to regain its legitimacy, with parents and the public alike.[45] Overall, a review of play literature 'highlights the importance of children having access to play spaces in their local communities, and the importance of adults having positive attitudes towards children playing freely outside, to a wider sense of well-being'.[46]

Australian research confirms the link between public environments and children's well-being. As part of their longitudinal study of over 10 000 children nationally,[47] the Australian Institute of Family Studies investigated the link between neighbourhood socio-economic status and the conduct problems and pro-social behaviour of almost 5000 four-to-five year olds.[48] One part of the survey assessed 'neighbourhood belonging', by asking parents to assess trust of neighbours, a sense of identity with the neighbourhood, how well informed they were about local affairs, and knowledge about where to find information about local services. The study found that *conduct problems* associated with lower socio-economic neighbourhoods were mediated by neighbourhood safety and neighbourhood belonging. The only neighbourhood variables that were significantly associated with *pro-social* behaviour were neighbourhood cleanliness and neighbourhood belonging. Researchers noted that it was a 'particularly striking aspect of the current study … that neighbourhood belonging was such an influential factor' and that in light of this, 'initiatives that encourage social gatherings of families in their neighbourhood may be particularly effective as a mechanism to build social relationships and develop mutual trust as well as providing information about local services'.[49]

At the risk of stating the obvious, play, and all of its accompanying benefits, cannot occur unless children have *somewhere* to play. As we noted at the beginning of Chapter 2, we are inescapably *embodied* beings: everything we do, everything we value, enjoy or is important to our well-being, like play, has to happen in a physical space that we are (ideally) legally entitled to occupy. Play can occur in privately owned spaces, but in strata developments, these are likely to be small. This has a profound effect on children's ability to play and to play safely (discussed below). As a result of small private play spaces in strata developments and community title developments, particularly those inspired by the principles of New Urbanism[50] (smaller private spaces and larger common space), common property will be a vital resource for children. In community title developments, that common property might be roads, parks, sporting facilities and bushland; that is, large swathes of privately owned and regulated land that in ordinary residential subdivisions would be public space. A child who lives in a community title development, that contains a public school and shops, may in fact rarely leave privately owned property on a day-to-day basis.

In light of the research from the Australian Institute of Family Studies, it is possible that strata, but particularly community title, could have a *positive* effect on children's well-being. If the proponents of private communities, who argue that they increase social capital and community engagement, are correct,[51] then community title should increase the 'neighbourhood belonging' that researchers identified as a significant factor in children's problem conduct or pro-social behaviour. Bodies corporate can use common property to organise 'social gatherings of families in their neighbourhood [which] may be particularly effective as a mechanism to build social relationships and develop mutual trust as well as providing information about local services'. Most of the larger developers organise community events in initial years and some communities continue these. However, the limited research in Australia to date suggests that such activities may drop off over time.[52] This is arguably inevitable if a sense of community has not developed organically.

One significant advantage private common property may have over genuinely public space is safer streets, as privately owned roads can have lower, by-law mandated speed limits. Liberty Grove in Sydney's inner west, has a speed limit of 20 km per hour and L-plate drivers are not permitted to drive on the community's roads.[53] If communities are gated (in fact, a rare feature in Australia), they will also have no through traffic. As noted above, concern about traffic is a major contributor to parents restricting vital outdoor play for children and privately owned and regulated community title roads can potentially mitigate these fears. Hopefully, however, Australia will never see the equivalent of attempts by United States communities to render their streets safer by creating 'sex offender covenants', which prohibit ownership and occupancy of houses in HOAs by registered sex offenders.[54]

As community title developments are squarely marketed to families, they often contain facilities for recreation, typically tennis courts, pools and bike paths. While valuable assets, they provide relatively static play spaces, with pre-determined uses. None allow for the kind of creative adaption research suggests is necessary for optimal play. Tennis courts can only be used for playing tennis; pools can only be used for swimming and of necessity, must be supervised by adults. Static, adult-focused leisure facilities are potentially concerning because '[a]ggressive behaviour has been linked to a lack of interesting and engaging environments and destructive behaviour is most common in boring spaces without trees, bushes or other natural boundaries'.[55] Lack of appropriate recreational facilities has been linked to problem behaviour in teenagers, with some research suggesting that the best service for teenagers in master planned communities is transport to get them out.[56]

Creative play spaces are expensive to construct and not unreasonably, developers are unlikely to pay for them if it is not certain they will see a return in sales. Further, as common property, play spaces and equipment will need to be maintained by the ultimate community through levies, and that requires a preparedness to pay.[57] Research reveals that even in relatively new communities, there is resident concern about the ongoing costs of services and facilities, as well as disagreement about which services and facilities should be maintained and to what level.[58] Unlike a local council, the decision-makers in strata and community title are the direct beneficiaries of services and facilities, as well as the direct funders. As noted at the beginning of Chapter 5, lot owners are not required to vote objectively or with any professional expertise at body corporate meetings; when voting, they are exercising a private right and are entitled to act in their own interest.[59] As a result, if people do not want to pay for play facilities because they do not personally benefit from them, they are entitled to vote in that regard.

Insurance and potential liability for injury on common property are fundamental considerations in the provision of any facilities, and a factor in the decision not to install any play equipment, let alone novel or challenging equipment in schemes.[60] The mere suggestion of a skate park for teenagers would be enough to give a community title lawyer or executive committee member heart palpitations. This is not unreasonable because as the occupier of common property, bodies corporate can be liable at common law for injury sustained.[61] While all bodies corporate are required to maintain insurance cover for death or bodily injury, if insurance is insufficient or unavailable, bodies corporate do not have limited liability and individual owners are liable in proportion to their unit entitlements. In the United States, judgments against homeowner associations for injuries sustained on common property have tipped US$30 million.[62]

The role of insurance and potential civil liability, along with legislatively sanctioned self-interest and a lack of professional expertise, are all good illustrations of the inherent problems in the privatisation of community facilities. With good reason, private citizens may be neither willing nor able to carry out functions typically performed by government. This reality needs to be weighed against the benefits of private communities and facilities before they are too enthusiastically adopted by developers, local councils and state governments.

Finally, while children can turn dull play spaces into more interesting ones with their own ingenuity (think of 8-year-old boys climbing on the roof of standard play equipment, or children gathering stones or bark from gardens to represent money in imaginary shops), strata and community title present a barrier to this kind of adaptation. This is because unlike public spaces, common property is regulated by detailed by-laws, written and enforced by the private citizens who own that space. It is to these by-laws that we now turn.

By-laws regulating children's play on common property

The preceding section went to some lengths to establish that children's play is a serious matter, fundamental to their well-being and that children need space to play. This section then considers the question, if play is essential to children, should private citizens be able to infringe a child's 'right' to play through the creation of by-laws?[63] The question will be explored through an examination of real by-laws in a range of communities.

At the most extreme end is the by-law considered in the Western Australian case of *The Owners of 111 The Broadview Landsdale – Survey Strata Plan 38894* v. *Colavecchio*.[64] The site of the dispute was Kingsdene Mews, a low-rise strata title estate in Perth.[65] The common property consists of grassed pavements and a circular cul-de-sac road around a central lawn and garden. The grassed pavements and centre lawn and garden would be ideal public play space for children, and the cul-de-sac road, safe to ride bikes or skateboards.

The by-laws for Kingsdene Mews included a blanket ban on children playing on *any* common property. They stated that a proprietor, occupier or other resident shall not

> permit any child of whom he has control *to play upon common property within the scheme*, or unless accompanied by an adult exercising effective control, to be or remain upon common property comprising a road, car parking area or any other area of possible danger or hazard to children.
>
> (emphasis added)

While model by-laws often prohibit children playing in areas that might be a danger to them, such as laundries or car parks, this by-law prohibited children even playing on the grassed pavements or central lawn. Further, while roads are generally a danger to children, this cul-de-sac is so quiet that even the most cautious parents would allow their children to ride bikes on it.

The applicant had sought the imposition of a fine on lot owners for a breach of by-laws by their tenants, alleging that the eight- and 10-year-old children of the tenants had damaged gardens and made undue noise on common property or within a lot. Significantly, it seems that the applicants did not realise that the by-laws banned *all* play on common property; had the children been playing tea parties on the lawn with dolls, it would have been a breach of the by-law.

In a 27-lot scheme, 10 proprietors of lots supported the application and 10 opposed it. The evidence of each camp was telling, revealing conflicting perceptions of appropriate property use, which in turn reflected people's own sometimes 'incommensurable' values.[66] Those supporting the application were primarily concerned with the noise of basketballs and the aesthetics of the hoop. They claimed to have been:

> plagued with the bouncing of basketballs, shouting and screaming from the tenants [*sic*] offspring and their visitors, along with the playing of tennis on and around our grassed areas, resulting to damaged trees and plants … both my wife and I the owners of 17 Kingsdene Mews neither want or need to see unsightly basketball hoops in our complex, nor hear loud music occasionally i.e. (our house is about ninety metres away from number six and in our back yard we can still hear the song What about me it isn't fair played up to six times in a row) or foul filthy language from these people from number six.[67]

Other supporters of the application said,

> [a]bove all this the children ride their bikes, bounce the ball, chase each other around the Estate and even play tennis. So as you can imagine there is a lot of screaming, loud laughing and a lot of loud talk.[68]

Opponents of the application argued that the children of the tenants:

> did nothing that any children their age could and should not have been doing. They have always been well behaved and courteous including the occasions on which Mr Baxter [their immediate neighbour and principal instigator of the application] intimidated them.[69]

It was suggested that there were owners in the complex who did not want any properties rented and had expressed the view that children should be 'banned'. Another opponent of the application said:

> It is my belief and understanding that Kingsdene Mews was built and designed as small family homes for a young growing Landsdale community. It is therefore reasonable to suggest that families with young children would likely reside within the Kingsdene Mews strata group and <u>therefore</u> play within the area.
>
> …

> It would be unreasonable to expect children not to play in and around their home. The road within Kingsdene is not a through road and therefore generally only receives vehicles belonging to the tenants or visitors.
>
> The lots within Kingsdene are very small and therefore the Common ground within Kingsdene is going to be used as a play area for those that have children. It would be unreasonable to think that during those times of play, children's balls etc. would not go outside the common area and onto the road or garden for example. One would be naïve to think that this wouldn't happen. One would also be naïve to think that there wouldn't be some noise associated with children's healthy play.[70]

A resident schoolteacher said that:

> I have no problems at all to the fact that they only play ball for short periods at a time and not at all in the early hours or late evening.
>
> As there have been no problems in allowing families with children to move into the estate, I feel that we need to allow them some space in which to play. The use of the common areas has been opposed therefore playing ball outside their own home at reasonable times during the day should be permitted.
>
> . . .
>
> I do not feel that the application is correct as it stated that the children play ball for four hours at a time during the school holidays. I am a school teacher on holidays at the same time as the children and never once did I hear them outside for more than one hour. On extremely hot days they were not out at all and I am sure during the winter months they will be out very little.[71]

The tenant had actually moved the basketball hoop a number of times in an effort to reduce noise and the principal complainant had refused to meet with the tenants to discuss the problem. The Strata Titles Referee held that before an application was made for settlement of a dispute, reasonable attempts should be made to resolve the dispute amicably.[72] He further held that the evidence was conflicting, it was not specified on which by-laws the order should be based, and that section 103I of the *Strata Titles Act 1985* (WA) only allowed for the fining of a lot owner, while in this case, it was the actions of the occupier that were in issue.[73] The application was dismissed.

If the houses had been in an ordinary residential subdivision, this dispute would not exist. There is nothing a resident can do to prevent children playing on a public footpath, bouncing balls or shooting hoops, nor playing audible music. By-laws prohibiting noise were first created for apartment blocks, in which residents have a greater capacity to affect others, but Kingsdene Mews is a subdivision of freestanding houses. It is higher density than many Australian suburbs but lower density than inner city areas. This prompts the question, why did the development need a noise by-law at all? Residents would have been protected by local council regulations and it is hard to see why they needed any higher level of protection. Most importantly, it is difficult to see why they should be empowered to create this higher level of protection at the expense of children's physical, social and cognitive well-being.

Section 35(1)(b) of the *Strata Title Act 1985* (WA) requires the body corporate (strata company) to 'control and manage the common property *for the benefit of all the proprietors*' (emphasis added). The Western Australian Tribunal has used this phrase when assessing the validity of by-laws in the past,[74] and thus it is possible that had *Colavecchio* been argued on the basis that the by-law banned all play on common property (a fact that no one seemed to have noticed), the by-law might have been found to be invalid, not being for the benefit of proprietors who were parents or who leased their property to parents. The phrase 'for the benefit of all proprietors' (which would be preferable if it read 'all residents'), recognises the potentially harmful effect of by-laws, and that the majority do not have an automatic right to create rules without reference to their consequences.

Slightly less extreme than the by-law in *Colavecchio*, are the ubiquitous 'no skateboards, no ball games' by-laws. If the outdoor common property consists of only a car park, as is the case with many 1960s and 1970s strata schemes (see Figure 6.1 above), these by-laws are necessary. However, new schemes are often built with some, albeit inadequate, open space, particularly if they are medium density. For example, a 2–3 bedroom townhouse development in Maroubra in Sydney's east, has a circular common property garden in its centre (see Figure 6.2). The garden would seem to be ideal for children playing with balls (particularly if the unprepossessing plants were removed), and the brick path that runs around the lawn would be good for scooters, if not bikes. However, the sign attached to the fence reveals that both are banned by by-laws (see Figure 6.3).

While this townhouse development is next to a large playing field and close to the beach, younger children could not play in these areas unsupervised. In an ordinary residential subdivision, they would play in their private yard, but in this townhouse development the private yards that exist for some, but not all lots, are tiny courtyards. Presumably, consistent with principles inspired by New Urbanism, the common property garden was meant to substitute for private yards; individually owned open space was consolidated into collectively owned open space. However, the purpose of this was defeated by the creation of a by-law that prohibits children using the common space as they would a private yard.

One final example of by-laws limiting children's play on common property can be seen at Liberty Grove. Liberty Grove is the community title redevelopment of the former Olympic Village site on the Rhodes peninsular in Sydney's inner west, the area, as noted above, that has needed a new primary school to accommodate the number of children living in new high-density developments. Liberty Grove has 12 strata schemes, with extensive common property. This includes two swimming pools, two tennis courts, a gymnasium, a basketball court, two playgrounds, two large parks, four small parks and a community hall.

Figure 6.2 Modest common property in medium density contemporary development in Sydney's eastern beaches.[75]

Figure 6.3 Sign banning children's play on common property lawn and paths.[76]

Much of the common property is accessible by the general public, although there has been some confusion and tension about this.[77]

Although it was initially assumed that children would not make up a significant proportion of the estate, this has not proved to be the case. People who have moved into the area have been disproportionately families with children,[78] including a significant proportion of Asian Australian families.[79] In 2011, there were 461 children living in Liberty Grove,[80] with households with children making up almost 50 per cent of the estate.[81] Liberty Grove has 295 courtyard homes and townhouses, 216 low-rise apartments and 284 high-rise apartments,[82] making it clear that the only outdoor space children are likely to have access to is the common property.

Liberty Grove's management statement includes limited regulation of common property open space. It reasonably bans 'organised' sports and games without written approval of the Executive Committee,[83] and requires children under 12 to be accompanied by an adult in the pools.[84] By-law 12 requires the community association to allow members of the public to use the 'public open space' and 'communal open space' for 'passive recreation', without restriction. This by-law was included as a condition of development consent and cannot be altered without the local council's consent. It indicates that at least in the planning stage and in relation to non-residents, the Council was mindful of the potential effect of by-laws on the use of recreation space, although it only thought to protect passive, and not active recreation.

In 2010, the management statement was amended to include new by-law 9.16. This states that the 'Liberty Grove Community Policies Handbook adopted at the 2010 Special General Meeting is binding on all proprietors in the same manner as the Community Management Statement'. Although there is a strong possibility that this provision would be invalid as a result of inconsistency with community land and Torrens requirements of registration,[85] if the Handbook is being implemented and observed, it will have a significant impact on children's play.[86]

The Handbook provides that '[c]hildren under 13 must be accompanied and supervised by an adult at all times while using any of the community facilities on the estate' and '[s]upervision of children must be carried out by an adult resident of age 21 or older'.[87] As noted above, the community facilities include parks, a basketball court and tennis courts. It is likely that the change was inspired by liability concerns, but as a result, a group of 10 or 12 year olds are not allowed to meet at the basketball court and shoot hoops without supervision. If their 20-year-old sister offered to watch them, this would not be sufficient, nor would their non-resident grandfather. If children want to kick or throw a ball in one of the small parks, this is not permitted at all, nor are they entitled to take up more than half of the green space playing ball games in the larger parks, regardless of who else is around. The children in Liberty Grove are in high-risk groups for childhood obesity, living in high-density urban housing and many having Asian family backgrounds.[88]

The by-laws at Liberty Grove, the townhouse development, and Kingsdene Mews all demonstrate the capacity of by-laws to compromise children's play, in particular active, outdoor play in 'public' spaces. As we saw in the beginning of the chapter, play is not optional for children but is necessary for their physical, social and cognitive development. Children in strata and community schemes need to engage in active play, ball sports, running and jumping to minimise their risk of childhood obesity, for their own benefit and the benefit of the rest of us who will bear the financial burden of obesity-related health care. Children in strata schemes, by virtue of their small private space and their cultural backgrounds, are at a higher risk of childhood obesity, and thus most in need of facilities to engage in active play. By-laws that ban ball games, skateboards and scooters all eliminate important modes of active play. There could be little doubt that the children in Kingsdene Mews, as a result of a dispute supported by the existence of by-laws, reduced their outdoor exercise, possibly preferring computer games and television, in an attempt to placate their neighbour.

Children in strata and community schemes need 'community play', the opportunity to meet friends in public spaces, to hang out, form friendships and networks. Children can develop social networks with their parents, but research is clear that they need time away from parents, time to hang out unsupervised, in spaces away from adult eyes. In a community like Liberty Grove, this opportunity may not come until children reach high school, despite the fact that most parents would be comfortable with children in mid-to-late primary school playing unsupervised in local parks. Most parents know instinctively what research confirms: allowing children to explore public space teaches them to navigate unfamiliar environments and to develop public trust. It is an important part of growing up. The irony is that Liberty Grove, as a relatively closed community, with no through streets,[89] and a 20 km per hour speed limit, is considerably safer than most ordinary residential subdivisions.

Research on play demonstrates that:

> Adults need to ensure that the physical and social environments in which children live are supportive of their play; otherwise their survival, well-being and development may be compromised. This does not necessarily mean providing specific services, although there may be circumstances where this is appropriate. But it does mean avoiding the temptation to dismiss play as frivolous, restrict it through fear for and of children, or control and appropriate it for more instrumental purposes.[90]

Strata and community title legislation allows adults to do exactly that. It empowers private citizens, with no planning or other expertise, to write rules regulating children's play spaces with no reference to children's well-being.

By-laws limiting children's play in lots

To suggest that this final section is solely about well-being would be misleading; as it relates to children's falls from high-rise windows or balconies, it is no exaggeration to say that the matter discussed is one of life and death. Like the preceding section on play generally, this section will include substantial background information, primarily to establish that what is being discussed is an authentic issue in strata living to which the law must be responsive. If the legitimacy of property rights is partially dependent on the acceptability of the externalities they may cause, then we need to actually understand what those externalities might be.

Background to children's falls from high-rise buildings

Children's falls from high-rise buildings are not random, freak events; they have been a well-documented phenomenon in urban areas around the world for the past 40 years. A New York City Department of Health study of child mortality due to falls from building heights found that falls represented 12 per cent of all accidental deaths in the City between 1965 and 1969. Window falls were responsible for 123 deaths.[91] Twenty-three fatal falls occurred in the South Bronx in the summer of 1971 alone, with one precinct patrol man reporting that he had personally picked up nine dead children.[92]

Not all falls are fatal and children suffer a range of serious injuries, including fractures, internal and brain injuries.[93] Owing to their different centre of balance from adults, children tend to topple head first, using their arms and hands to break their fall. In a United States study on 70 children's falls from one to three or more storeys, only 15 children suffered minor soft tissue trauma. Fifty-four per cent suffered head trauma and 33 per cent skeletal trauma.[94] The likelihood of a child requiring life-long medical care and assistance following a fall is significant.

In response to the alarming numbers of falls in the 1960s and 1970s, New York initiated a 'Kids Can't Fly' program. The program relied on intensive public education, including home visits, distribution of multi-lingual posters and flyers at health clinics, schools, preschools, churches, supermarkets and welfare offices, as well as the provision of 16 000 free window guards by the City. But, in 1976, the Health Code was changed to require landlords to install window guards in all premises housing children under 10. Window guards, like fire sprinklers, remedy inherent building risks, saving lives regardless of parenting skills, income or education.[95] Legislatively mandated window guards raised the initial reduction in falls produced by education from 50 per cent to 96 per cent.[96] This demonstrates the importance of public regulation of buildings in which private property rights are split between multiple parties, who, driven by diverse self-interest, may not be able to agree on optimal or even essential action. Thirty years later those provisions remain in place.[97]

In the past decade, trauma doctors at the Children's Hospital, Westmead in Sydney, noticed a sharp spike in the numbers of children they were treating as a result of falls from windows and balconies. In the period 1998–2006, on average, the Hospital treated seven children a year who had fallen from windows. In 2007, the Hospital treated 17. Balcony falls saw a similar change, increasing from an average of six, between 1998–2006, to a high of 11 in 2007.[98] While

Westmead is the largest children hospital in New South Wales, it does not treat all seriously injured children. The Sydney Children's Hospital, Randwick and John Hunter Children's Hospital, Newcastle, are also large paediatric hospitals which treat children who have fallen from windows and balconies. Unfortunately, centralised records on children's falls are not maintained and thus it is difficult to know exactly how many children have fallen from windows and balconies in the Sydney metropolitan region, let alone nationwide, nor is it possible to know the precise division between houses and apartments. However, increases were sufficient for staff at Westmead to be seriously concerned.

The circumstances of Iman Akter Mostafa's fatal fall on 18 October 2009 were typical of the kind of accident confronting doctors.[99] The three-year-old boy[100] was playing in a bedroom of an apartment in Kogarah, in the local government area of Sydney with the fourth largest number of children living in apartments.[101] This is older apartment stock, as noted at the beginning of the chapter, often rented by recent migrant families with young children.

While Iman was a visitor at the flat, the tenants had a three-year-old of their own and were expecting their second child. The two three-year-olds were playing in a tiny bedroom, barely wider than the single bed that was placed under the window. The aluminium sliding window opened roughly a metre and despite the tenant's requests to his landlord, was not fitted with a lock. Like many other children, Iman fell through the flyscreen, 15 metres to the concrete below.

Iman's fatal accident illustrates a number of common features in children's falls. First and most obviously, the apartment buildings from which children fall are high. The further the fall, the greater the likelihood of serious or fatal injury.[102] Second, when a child falls from an apartment building they are unlikely to fall on anything other than concrete in the form of paths or driveways. Children who fall from freestanding homes may fall into garden beds or on grass, which greatly reduces the severity of injuries.[103] Third, owing to the lack of play space around apartment blocks, in particular those dating from the 1960s and 1970s, as well as parents' inability to supervise children when several storeys up, children are more likely to be confined to bedrooms to play.[104] Fourth, children living in apartments overwhelmingly live in those with two or less bedrooms.[105] Limited floor space means that furniture is often placed under windows allowing children to climb up to sill level and fall. Finally, many children, Iman included, fall through flyscreens in windows that can only be locked shut or left entirely open. No ordinary flyscreen can support a child's weight and without a lock or limiting device,[106] parents cannot partially open a window, locking it to an opening through which a child cannot fall.

As a result of the sharp increase in falls like Iman's, in 2009 The Children's Hospital at Westmead formed a *Working Party for the Prevention of Children Falling from Residential Buildings*, made up of experts in medicine, law, planning, building and children's safety.[107] The Hospital also began an ongoing education campaign, *Kids Can't Fly*, to increase public awareness of the risks windows and balconies pose for young children.[108]

The Working Party produced a Report outlining a number of practical measures that can be taken to render windows and balconies in existing buildings safer for children. The most important measure is ensuring that windows can be secured to an opening of not more than 10 centimetres to prevent a baby or toddler's head passing through the opening.[109] Windows can be fitted with locks or limiting devices, many of which are inexpensive and easy to install. Retrofitting balconies is more expensive, but the addition of Perspex panels, or wrapping shade cloth around the balustrade, can eliminate climbable elements and raise balustrade height.[110]

Using the New York experience as a guide, rather than relying solely on community education, the Report recommended legislative reform. This included amendments to the *Strata Schemes Management Act 1996* (NSW) to compel bodies corporate to retrofit windows more than three metres from the ground so that they can be locked to an opening of not more than 10 cm.[111] The rationale for this recommendation was that bodies corporate already have responsibility for windows, which contrary to lay assumption, are invariably drawn on strata plans as common property. Bodies corporate have a statutory obligation to maintain and repair common property, which can include a requirement to add features to common property that were not originally included in the building.[112] Safety is an aspect of repair. In New South Wales, while a body corporate can decide not to repair some parts of common property (for example, an obsolete air-conditioner), it cannot so decide if this would compromise the safety of the building.

In response to the Westmead Report, the Building Codes Board of Australia announced changes to the Building Code of Australia (BCA) to require all windows in early childhood centres and habitable rooms in new residential buildings to be fitted with a secure screen or a device to restrict the window opening to less than 125 mm. The Board 'weighed its responsibility for child safety as paramount'.[113] The changes came into force in early 2013,[114] and are applicable nationwide. However, as the BCA is not retrospective, the provisions only affect new buildings, not existing stock.

As a result, in late 2013, the New South Wales government passed the *Strata Schemes Management Amendment (Child Window Safety Devices) Act 2013* (NSW) implementing the recommendations of the Children's Hospital, Westmead,[115] and requiring bodies corporate to retrofit existing windows so that they do not present a safety risk to children. Screens, locks or other devices can be used to ensure that any opening on a window two metres or more above any external surface can be restricted to an opening of 125 mm.[116] Lot owners are also expressly authorised to install window safety devices,[117] and while tenants will usually be unable to do so as well, an 'interested person' (which would include a lot owner or tenant) can make an application to an adjudicator for an order that the body corporate install window safety devices if it has not done so within two months of being asked.[118]

The window safety amendments effected two important changes which address problems highlighted in this book. First, the amendment recognises the problematic nature of multiple private citizens controlling collectively owned property through majoritarian vote, particularly when votes can legitimately be exercised in self-interest. The majority may act in ways that are harmful to others, including children, even to the point of making decisions that are inimical to their physical safety. The safety of children in strata schemes has clearly been an issue in New South Wales since the 1996 amendments,[119] if not earlier, and yet almost 20 years later, there are still thousands of strata schemes all across Sydney with profoundly unsafe common property windows. Left to their own devices, bodies corporate have not protected children and so the legislature needed to step in.

Second, the amendment recognises that while most windows are the property of the collective, and it is preferable for the body corporate to take responsibility for them, that collective property right cannot trump a parent or carer's right to protect their child from serious injury or death. If the body corporate fails to act, a parent or carer can fix the windows themselves. As Singers says, the 'mix of entitlements and obligations we can legitimately claim [through property law] depends on the kinds of human relationships we can defend, nothing more and nothing less'.[120] A collective property right that prevents a parent from protecting their own child from serious injury or death, inside their own home, is difficult to defend.

Unfortunately, because balconies were not covered by the amendments, the conviction that a body corporate's property rights trump a parent's right to protect their child, remains common in the strata community. Balconies are still vulnerable to the body corporate's by-law-making power.

Balconies and the impact of by-laws on children's safety and play

In the last six years of community discussion about window and balcony safety, it has become evident that there is an unresolved tension between an individual lot owner's right to remedy their balconies and the right of bodies corporate to regulate common property. This tension is a direct result of the expansive nature of the by-law-making power, and the culture that this expansive power has fostered among executive committee members, strata managers and lot owners.

Although only accessible via private lots, as noted above, balconies and windows are invariably drawn on strata plans as common property; this is so bodies corporate can maintain the fabric of the building and ensure a uniform façade. A particularly contentious issue in relation to balconies has been the installation of high-tensile nets to prevent children (or pets) falling. These nets are apparently common in high-rise buildings in South America, and are being imported into Australia.[121]

In media discussions of children's falls, it came to light that some bodies corporate were actively preventing parents from taking measures to protect their children from falls.[122] *At their own expense*, parents had acquired protective nets. They had then asked strata managers or executive committees if they could install the nets and had been denied permission on the grounds that they impaired the aesthetics of the building. While not necessarily representative of the bodies corporate involved, many of the comments in online discussion of nets revealed an extraordinary lingering antipathy to children in apartments.[123]

There seems to be some confusion among strata managers and executive committees about the operation of the law in this area, which is indicative of a wider confusion about the appropriate function and ambit of by-laws. Many residential schemes in Sydney use the New South Wales' model by-laws which prohibit the driving of screws through common property without body corporate consent. However, the model by-law exempts an owner, or person authorised by the owner, installing 'any structure or device to prevent harm to children'.[124] As a result, an owner can install barriers or nets to protect children, *without* body corporate permission. In the Second Reading Speech for the Strata Schemes Management Bill 1996 (NSW), the Minister for Fair Trading, Mrs Lo Po said that:

> The bill takes special account of circumstances where children live in strata schemes … security measures taken to ensure the safety of children, for instance falling from balconies, will be automatically permitted. The body corporate will not be able to prevent an owner or someone authorised by the owner from taking these measures [if] an appropriate standard and in keeping with the appearance of the building.[125]

It seems clear that the model by-law was intended to remove safety devices for children from the general authority of the collective, and to vest it in individual owners, that is, parents.

However, section 108 of the *SSMA* 2015 (like s 65A of the old SSMA 1996) requires a special resolution of the body corporate for additions to common property and under section 136(2) of the *SSMA* 2015, by-laws cannot be inconsistent with the *SSMA*. As a result, it is possible that the model by-law cannot be read to circumscribe the *Act*, as a logical reading of the by-law would suggest; that is, a special resolution of the body corporate is needed for additions to common property generally, but safety devices for children are exempted. In any event, some schemes have never had model by-laws or have altered them as they are empowered to do. Further, under the ordinary by-law-making power it would be possible for a body corporate to write a by-law banning owners from making *any* alteration to balconies, regardless of the need for the alteration. This would clearly be a by-law that relates to the use and enjoyment of common property and would be presumptively valid.

It is possible that a by-law banning parents rendering their balcony safer might be considered 'unconscionable', contrary to new section 139(1), but that bar is quite high. A requirement that by-laws be 'reasonable' might be easier to argue, although that may depend on the strength of a decision-maker's adherence to notions of freedom of contract. However, a provision such as Tasmania's prohibiting by-laws that adversely affect 'the health, welfare or safety of any person'[126] would clearly invalidate any prohibition on rendering balconies safer.

This analysis viscerally demonstrates the central problem highlighted in this book – an insufficient legislative recognition of the inappropriateness of freedom of contract norms in the property context. If private parties are free to create by-laws that prevent their neighbours from safeguarding their children's lives, something is very wrong. Attention is not being paid to the harmful effects of the exercise of property rights and as a result, our property law is not consistent with basic morality or the norms of democratic society.

Conclusion

In the course of writing this chapter, I have tried hard to think of other groups whose well-being might be compromised through by-laws. As is often the case in private communities, imagination can run a poor second to real world practice. This was brought home to me when I found an article about a New York co-operative which created a rule (analogous to a strata by-law) that limited the use of common property to two hours. What was the unreasonable use of common property that prompted this attempted regulation? An elderly lady, who liked to sit in the foyer of the building so that she could greet her neighbours as they came in the door, in particular school children on their way home from school. The lawyer who represented the co-op said, '[s]he was lonely is what she was. She wasn't doing anything destructive, but the lobby wasn't sprawling'.[127] The rule was passed, but fortunately some owners had the humanity to insist on its rescission. Like children, public space is important for elderly people's well-being and ability to socialise, and in strata and community schemes that 'public space' will be private common property.

When we grant private citizens the entitlement to regulate common property, we must be attentive to the externalities caused by the exercise of that right. If its exercise conflicts with values we care about – health, happiness, human dignity, companionship and well-being – then that particular manifestation of the right cannot be justified. How we guard against such unjustifiable property practice, as well as the practices considered in the previous two chapters, will be the subject of the book's conclusion.

Notes

1 Joseph William Singer, 'Democratic Estates: Property Law in a Free and Democratic Society Essay' (2008) 94 *Cornell Law Review* 1009, 1059.
2 Gregory S Alexander *et al.*, 'A Statement of Progressive Property' (2008) 94 *Cornell Law Review* 743, 744.
3 Gregory S Alexander, 'The Social-Obligation Norm in American Property Law' (2008) 94 *Cornell Law Review* 745.
4 Thomas W Merrill and Henry E Smith, 'The Morality of Property' (2007) 48 *William and Mary Law Review* 1849; Carol M Rose, 'The Moral Subject of Property' (2007) 48 *William and Mary Law Review* 1897.
5 Alexander *et al.*, 'A Statement of Progressive Property', above n 2, 744.
6 Parents around the world apparently share this antipathy: W van Vliet, 'Families in Apartment Buildings: "Sad Storey for Children?"' (1983) 15(2) *Environment and Behaviour* 211, 217.

7 Mrs Dwyer, Trades Hall Council, submission to *Royal Commission for the Improvement of the City of Sydney and Its Suburbs* (NSWPP, 1909), cited in Caroline Butler-Bowdon *et al.*, *Homes in the Sky* (Melbourne University Publishing, 2007) 4.
8 Butler-Bowdon *et al.*, above n 7, 5.
9 Ibid. 5.
10 Robert Freestone, 'The Great Lever of Social Reform: The Garden Suburb 1900–30' in Max Kelly (ed.), *Sydney: City of Suburbs* (New South Wales University Press, 1987) 53, 59.
11 Harry Seidler, quoted in Robin Boyd, *The Australian Ugliness* (Penguin, 1980) 255, cited in Butler-Bowdon *et al.*, above n 7, 160.
12 Butler-Bowdon *et al.*, above n 7, 158–66.
13 Photo by Lola Hubner.
14 The residential and commercial development of disused industrial sites on once-working harbours and rivers is a global phenomenon: Brian Hoyle, 'Global and Local Change on the Port-City Waterfront' (2000) 90(3) *Geographical Review* 395.
15 See Jacksons Landing www.jacksonslanding.net.au/.
16 Balmain Cove was the subject of *Community Assoc DP No 270180* v. *Arrow Asset Management Pty Ltd* [2007] NSWSC 527; see Chapter 4, n 55.
17 AGL still own a stratum parcel of Breakfast Point, underneath the Village Green. It is a cubic parcel of earth, capped with concrete to contain contamination.
18 G Woolcock, B Gleeson and B Randolph, 'Urban Research and Child-friendly Cities: A New Australian Outline' (2010) 8(2) *Children's Geographies* 183.
19 M Wulff, E Healy and M Reynolds, 'Why Don't Small Households Live in Small Dwellings?' (2004) 12(1) *People and Place* 56–71; J Yates, 'The Rhetoric and Reality of Housing Choice: The Role of Urban Consolidation' (2001) 19(4) *Urban Policy and Research* 491–527.
20 B Judd, D Olsberg, J Quinn, L Groenhart and O Demirbilek, *Dwelling, Land and Neighbourhood Use by Older Home Owners, AHURI Final Report* (Australian Housing and Urban Research Institute, 2010).
21 H Easthope and A Tice, 'Children in Apartments: Implications for the Compact City' 2011 29(4) *Urban Policy and Research*, 431.
22 In 2011, 28.0 per cent of all households living in a flat, unit or apartment were family households with children while 55.5 per cent of households in other dwellings (including detached houses and townhouses) were family households with children: L Troy, H Easthope, B Randolph and S Pinnegar, *Renewing the Compact City: Interim Report* (City Futures Research Centre, UNSW, 2015) 6.
23 B Randolph, *Children in the Compact City: Fairfield as a Suburban Case Study* (City Futures Research Centre, UNSW, 2006). There are some families living in high-rise public housing in inner Sydney, and more in inner Melbourne, where Housing Victoria built 45 public housing towers in the 1960s: Peter McDonald and Helen Brownlee, 'High-rise Parenting: Raising Children in Melbourne's High-rise Estates' in *Parenting in the 90s*, Family Matters No. 36, Australian Institute of Family Studies, 1993. However, as public housing is not regulated by strata legislation, these apartments and families are beyond the scope of this book.
24 H Easthope and A Tice, 2011, n 21; C Whitzman and D Mizrachi, 'Creating Child-friendly High-rise Environments: Beyond Wastelands and Glasshouses' (2012) 30(3) *Urban Policy and Research*, 233–49.
25 Cathy Sherry and Hazel Easthope, 'Under-supply of Schooling in the Gentrified and Regenerated Inner City' (2016) 56 *Cities*, 16–23.
26 Ibid.
27 Henrietta Cook and Bianca Hall, 'Wait Goes On For Docklands School As Government Begins Another Study' *The Age*, 25 June 2015, www.theage.com.au/victoria/wait-goes-on-for-docklands-school-as-government-begins-another-study-20150624-ghws7m.html#ixzz3zZWZN8gI.

28 Bingara Gorge Online, *Wilton Public School Opens at Bingara Gorge!* www.bingaragorgeonline.com.au/Wilton-Public-School-at-Bingara-Gorge-Now-Open-/default.aspx.

29 Marissa Calligeros, 'Gold Coast: From Glitter Strip to Nappy Valley', 13 August 2013, Brisbane Times, www.brisbanetimes.com.au/queensland/gold-coast-from-glitter-strip-to-nappy-valley-20130812–2rs4e.html.

30 Josie Gleave and Issy Cole-Hamilton, 'A World Without Play: A Literature Review' (Play England, 2012).

31 James F Christie and E P Johnsen, 'The Role of Play in Social-Intellectual Development' (1983) 53 *Review of Educational Research* 93; Stuart Lester and Wendy Russell, 'Children's Right to Play: An Examination of the Importance of Play in the Lives of Children Worldwide' (Working Paper No. 57, Bernard van Leer Foundation, December 2010).

32 Gleave and Cole-Hamilton, above n 35, 6.

33 See Randolph, above n 26.

34 Louise Hardy, 'NSW Schools Physical Activity and Nutrition Survey: Executive Summary' (University of Sydney, 2010).

35 J J Reilly and J Kelly, 'Long-Term Impact of Overweight and Obesity in Childhood and Adolescence on Morbidity and Premature Mortality in Adulthood: Systemic Review' (2011) 35 *International Journal of Obesity* 891.

36 Gleave and Cole-Hamilton, above n 35, 9.

37 Ibid. 11.

38 Ibid. 12.

39 Ibid. 12.

40 Lester and Russell, above n 36, 30–2.

41 Lester and Russell, above n 36, 50 (emphasis added).

42 Gleave and Cole-Hamilton, above n 35, 14.

43 Ibid. 15.

44 Lester and Russell, above n 36, 44–5.

45 Gleave and Cole-Hamilton, above n 35, 17.

46 Ibid. 21.

47 Growing Up in Australia, *Growing Up in Australia: The Longitudinal Study of Australian Children* www.growingupinaustralia.gov.au/.

48 Benjamin Edwards and Leah M Bromfield, 'Neighborhood Influences on Young Children's Conduct Problems and Pro-Social Behavior: Evidence from an Australian National Sample' (2009) 31 *Children and Youth Services Review* 317.

49 Ibid. 322–3.

50 See Chapter 5, n 106.

51 Robert H Nelson, *Private Neighbourhoods and the Transformation of Local Government* (Urban Institute Press, 2005); Fred Foldvary, 'Proprietary Communities and Homeowner Associations' in David T Beito, Peter Gordon and Alexander Tabarrok (eds), *The Voluntary City: Choice, Community, and Civil Society* (University of Michigan Press, 2002) 258; Robert C Ellickson, 'Cities and Homeowners Associations' (1981) 130 *University of Pennsylvania Law Review* 1519.

52 Robin Goodman and Kathy Douglas, 'Life in a Master Planned Estate – Community and Lifestyle or Conflict and Liability?' (2010) 28(4) *Urban Policy and Research* 451.

53 Liberty Grove Management Statement DP270137, amended by-law 4.1 and 'Liberty Grove Community Policies Handbook' (12 May 2011) https://docs.google.com/viewer?a=v&pid=sites&srcid=bGliZXJ0eWdyb3ZlLm5ldC5hdXxob21lfGd4OjdjOThmYzhmYjgxZTljODg. However, this provision cannot bind members of the general public who are entitled to use the roads of the estate.

54 Asmara M Tekle, 'Safe: Restrictive Covenants and the Next Wave of Sex Offender Legislation' (2009) 62 *Southern Methodist Law Review* 1817.

55 Gleave and Cole-Hamilton, above n 35, 9.

56 Philippa Williams and Barbara Pocock, 'Building "Community" for Different Stages of Life: Physical and Social Infrastructure in Master Planned Communities' (2010) 13(1) *Community, Work and Family* 71.

57 *SSMA 1996* s 62; *Community Land Management Act 1989 (NSW)* sch 1 cl 4.

58 Therese Kenna and Deborah Stevenson, 'Negotiating Community Title: Residents' Lived Experiences of Private Governance Arrangements in a Master Planned Estate' (2010) 28(4) *Urban Policy and Research* 435, 444; City Futures, 'Common Ground: The Costs and Provision of Community Infrastructure in Community Title Schemes in NSW' (UNSW, 2008) 16.

59 *Houghton* v. *Immer (No 155) Pty Ltd* (1997) 44 NSWLR 46, 52–3.

60 Email from Carolyn Chudleigh, Partner, Holding Redlich, to Cathy Sherry, 21 May 2013. Ms Chudleigh created the legal structure for a number of significant community title developments in Sydney, including Breakfast Point, The New Rouse Hill and Bingara Gorge.

61 E.g. *Ridis* v. *Strata Plan 10308* (2005) 63 NSWLR 449; *Morgan* v. *Owners Strata Plan 13937* [2006] NSWSC 1019. See Cathy Sherry, 'Kids Can't Fly: The Legal Issues in Children's Falls from High-Rise Buildings' (2013) 2(1) *Property Law Review* 22.

62 When brothers Andrew and James Kennedy fell into an ice covered lake on HOA property, a jury awarded $30.7 million against the HOA. The boys had gone to help Andrew's twin, Christopher, who had walked on the ice, causing it to crack. Andrew drowned, James suffered permanent brain damage and Christopher survived, but suffered psychological trauma. It was alleged that the HOA had not restricted access to the area near an over-flow crib, and had not posted signs or installed safety equipment: *Wolsieffer* v. *Lakes of the Four Seasons Prop Owners Assn* (Ind Lake Co Cir, No 45C01–0303-CT-00045, 25 February 2011) ('*Wolsieffer*').

63 While Australia is a party to the *Convention on the Rights of the Child* (opened for signature, 20 November 1989, 1557 UNTS 3 (entered into force 2 September 1990)) which has recognised children's rights under Art 31 to 'engage in play and recreational activities appropriate to [their] age', and Australia has agreed to 'encourage the provision of appropriate and equal opportunities for … recreational and leisure activity', the *Convention* creates no legally enforceable rights for individual Australian children.

64 *Owners of 111 The Broadview Landsdale – Survey Strata Plan 38894* v. *Colavecchio* [2004] WASTR 15 ('*Colavecchio*').

65 In Western Australia the term 'strata title' is used for both high- and low-rise legislatively created developments under the *Strata Titles Act 1985* (WA). While there are some differences between the Western Australian and New South Wales legislation, the by-law in question could just as easily have been written in New South Wales.

66 Alexander *et al.*, above n 2.

67 *Colavecchio* [2004] WASTR 15, [32.2].

68 Ibid. [32.3].

69 Ibid. [38(a)].

70 Ibid. [39(b)].

71 Ibid. [41(d)].

72 Ibid. [44].

73 Ibid. [46]–[50].

74 In *Grant and Owners of Rosneath Farm – Strata Plan 35452* [2006] WASAT 162, the Tribunal Member held that while the detailed by-laws based on permaculture and pattern language principles restricted use of common property, they did so for all owners. The Tribunal Member held at [104] that

> [t]o the extent that a proprietor accepts a by-law which restricts his rights in the common property, a corresponding benefit is received in the sense that others may not act in the same way so as to affect the amenity of the first proprietor.

75 Photo by Lola Hubner.
76 Photo by Lola Hubner.
77 City Futures, 'Common Ground', above n 58, 8.
78 Bill Randolph, Darren Holloway and Kristian Ruming, 'Social Outcomes of Residential Development, Sydney Olympic Park Stage 1: Local Area Analysis' (City Futures, 2005) 27.
79 Ibid. 30.
80 Profile.id, *Liberty Grove: Service Aged Groups* http://profile.id.com.au/canada-bay/service-age-groups?WebID=180.
81 Profile.id, *Liberty Grove: Households with Children* http://profile.id.com.au/canada-bay/households-with-children?WebID=180.
82 City Futures, 'Common Ground', above n 58.
83 Canada Bay Council had envisaged that the large parks would be used for school and junior sporting competitions. This is still possible, but the ultimate decision has been ceded to a private body corporate: ibid. 24. As the Council requires the private community to pay for the space and makes no contribution to its upkeep, this is not necessarily inappropriate.
84 Management Statement, DP270137, cls 4.2, 6.2(c)(2).
85 The practice of using unregistered 'handbooks' or architectural guidelines was probably copied from the United States where by-laws are often unregistered. Handbooks or guidelines can be altered more easily than by-laws, precisely because they are not registered. However, strata and community title Acts prohibit by-laws that are inconsistent with the parent Act and parent Acts require by-laws to be registered to be effective. An attempt to create a binding rule without registration is arguably inconsistent with the requirement of registration. Avoidance of registration is also inconsistent with the Torrens principle of indefeasibility: that landowners can only be bound by interests recorded on the Torrens register. This is not a minor technicality, but a fundamental part of our land system, designed to eradicate the uncertainty and economic risks that flowed from purchasers being bound by unknown or unknowable prior interests in land. Unregistered community 'handbooks' and architectural guidelines constitute restrictions and obligations on land, and if people have not been made aware of these prior to purchase, their title is undermined and frustration will inevitably result. This reasoning is consistent with the High Court's decision in *Westfield Management Ltd* v. *Perpetual Trustee Co Ltd* (2007) 233 CLR 528 which held that it was impermissible to require purchasers to look beyond the register to determine easement rights.
86 The phenomenon of registered but arguably invalid by-laws is not uncommon, just as there are registered but invalid easements and covenants, e.g. *Clos Farming Estates* v. *Easton* [2002] NSWCA 389. Australian land registries do not check by-laws, easements or covenants for validity prior to registration. Registration does not cure invalidity when an easement or covenant does not comply with the common law, or a by-law was created beyond power.
87 'Liberty Grove Community Policies Handbook', above n 63.
88 Reilly and Kelly, above n 40.
89 Liberty Grove is not gated, but it is enclosed between Homebush Bay Drive, a major freeway, and the Northern rail line.
90 Lester and Russell, above n 36, 52.
91 L Bergner, S Mayer and D Harris, 'Falls From Heights: A Childhood Epidemic in an Urban Age' (1971) 61(1) *American Journal of Public Health* 90, 92.

92 C Spiegel and F Lindaman, 'Children Can't Fly: A Program to Prevent Childhood Morbidity and Mortality from Window Falls' (1977) 67(12) *American Journal of Public Health* 1143.

93 M Smith, J Burrington and A Woolf, 'Injuries in Children Sustained in Free Falls: An Analysis of 66 Cases' (1975) 15(11) *Journal of Trauma* 987.

94 C Musemeche *et al.*, 'Pediatric Falls from Heights' (1991) 31(10) *Journal of Trauma* 1347, 1348.

95 E Shenassa, A Stubbendick and M Brown, 'Social Disparities in Housing and Related Pediatric Injury: A Multilevel Study' (2004) 94(4) *American Journal of Public Health* 633.

96 American Academy of Pediatrics, Committee on Injury and Poison Prevention, 'Falls from Heights: Windows, Roofs, and Balconies' (2001) 107(5) *Pediatrics* 1188, 1189.

97 New York City Health Code, §131.15. Window guards were only mandatory for landlords to install, not owner-occupiers, but in 2011, the law was amended to require owners of condominiums to install window guards on their personally occupied, as well as tenanted, apartments: Local Law of the City of New York, No 57, 2011.

98 The Children's Hospital at Westmead, *Working Party for the Prevention of Children Falling from Residential Buildings Outcomes Report* (February 2011) 10.

99 Arjun Ramachandran, 'Man Describes Horrific Moment when his Friends' Three-Year-Old Son Fell from Sydney Unit Window' *Sydney Morning Herald* (online), 19 October 2009 www.smh.com.au/national/man-describes-horrific-moment-when-his-friends-threeyearold-son-fell-from-sydney-unit-window-20091019-h3mr.html.

100 Most studies show that boys outnumber girls roughly 2 : 1 sustaining injuries from falls: Monash University Accident Research Centre, *The Relationship Between Slips, Trips and Falls and the Design and Construction of Buildings* (Report No 281, 2008) 33.

101 Randolph, above n 26, 11.

102 Monash University Accident Research Centre, above n 124, 34; Smith, Burrington and Woolf, above n 118; Musemeche *et al.*, above n 118.

103 Monash University Accident Research Centre, above n 124, 34.

104 Randolph, above n 26, 25. International research confirms that 'easy parental supervision and easy access to the outdoor environment' are the two principle obstacles to children's play when they live in apartments: van Vliet, above n 6, 221. Bizarrely, a 1967 report on families living in high-rise public housing in Melbourne criticised parents who were reluctant to let their 3–5 year olds play outside unsupervised: Peter McDonald and Helen Brownlee, 'High-rise parenting: raising children in Melbourne's high-rise estates' in *Parenting in the 90s*, Family Matters No. 36, Australian Institute of Family Studies, 1993.

105 Some 76.2 per cent of children living in apartments in Sydney live in those with two or less bedrooms: Randolph, above n 26, 17.

106 A window-limiting device is a mechanism that prevents a window opening further than a defined aperture.

107 The Children's Hospital at Westmead, above n 98. I have been on the Working Party since its inception and wrote the legal recommendations in this Report.

108 Kids Health: The Children's Hospital at Westmead, *Kids Can't Fly: Window and Balcony Safety for Children* 2011 http://kidshealth.chw.edu.au/projects/falls-windows-and-balconies.

109 This conforms to overseas practice and policy. In its policy statement on paediatric falls, the American Academy of Pediatrics' Committee on Injury and Poison Prevention recommended the installation of 'locks on windows to prevent sliding windows not intended for egress from opening more than 4 in', which is roughly 10 cm, as well as the installation of operable window guards where fire regulations permit:

American Academy of Pediatrics Committee on Injury and Poison Prevention, above n 96, reaffirmed on 1 September 2007.

110 The Children's Hospital at Westmead, above n 98, 32–43.

111 Ibid. 49–53. I wrote these suggested amendments to the *SSMA*, partly informed by research done for this book.

112 *Proprietors Strata Plan No 6522* v. *Furney* [1976] 1 NSWLR 412; *Lin* v. *Owners – Strata Plan No 50276* (2004) 11 BPR 21,463. In relation to the obligation to add to or update common property, now see *Owners Strata Plan 50276* v. *Thoo* [2013] NSWCA 270 ('*Thoo*'); it should be noted that *Thoo* concerned an $800 000 proposed update of a ventilation system, not a minor updating or addition to common property.

113 Letter from Ivan Donaldson, General Manager, Australian Building Codes Board, to Stakeholders, 22 November 2011.

114 National Construction Code, BCA2013, Vol 1, D2 24.

115 NSW Fair Trading, 'Children and Window Safety: Consultation Paper' (13 March 2013).

116 *SSMA 1996* s 64A. See also *Strata Schemes Management Regulation 2010* (NSW) regs 31–2.

117 *SSMA 1996* s 64A(3).

118 *SSMA 1996* s 140A.

119 See Parliament of New South Wales, *Parliamentary Debates*, Legislative Assembly, 13 November 1996, 5922 (Lo Po).

120 Joseph William Singer, *Entitlement: The Paradoxes of Property* (Yale University Press, 2000) 215–16.

121 Safety Nets for Life www.safetynetsforlife.com.au/.

122 Kelsey Munro, 'Parents Denied Safety Nets to Protect Kids in Apartments', *Sydney Morning Herald* (online), 24 January 2012 http://smh.domain.com.au/real-estate-news/parents-denied-safety-nets-to-protect-children-in-apartments-20120123–1qe3r.html; Flat Chat Forum, 'Safety Nets on Windows or Balconies – Should They be Allowed?' www.flat-chat.com.au/forum/common-property/safety-nets-in-windows-or-balconies-to-protect-children-from-falls-should-they-be-allowed.

123 Comments included the following:

> Why should it be the government that has to look after children? Why can't parents do it and find a safe place for their family to live? Mandatory netting is the most patently absurd measure I can think of. It removes all parental responsibility, it would be an eyesore on an already ugly city and you'd have older kids using them as play apparatus. i.e. You'd just be seeing dead 12 year-olds instead of dead toddlers.

and

> You idiot !!!!!!!!! What should be required by law is that, just for once, just for a few seconds, parents take some responsibility for their damn kids. Make your balconies safe so that your idiot children CAN'T climb over the edge. NOT MY PROBLEM.
>
> Munro, above n 122

These comments were made despite the fact that the article made it clear that parents were paying for the safety nets themselves.

124 Model by-laws are found in *Strata Schemes Management Regulation 2010* (NSW) sch 2(5)(3)(c). The latter regulation is the newest form of the model by-law and includes a requirement that additions must not compromise fire safety. Updated model by-laws only apply to existing schemes if they are specifically adopted.

125 New South Wales, *Parliamentary Debates*, Legislative Assembly, 13 November 1996, 5922 (Lo Po).

126 *Strata Titles Act 1998* (Tas) s 91(3)(c).
127 Joanne Kaufman, 'Because the Board Says So', *New York Times* (online), 24 May 2013 www.nytimes.com/2013/05/26/realestate/quirky-building-rules-because-the-board-says-so.html?pagewanted=1&_r=0&adxnnl=1&adxnnlx=1369784026-g Ts8m MYCyHPlbj6vMdeUA.

Conclusion

This book began with an historical, social and legal explanation of strata and community title. Chapter 1 recounted the resistance to high-rise development in Australia, as well as the growing pressure in the post-war period to accept apartments as homes and facilitate their development with legislation. Legislation was necessary as a result of property law's refusal to allow positive obligations to attach to freehold land, preventing the enduring maintenance obligations necessary in a collectively owned building. While company title presented a partial solution to this problem, owing to limitations on transfer and leasing in articles of association, banks were not prepared to lend money for the purchase of a company title apartment. The *Conveyancing (Strata Titles) Act 1961* (NSW), followed quickly by other states' Acts, facilitated the vertical subdivision of land into registered freehold titles, satisfying banks' concerns, and imposing positive obligations of maintenance in the form of levies, overcoming the limitations of the common law.

In addition, the legislation gave bodies corporate the power to write by-laws empowering private citizens to create rules for their neighbours with reference to both collectively owned common property and privately owned lots. Over the past 50 years, the by-law-making power has remained consistently broad in all states, generally requiring by-laws to relate 'the control, management, administration, use or enjoyment' of lots or common property. Few by-laws fail to meet that description. With the passage of the Queensland *Group Titles Act 1973*, and its subsequent counterparts in other states, this power was extended to horizontal subdivisions, allowing developers to copy the kinds of privately regulated communities that had been built in the United States throughout the twentieth century. As the number and size of strata and community title developments have grown, so too have the number of citizens both invested with and subject to the by-law-making power. Despite this very significant change in private property law, the by-law-making power has attracted little judicial or academic scrutiny. This book set out to fill some of that gap.

Strata and community title are a subset of property law and as a result, an effective analysis requires a theoretical framework for property. The theoretical framework employed in the book was established in Chapter 2, which argued that land has a significance that other forms of property lack: everything we do, everything we value, our relationships, work and recreation, must occur on some piece of land which we are legally entitled to occupy.[1] Further, we can never take land with us when we move on or die; it remains to be used, of necessity, by those who come after us. Land law, as distinct from more generic property law, must, (and typically does), account for this special role of land.

Chapter 2 argued that land law, like all property law, is a contested concept. We must argue for what is in and what is out of the sphere of protected property rights and this is invariably and *unavoidably* done by reference to basic morality. There is 'no baseline free from judgment',[2] and thus '[n]o arbitration among the different available conceptions of property is possible without some normative apparatus'.[3] Ultimately, property law should amount to 'justificatory practice'.[4] The normative apparatus advocated in the book was democracy; that is, property entitlements can be justified so long as they accord with the values of a modern, pluralist, democratic community; values which include economic liberty and personal autonomy, mobility, widespread distribution of property, and fair social relations.[5]

By using 'boundary rules', the *numerus clausus* principle or the estates system, property law has already done some of the work of determining justifiable practice for us. Unlike contract law, which allows parties to negotiate almost any entitlement they please, property law uses doctrine to prefigure a number of acceptable boxes into which property rights must fall. Those boxes were largely developed in the period in which English society moved from feudalism to democracy,[6] and not surprisingly, most of those boxes promote democratic norms, both economic and personal. In particular, by imposing a range of prohibitions on individually created entitlements, they promote the free alienability of land and unfettered use by current owners.

Modern property law does not just reflect modern, pluralist democracy; it was, and remains, integral to its creation. This is because property entitlements multiplied thousands of times, through time and across a physical landscape, create a political and social system. This was demonstrated in the democratic aberration of racially restrictive covenants – entitlements between individual vendors and purchasers that multiplied across the United States' suburban landscape to create a system of residential apartheid, which, because of the permanence of land and rights attached to it, endures to this day.

Racially restrictive covenants demonstrate that while the core of property law might be democratic, property law can still go awry and we must be vigilant in this regard. Within 'boundary rules' parties have some freedom to tailor rights to their own needs and this tailoring can produce externalities to which property must be attentive. A system of property that ignores the harmful effects of the exercise of particular property rights by some citizens on others cannot claim to be democratic. The process of attending to these externalities is necessarily complex and messy. As Singer argues, while the *numerus clausus* principle or the estates system might provide us with some clear answers about legitimate property entitlements, they will not provide all the answers.[7] On the edges of 'boundary rules', we must do the work ourselves through a process of 'reasoned deliberation'[8] with reference to 'the full range of human values we care about'.[9]

Chapter 3 argued that the extraordinarily complex 'quagmire' of United States servitude law was an example of this reasoned deliberation. In an anomalous nineteenth-century breach of the *numerus clausus* principle, United States law allowed contractual agreements between individual citizens to transform into property rights and run with the fee.[10] This meant that future owners would be burdened by positive obligations agreed by predecessors in title, which unlike lease covenants would not expire with time. Courts recognised that even with notice, these interests could impose unacceptable externalities on successors in title, as well as on the community at large. Positive covenants, along with their restrictive cousins, adopted after the English decision in *Tulk v. Moxhay*, had the potential to undermine the democratic gains made by property law over centuries: by allowing freedom of contract to invade property law, they could limit the current owner's economic and social freedom to use land, 'creating modern variations of feudal serfdom'.[11]

As a result, United States courts developed a doctrinal requirement that 'real covenants' had to 'touch and concern' the land. Essentially, this was a mini-boundary rule that limited covenants to 'an objective purpose of land planning',[12] allowing courts to invalidate covenants that were harmful to current individual owners and the wider community. We saw that, in attempting to modernise the 'touch and concern' doctrine, the *Restatement (Third) of Property: Servitudes* (2000) did not intend:

> to remove courts from their historic role of safeguarding the public interest in maintaining the social utility of land resources. By reformulating the inquiry to ask the question directly – whether the servitude in issue poses such a threat to the public welfare that the rights or obligations it creates should not be allowed to run with land – the law will encourage clearer identification of the issues and clearer explanations of the reasons why servitudes may not be used to implement particular arrangements than was ever possible using the 'touch and concern' doctrine.[13]

The *Restatement* invalidates servitudes that are illegal, unconstitutional, unconscionable, against public policy or an unreasonable restraint on trade or alienation. In addition, the decision in *Hidden Harbour Estates, Inc* v. *Norman*,[14] replicated in the *Restatement*, invalidates rules of an association that are 'unreasonable'. None of these terms can be predefined with any certainty but must be assessed in the context of a servitude's operation on a case-by-case basis. They require courts to consider 'not only choices about initial acquisition and transfer but also the proper moral and legal response to exercises of rights as they happen over time'.[15] They maintain the 'reasoned deliberation' that is a necessary part of determining legitimate property entitlements.

In addition to considering the technical detail of servitude law, Chapter 3 examined the way in which United States academics and courts have recognised the role that servitudes have played in creating substantial swathes of American cities and suburbia. In doing so, they are acknowledging that property law 'is part of the way we define a legitimate social order'[16] and that '[e]very legal right should be understood … by reference to … the shape and character of the social relationships engendered by those rights and powers'.[17] While not in agreement with one another, Ellickson[18] and Alexander[19] both recognised that property rules create our community: Ellickson wanted to see them used to create a liberal society in which people are free to form associations governed by the association's internal values, while Alexander argued that 'groups must be held to an obligation more demanding than mere compliance with their own internal aspirations' and that '[t]hrough the dialogic process of adjudication, courts can serve as a bridge between communities and society, preventing the former from lapsing into a solipsistic perspective that would pervert the ideal of community'.[20] Chapter 3 noted that while Ellickson and Alexander diverge on the point at which they would expect courts to adjudicate on community rules – that is, the line in the sand beyond which small groups are not permitted to step – both agree that *at some point*, courts must act and draw that line.

Chapter 3 looked at the work of other commentators who brought together these two strands of scholarship – the technical property law and the political theory – examining the real world operation of HOAs.[21] McKenzie's research demonstrated that the exponential rise of HOAs in the United States was not simply a result of consumer demand for private communities but rather the result of complex forces of land economics and trends towards privatisation. McKenzie argued that stakeholders did not anticipate the problems associated with giving private citizens responsibility for substantial infrastructure, as well as governing power in relation to their neighbours.

All of the United States research revealed that the legal ability to attach negative and positive obligations to freehold land throws up questions about legitimate property entitlements, on both an individual and a community level, which cannot be answered easily. Considerable attention was paid to United States research because strata and community title by-laws and management statements are the legislative equivalent of United States positive and restrictive covenants and association rules. We now have the corresponding law, and are rapidly acquiring corresponding developments. Rather than re-invent the wheel, we should pay heed to United States judicial deliberations and academic debate.

Chapter 4 returned the discussion to Australian law and demonstrated, that unlike United States courts which have exercised caution in relation to private regulation of land, higher courts, particularly in New South Wales, have consistently confirmed a broad interpretation of the by-law-making power. While Young JA's judgment in *Casuarina* contained a promising analogy between by-laws and easements, as well as a reference to the need for by-laws to be 'reasonable', ultimately, the judiciary has declined to craft any limitation or 'boundary rule' on the kinds of rights private citizens can dream up and attach to freehold land via by-laws. This is both anomalous and risky because, as Heller says:

> The pervasive presence of boundary rules [in property law] challenges legal and economic theories that suggest unstructured fluidity to private property.

> Instead, the overwhelming evidence suggests that the notion of an open-ended bundle of property rights is wrong.[22]

There is already evidence in Australia that Heller's warning is correct. The 'open-ended bundle of property rights' created by the almost unlimited by-law-making power has spawned developments in which initial owners (developers) have been able to use by-laws to create long-term, financially burdensome contracts, which compel owners to pay for services that they do not want, either because they do not need them or they do not represent value for money. Some case law now supports the use of by-laws to extract sums of money from owners, which, unlike the levying provisions, do not directly relate to the genuine costs of maintaining buildings in accordance with the initial rationale for the strata legislation.[23] While these arrangements will have been disclosed to owners in contracts of sale, notice has never operated as a cure-all in property law. This was arguably the error of Lord Cottenham in *Tulk* and of Epstein.[24] The collective judicial retreat from *Tulk*, and the consistent invalidation of servitudes by United States judges in accordance with the 'touch and concern' doctrine, suggests that in the opinion of the judiciary, in addition to most academic commentators, notice is not sufficient to bind successors in title. Purchasers, like covenant or by-law creators, cannot have perfect foresight in relation to the ultimate effect of these burdens on themselves, let alone the land market or community as a whole.

Consistent with the traditional concerns of democratic property law, Chapter 5 expanded the analysis beyond economics to consider the social or personal impact of by-laws. The chapter noted that all Australian states have sensibly prohibited a range of by-laws that are clearly harmful, but these prohibitions may not be sufficient. In particular, the unanswered question of whether discrimination legislation applies to bodies corporate, highlights the problematic blurring of the public/private divide created by strata and community schemes.

Chapter 5 looked at the concerning absence of the fundamental democratic principle of 'negative liberty' in strata and community legislation, using the example of by-laws that create bans or restrictions on pets. We do not normally expect to control what our neighbours do inside their own homes, 'the scene of [their] history and future, [their] life and growth',[25] and to grant people this right of control, *without adequate justification*, creates undemocratic property entitlements. Multiplied thousands of times across our urban landscape and through time, these entitlements then have the potential to undermine the fabric of our democratic community. The chapter then looked at the more contentious issue of by-laws that regulate behaviour that falls on the border of self-regarding and other regarding acts, and noted that these by-laws have the capacity to foster communities of intolerance, where activity that we would ordinarily be expected to accept, like a shade of house paint or blinds we dislike, or the sight of a commercial vehicle, can be controlled. It was argued that we must not be blind to the class, personal autonomy or free speech implications of these privately created rules, or to the consequences of fostering a conviction in citizens that they have a legitimate interest in these issues and a justifiable power to regulate them.

Finally, in Chapter 6, consistent with our need to account for 'the full range of human values we care about' when determining property entitlements,[26] the analysis was broadened again to look at the effect that by-laws can have on 'well-being'. The chapter used a case study on children's safety and play to illustrate the way by-laws can compromise children's ability to engage in activities that are fundamental to their development, and in worst case scenarios, compromise their safety, presenting very real risks to their lives. The chapter argued that allowing private citizens to compromise others' well-being in this way, through by-laws, amounted to unjustifiable property practice.

What then is the solution to these problems that have arisen in Australian strata and community title law and practice? The solution can be found back where the book began with property theory and the United States law of servitudes. First, we must accept that the content of property law is not inevitable; it is contested, it is moral and it must amount to justificatory practice judged with reference to a wide range of values. Second, the United States has had over a century of experience with private communities created through property law, which can help us to determine justificatory practice in our own private communities. United States judges and academics have consistently argued that while the ability to attach positive and negative obligations to freehold land should be accepted in property law, this must be done with an awareness of the potential of servitudes to create 'modern variations of feudal serfdom'.[27] Obligations on freehold land, particularly those that impose monetary burdens and which control owners' behaviour, can compromise the hard-won gains of property law in the post-feudal period, and it is the proper role of legislatures *and courts* to prevent this.

Our courts have always done this with the application of property 'boundary rules' to ordinary land and there is no valid reason for strata and community title to be excepted. Strata and community title are simply a form of land subdivision and irrefutably a subset of property law. Using the United States case law as our guide, the power to write by-laws should be expressly limited by a judicial or legislative requirement of reasonableness, judged with reference to the interests and well-being of all residents, not just owners. Queensland and Tasmania already have express legislative requirements of reasonableness, and in New South Wales, Young JA has stated *in obiter* comments that by-laws are required to be 'reasonable'.[28] New South Wales now has a legislative provision prohibiting by-laws that are harsh, unconscionable or oppressive, but this is an unnecessarily high bar. It is difficult to see why any community should be entitled to create rules that, while not harsh, unconscionable or oppressive, are objectively unreasonable.

However, a requirement of reasonableness will have limited effect if its function is not properly appreciated. Although Queensland adjudicators have long had the power to invalidate by-laws that are unreasonable, some still felt compelled to uphold blanket pet bans, even though continued litigation (and subsequent adjudicator decisions) indicated that many people believed that blanket bans were deeply *unreasonable*.[29] If the cultural and theoretical framework within which by-laws are being evaluated is freedom of contract, almost any regulation that seems to be freely agreed, either by special majority vote or by voluntary purchase, will seem 'reasonable'. If, consistent with the reasoning of Ellickson and Epstein, we believe that it is acceptable for private communities to regulate their own land and lives with minimal public supervision, whatever communities freely agree will in turn be acceptable. If, however, we demand more of property law, and insist that its legitimacy also be judged by reference to its effect on parties *outside* the voluntary transaction (the minority in the scheme, people excluded from the scheme by the by-laws now and in the future, the wider community who inevitably have to interact with people enculturated by strata and community living, etc.), then we may conclude that even majority, voluntarily agreed by-laws can be unreasonable. In short, we must understand the word 'reasonable' within the framework of property law, not contract, and in property law, the ethos 'I/we want, therefore I/we should have', has never been accepted.

A requirement of 'reasonableness' must also be interpreted with reference to the quasi-legislative nature of the by-law-making power, particularly when the power is exercised over a large geographical area and/or population. There must be an acknowledgement that such powers are normally exercised by accountable officials, whose actions are constrained and reviewable through public law. This power in the hands of private citizens is unusual and inherently dangerous. As McKenzie cautions:

> people cannot be trusted with unlimited power because we are naturally selfish creatures and our emotions override our intellect. We are easily convinced that, by an amazing coincidence, the very course of action that suits our own self-interest just happens to be … morally correct and wise.[30]

As a result, the power to write by-laws must be subject to robust judicial review. As Alexander argues:

> No single group within a pluralistic communitarian society can be permitted to arrogate to itself unilateral control over the inside-outside problem. There must be some agency that is responsible for maintaining the framework within which a variety of groups can engage in a dynamic process of formulating and reformulating the good.… Courts are an appropriate institution to maintain that framework.[31]

Unless decision-makers have a clear understanding of the crucial role that judicial review of by-laws is performing, a legislative stipulation of 'reasonableness' will have limited effect. In other words, it is not enough to create the power to review; those invested with that power must appreciate why the power exists and be prepared to use it. The recent Queensland Court of Appeal decision in *Albrecht* v. *Ainsworth*[32] confirmed that when empowered to review a body corporate's actions, a decision-maker can substitute their own objective conclusion on reasonableness based on all of the evidence and does not have to defer to the preferences of the private community, as expressed through voting.

The task of determining reasonableness is not always easy, but it is hardly novel. Literally tens of thousands of Acts and precedents require tribunals and courts to make judgments about reasonableness. It is untenable to argue that the term is too hard to define so we should avoid the question altogether. However, in order to assist judges, lawyers and members of strata and community title schemes, the legislation could provide some guiding principles, based on particular values upon which we might safely suggest we all agree.

The first would be based on negative liberty and would hold that any regulation of activity within private lots is presumptively invalid, unless the activity in question affects other people in some meaningful way. Whether inconsistent blind or house colours unreasonably interfere with others' enjoyment of common property could be determined on a case-by-case basis. A shade of fluorescent pink may well interfere with others' enjoyment of their common master planned estate to an extent that a court may find unreasonable, however a shade of beige that varies from the shade of beige chosen by the architectural committee may not.

Second, by-laws which impose monetary obligations on owners should be prohibited in all states. Monetary obligations on freehold land have been prohibited for centuries, in common and civil law systems. An exception to this general rule was needed to compel maintenance of commonly owned property and for the administration of a community of owners; that is, the *genuine and necessary* costs of collectively owned buildings and land. The exception was created in the levying provisions in the legislation, with appropriately crafted safeguards for the collection and spending of money by private citizens. If money can be collected via the by-law provisions, either directly or through contracts supported by by-laws, there is almost no limit to the monetary obligations developers and subsequent majorities can impose on owners of land.

Chapter 4 gave the example of a by-law to fund rock bands playing on common property every Friday night. Such a by-law relates to the use and enjoyment of common property, and in New South Wales at least, is currently prima facie valid. However, it is also a monetary obligation that would reduce the value of property to all but a small proportion of potential owners and would compel minority owners to pay for a service they never wanted, and could not have reasonably anticipated. It is the proper subject of individual contractual agreements, not a binding interest in land, and its prima facie validity demonstrates the inadequacy of the current legislation. A prohibition on by-laws that effectively impose ongoing monetary obligations on owners may have implications for tourist and resort-style schemes, but extensive litigation and dispute suggests that these are far from functional forms of property.

Third, like the United States, prohibitions should be imposed on by-laws and contracts that are unconscionable or unreasonable restraints on trade. Broad concepts like 'unconscionability' and 'unreasonableness' would allow courts to assess by-laws and positive obligations on land in the form of body corporate contracts, not just at the point of formation, but over time, as all property entitlements should be assessed.[33] As the litigation on management rights demonstrates, attempts to control developer practice via narrow legislative provisions and disclosure requirements have manifestly failed to prevent unjustifiable externalities or to safeguard the rational and efficient use of land.

Fourth, provisions that the legislature considers important or essential (for example, the power of individual lot owners to make additions to common property for the safety of children) must be taken out of the 'model' by-laws. If provisions are important or essential, the collective should not have the power to alter or remove them and thus, by definition, they should not be in by-law sections. This current practice illustrates the confusion over the proper, subordinate place of freedom of contract in strata and community title law, which dates back to the initial *Conveyancing (Strata Titles) Act 1961*.[34] As a subset of property law, freedom of contract should not be the dominant ethos, and ultimately the power of individual owners must be constrained.

Finally, a provision such as Tasmania's prohibition on by-laws that unreasonably 'adversely affect the health, welfare or safety of any person' would prompt both drafters and reviewers of by-laws to consider their effects on people, both inside and outside the scheme. It would require communities like Liberty Grove to think about the effect of a by-law limiting children's freedom of movement and play, instead of allowing the community to pretend that detrimental effects of the by-law do not exist.

A by-law-making power limited in these ways is more complex than existing powers for good reason. As this book has demonstrated, when 'devising new strategies in an old war among generations of land owners … we should not be overly surprised if we wind up in some of the same old trenches'.[35] We can try to rationalise property law to ensure that it is comprehensible and useable for contemporary landowners, but there is no way that we can eradicate the tensions that will always exist between past, current and future owners of the limited resource that is land. There is also no way of escaping the fact that the power to create enduring positive and negative obligations on land amounts to a quasi-legislative power in the hands of private citizens, which must be controlled by the state via the legislation creating the power and via later judicial review. By giving citizens a broad blanket power to write by-laws we have not avoided the messy, complicated questions about what constitutes a legitimate property entitlement in a democratic community (for example, is the power to regulate clothing or pet ownership legitimate?); we have just given those questions a blanket affirmative answer without engaging with their content. As 'property law is part of the way we define a legitimate social order',[36] by failing to engage with those questions we are failing to protect our current, democratic social order. When assessing strata and community title by-laws, like all property law, we must continually ask ourselves, '[i]n which world would we rather live?'[37]

Notes

1 Jeremy Waldron, 'Homelessness and the Issue of Freedom' (1991) 39 *UCLA Law Review* 295.
2 Joseph William Singer, *Entitlement: The Paradoxes of Property* (Yale University Press, 2000) 215.
3 Hanoch Dagan, 'The Limited Autonomy of Private Law' (2008) 56 *American Journal of Comparative Law* 809, 814.
4 Ibid.
5 Joseph William Singer, 'Democratic Estates: Property Law in a Free and Democratic Society Essay' (2008) 94 *Cornell Law Review* 1009.
6 See A W B Simpson, *A History of the Land Law* (Clarendon Press, 2nd edn, 1986).
7 Singer, 'Democratic Estates', above n 5, 1027.
8 Gregory S Alexander *et al.*, 'A Statement of Progressive Property' (2008) 94 *Cornell Law Review* 743, 744.
9 Singer, 'Democratic Estates', above n 5, 1059.
10 Uriel Reichman, 'Toward a Unified Concept of Servitudes' (1981) 55 *Southern California Law Review* 1177; Susan F French, 'Toward a Modern Law of Servitudes: Reweaving the Ancient Strands' (1981) 55 *Southern California Law Review* 1261.
11 Reichman, above n 10, 1233.
12 Ibid.
13 *Restatement (Third) of Property: Servitudes* (2000) §3.2(a).
14 309 So 2d 180 (Fla Dist Ct App, 1975).
15 Singer, *Entitlement*, above n 2, 174.
16 Singer, 'Democratic Estates', above n 5, 1049.
17 Ibid. 1047.
18 Robert C Ellickson, 'Cities and Homeowners Associations' (1981) 130 *University of Pennsylvania Law Review* 1519.
19 Gregory S Alexander, 'Dilemmas of Group Autonomy: Residential Associations and Community' (1989) 75 *Cornell Law Review* 1.
20 Ibid. 60, 55–6.
21 Clayton P Gillette, 'Courts, Covenants, and Communities' (1994) 61 *University of Chicago Law Review* 1375; Gerald Korngold, 'Resolving the Flaws of Residential Servitudes and Owners Associations: For Reformation Not Termination Reply' [1990] *Wisconsin Law Review* 513; Stewart E Sterk, 'Foresight and the Law of Servitudes' (1987) 73 *Cornell Law Review* 956; Stewart E Sterk, 'Freedom from Freedom of Contract: The Enduring Value of Servitude Restrictions' (1984) 70 *Iowa Law Review* 615; James L Winokur, 'The Mixed Blessing of Promissory Servitudes: Toward Optimizing Economic Utility, Individual Liberty, and Personal Identity' [1989] *Wisconsin Law Review* 1.
22 Michael A Heller, 'The Boundaries of Private Property' (1999) 108 *Yale Law Journal* 1163, 1192.
23 *Italian Forum Ltd* v. *Owners – Strata Plan 60919* [2012] NSWSC 895 (White J); cf. *Owners Strata Plan No 60919* v. *Consumer Trader and Tenancy Tribunal* (2012) [2009] NSWSC 1158 (Patten AJ).
24 Richard A Epstein, 'Notice and Freedom of Contract in the Law of Servitudes Comments' (1981) 55 *Southern California Law Review* 1353.
25 Margaret Jane Radin, 'Property and Personhood' (1981–1982) 34(5) *Stanford Law Review* 957, 992.
26 Singer, 'Democratic Estates', above n 5, 1059.
27 Reichman, above n 10, 1233.
28 *Casuarina Rec Club Pty Ltd* v. *Owners – Strata Plan 77971* (2011) 80 NSWLR 711, 722 [89].
29 See Chapter 5, nn 91–4.

30 Evan McKenzie, *Beyond Privatopia: Rethinking Residential Private Government* (Urban Institute Press, 2011) 114.
31 Alexander, 'Dilemmas of Group Autonomy', above n 19, 55–6.
32 *Albrecht v. Ainsworth* [2015] QCA 220.
33 Some might object to this suggestion on the grounds that it will cause uncertainty and monetary loss for management and other contract rights holders, making them less inclined to invest in businesses in strata and community schemes. However, no matter what option we take, there are losers, both monetary and otherwise. Our current law gives certainty and commercial security to developers and management rights holders at the expense of strata owners. This in turn makes people disinclined to purchase strata apartments, particularly for their own home, reducing the value of current owners' investments. The cost of management rights litigation launched by dissatisfied owners would run to millions of dollars, not to mention the non-monetary costs of living and/or being a member of a body corporate or executive committee in a building with an ongoing intractable management dispute. If developers and management rights holders knew that courts would invalidate their by-laws and agreements that were unconscionable or unreasonable, they would have an incentive to ensure that they were fair to ultimate lot owners.
34 See Chapter 1, 26–7.
35 Carol M Rose, 'Servitudes, Security, and Assent: Some Comments on Professors French and Reichman' (1981) 55 *Southern California Law Review* 1403, 1404.
36 Singer, 'Democratic Estates', above n 5, 1049.
37 Singer, *Entitlement*, above n 2, 138.

Bibliography

A Articles/books/reports/other

ABC Radio National, 'Sanctuary', *360documentaries*, 30 September 2012 (Kirsti Melville) www.abc.net.au/radionational/programs/360/sanctuary/4268164.

Alexander, Gregory S, 'Freedom Coercion and the Law of Servitudes' (1987) 73 *Cornell Law Review* 883.

Alexander, Gregory S, 'Dilemmas of Group Autonomy: Residential Associations and Community' (1989) 75 *Cornell Law Review* 1.

Alexander, Gregory S, 'The Publicness of Private Land Use Controls' (1999) 3 *Edinburgh Law Review* 176.

Alexander, Gregory S, 'The Social-Obligation Norm in American Property Law' (2008) 94 *Cornell Law Review* 74.

Alexander, Gregory S, 'Pluralism and Property' (2011) 80 *Fordham Law Review* 101.

Alexander, Gregory S and Eduardo M Peñalver, 'Properties of Community' (2009) 10 *Theoretical Inquiries in Law* 127.

Alexander, Gregory S and Eduardo M Peñalver, *Property and Community* (Oxford University Press, 2010).

Alexander, Gregory S and Eduardo M Peñalver, *An Introduction to Property Theory* (Cambridge University Press, 2012).

Alexander, Gregory S, Eduardo M Peñalver, Joseph William Singer and Laura S Underkuffler, 'A Statement of Progressive Property' (2008) 94 *Cornell Law Review* 743.

Alexander, Larry, 'Illiberalism All the Way Down: Illiberal Groups and Two Conceptions of Liberalism' (2001–2002) 12(2) *Journal of Contemporary Legal Issues* 625.

Alford, Lindsey and Jasmine Sommer, 'Protection of Minority Owners in a Body Corporate' (2005) 11 *Australian Property Law Journal* 141.

American Academy of Pediatrics, Committee on Injury and Poison Prevention, 'Falls from Heights: Windows, Roofs, and Balconies' (2001) 107(5) *Pediatrics* 1188.

American Law Institute. *Restatement (Third) of Property: Servitudes* (2000).

Architectural Standards and Landscape Design Guidelines for Macquarie Links, DP270152, (Land and Property Information, NSW).

Asher, Stephen E, 'Interdistrict Remedies for Segregated Schools' (1979) 79 *Columbia Law Review* 1170.

Austen, Jane, *Pride and Prejudice* (T Egerton, 1813).

Australian Bureau of Statistics, 'Home Owners and Renters' (ABS Year Book Australia, 2012).

Australian Securities and Investment Commission, ASIC Management Rights Schemes, *CO 02/303, 02/304, 02/305*.

Balmain Cove, Management Statement DP270180, (Land and Property Information, NSW).

Batsel, G, 'Florida Condominiums – Developer Abuses and Securities Law Implications Create a Need for a State Regulatory Agency' (1972–1973) 25 *University of Florida Law Review* 350.

Batt, E McD, 'Housing Innovation in Brisbane' (1970) 8(3) *Royal Australian Planning Institute Journal* 80.

Bennett Moses, Lyria and Brendan Edgeworth, 'Taking it Personally: Ebb and Flow in the Torrens System's in Personam Exception to Indefeasibility' (2013) 35(1) *Sydney Law Review* 1.

Bennett Moses, Lyria, Brendan Edgeworth and Cathy Sherry (eds), *Property and Security: Selected Essays* (Lawbook Co, 2010).

Bergner, L, S Mayer and D Harris, 'Falls from Heights: A Childhood Epidemic in an Urban Age' (1971) 61(1) *American Journal of Public Health* 90.

Berlin, Isaiah, 'Two Concepts of Liberty' in Henry Hardy (ed.), *Liberty* (Oxford University Press, 2002).

Bingara Gorge Online, *Wilton Public School Opens at Bingara Gorge!* www.bingaragorgeonline.com.au/Wilton-Public-School-at-Bingara-Gorge-Now-Open-/default.aspx.

Blakely, Edward J and Mary Gail Snyder, *Fortress America: Gated Communities in the United States* (Brookings Institution Press, 1997).

Blandy, Sarah, Ann Dupuis and Jennifer Dixon, *Multi-Owned Housing: Law, Power and Practice* (Ashgate, 2010).

Booth, Philip, 'Speculative Housing and the Land Market in London 1660–1730: Four Case Studies' (1980) 51(4) *Town Planning Review* 379.

Booth, Philip, *Planning by Consent: The Origins and Nature of British Developmental Control* (Routledge, 2003).

Bounds, Michael, 'High-Rise Gentrification: The Redevelopment of Pyrmont Ultimo' (2005) 10 *Urban Design International* 179.

Boyd, Robin, *The Australian Ugliness* (Penguin, 1980).

Breakfast Point, www.breakfastpoint.com.

Breakfast Point Management Statement DP270347, (Land and Property Information, NSW).

Brussack, Robert D, 'Group Homes, Families, and Meaning in the Law of Subdivision Covenants' (1981–1982) 16 *Georgia Law Review* 33.

Buchanan, James M, 'An Economic Theory of Clubs' (1965) 32(125) *Economica* 1.

Bugden, Gary, *Strata Title Management Practice in New South Wales* (CCH Australia 3rd edn, 1981).

Bugden, Gary, 'Management Rights: Developments in Building Management Agreements for Strata Schemes' (2003) 41(3) *Law Society Journal* 42.

Bugden, Gary, *All About Strata and Building Management Statements and Building Management Committees: A Complete Guide to Working With Building Management Committees* (Institute of Strata Management, 2006) www.mystrata.com/doc-store/Building_Management_Statements_Paper_for_ISTM_Nov_06.pdf.

Bureau of Infrastructure, Transport and Regional Economics, *Cities: Population Growth, Job Growth and Commuting Flows in South East Queensland*, Research Report 134, (Australian Government, 2013).

Butler-Bowdon, Caroline and Charles Pickett, *Homes in the Sky: Apartment Living in Australia* (Melbourne University Publishing, 2007).

Butt, Peter, *Land Law* (LBC, 3rd edn, 1996).

Butt, Peter, *Land Law* (Thomson, 5th edn, 2006).

Byron Shire Council, *Multiple Occupancies to Community Title* (February 2009) www.byron.nsw.gov.au/multiple-occupancies-to-community-title.

Calabresi, Guido Melamed and A Douglas, 'Property Rules, Liability Rules, and Inalienability: One View of the Cathedral' (1971) 85 *Harvard Law Review* 108.

Cannadine, David and David Reeder (eds), *Exploring the Urban Past: Essays in Urban History by H J Dyos* (Cambridge University Press, 1982).

Casner, A James, W Barton Leach, Susan F French, Gerald Korngold and Lea VanderVelde, *Cases and Text on Property* (Aspen Publishers, 5th edn, 2004).

Cassidy, Kelly and Chris Guilding, 'Defining an Emerging Tourism Industry Sub-Sector: Who are the Strata Titled Tourism Accommodation Stakeholders?' (2010) 29 *International Journal of Hospitality Management* 421.

Castle, The (Directed by Rob Sitch) Village Road Show, 1997.

Christensen, Sharon and William D Duncan, 'Is it Time for a National Review of the Torrens' System? – The Eccentric Position of Private Restrictive Covenants' (2005) 12 *Australian Property Law Journal* 104.

Christensen, Sharon and William D Duncan, 'Aligning Sustainability and the Torrens Register: Challenges and Recommendations for Reform' (2012) 20 *Australian Property Law Journal* 112.

Christensen, Sharon and Anne Wallace, 'Links Between Physical and Legal Structures of Community Title Schemes and Disputes' (2006) 14 *Australian Property Law Journal* 90.

Christensen, Sharon A, William D Duncan and Amanda P Stickley, 'Evaluating Information Disclosure to Buyers of Real Estate – Useful or Merely Adding to the Confusion and Expense?' (2007) 7(2) *Queensland University of Technology Law and Justice Journal* 148.

Christensen, Sharon A, William D Duncan and Amanda P Stickley, 'Behavioural Biases and Information Disclosure Laws Relating to Residential Property Sales: Narrowing the Gap Between Existing Laws and Calls for Future Reforms' (2009) 9(2) *Queensland University of Technology Law and Justice Journal* 251.

Christie, James F and E P Johnsen, 'The Role of Play in Social-Intellectual Development' (1983) 53 *Review of Educational Research* 93.

Chudleigh, Carolyn, Partner, Holding Redlich: Email to Cathy Sherry, 21 May 2013.

City Futures Research Centre, *Common Ground: The Costs and Provision of Community Infrastructure in Community Title Schemes in NSW* (UNSW, 2008).

City Guide – Coto de Gaza, *The History* www.orangecounty.net/cities/CotoDeCaza.html.

Clarke, D N, 'Commonhold – A Prospect of Promise' (1995) 58(4) *Modern Law Review* 486.

Coase, R H, 'The Problem of Social Cost' (1960) 3 *Law and Economics* 1.

Coase, R H, 'Coase Theorem and the Empty Core: A Comment' (1981) 24 *Journal of Law and Economics* 18.

Cook, Henrietta and Bianca Hall, 'Wait Goes on for Docklands School as Government Begins Another Study' *The Age*, 25 June 2015, www.theage.com.au/victoria/wait-goes-on-for-docklands-school-as-government-begins-another-study-20150624-ghws7m.html#ixzz3zZWZN8gI.

Cowen, D V, 'The South African Sectional Titles Act in Historical Perspective: An Analysis and Evaluation' (1973) 6(1) *Comparative and International Law Journal of Southern Africa* 1.

Cruickshank, Dan and Neil Burton, *Life in the Georgian City* (Viking, 1990).

Dagan, Hanoch, 'The Limited Autonomy of Private Law' (2008) 56 *American Journal of Comparative Law* 809.

Davis, Kingsley, 'The Urbanization of the Human Population' (1965) 213(3) *Scientific American* 40.

Davison, Aidan, 'Stuck in a Cul-de-Sac? Suburban History and Urban Sustainability in Australia' (2006) 24(2) *Urban Policy and Research* 201.

Davison, Graeme, 'The Past and Future of the Australian Suburb' (1993) 31(2) *Australian Planner* 63.

Demsetz, Harold, 'Toward a Theory of Property Rights II: The Competition between Private and Collective Ownership The Evolution of Property Rights' (2002) 31 *Journal of Legal Studies* 65.

Depoorter, Ben W F and Francesco Parisi, 'Fragmentation of Property Rights: A Functional Interpretation of the Law of Servitudes' (Working Paper No 284, John M Olin Center for Studies in Law, Economics and Public Policy, 22 May 2003) http://digitalcommons.law.yale.edu/lepp_papers/284/.

Dilger, Robert, 'Residential Community Associations: Issues, Impacts, and Relevance for Local Government' (1991) 23(1) *State and Local Government Review* 17.

Douglas, Kathy and Rebecca Leshinsky, 'Pre-Action Dispute Resolution under the Owners Corporations Act 2006 (Vic): Teaching Conflict Resolution Strategies' (2012) 20 *Australian Property Law Journal* 224.

Duany, Andres, Elizabeth Plater-Zyberk and Jeff Speck, *Suburban Nation: The Rise of Sprawl and the Decline of the American Dream* (Farrar, Straus, and Giroux, 2000).

Dyos, H J, *Victorian Suburb: A Study of the Growth of Camberwell* (Leicester University Press, 1961).

Dyos, H J, 'The Speculative Builders and Developers of Victorian London' in David Cannadine and David Reeder (eds), *Exploring the Urban Past: Essays in Urban History by H J Dyos* (Cambridge University Press, 1982) 154.

Easthope, Hazel and Andrew Tice, 'Children in Apartments: Implications for the Compact City' (2011) 29(4) *Urban Policy and Research* 431.

Easthope, Hazel, Bill Randolph and Sarah Judd, 'Managing Major Repairs in Residential Strata Developments in NSW' (City Futures Research Centre, 2009).

Easthope, Hazel, Bill Randolph and Sarah Judd, 'Governing the Compact City: The Role and Effectiveness of Strata Management' (City Futures Research Centre, 2012).

Edgeworth, Brendan, 'The Numerus Clausus Principle in Contemporary Australian Property Law' (2006) 32 *Monash University Law Review* 387.

Edgeworth, Brendan, Christopher Rossiter, Margaret Stone and Pamela O'Connor, *Sackville & Neave Australian Property Law* (LexisNexis Butterworths, 9th edn, 2013).

Edwards, Benjamin and Leah M Bromfield, 'Neighborhood Influences on Young Children's Conduct Problems and Pro-Social Behavior: Evidence from an Australian National Sample' (2009) 31 *Children and Youth Services Review* 317.

Ellickson, Robert C, 'Alternatives to Zoning: Covenants, Nuisance Rules, and Fines As Land Use Controls' (1972) 40 *University of Chicago Law Review* 681.

Ellickson, Robert C, 'Cities and Homeowners Associations' (1981) 130 *University of Pennsylvania Law Review* 1519.

Epstein, Richard A, 'Notice and Freedom of Contract in the Law of Servitudes Comments' (1981) 55 *Southern California Law Review* 1353.

Epstein, Richard A, 'Covenants and Constitutions' (1987) 73 *Cornell Law Review* 906.

Epstein, Richard A, *Simple Rules for a Complex World* (Harvard University Press, 1995).

Federal Department of Housing and Urban Development, 'Condominium and Cooperative Study' (Washington, DC, 1975).

Fisher, Warwick, 'The Future of Rural Land Sharing Communities in Far North Coast New South Wales' (2004) 8 *Southern Cross University Law Review* 51.

Flat Chat Forum, 'Safety Nets on Windows or Balconies – Should They be Allowed?' www.flat-chat.com.au/forum/common-property/safety-nets-in-windows-or-balconies-to-protect-children-from-falls-should-they-be-allowed.

Foldvary, Fred, 'Proprietary Communities and Homeowner Associations' in David T Beito, Peter Gordon and Alexander Tabarrok (eds), *The Voluntary City: Choice, Community, and Civil Society* (University of Michigan Press, 2002) 258.

Franzese, Paula, 'Does it Take a Village? Privatization, Patterns of Restrictiveness and the Demise of Community' (2002) 47 *Villanova Law Review* 553.

Freestone, Robert, 'The Great Lever of Social Reform: The Garden Suburb 1900–30' in Max Kelly (ed.), *Sydney: City of Suburbs* (New South Wales University Press, 1987).

Freestone, Robert, *Model Communities: The Garden City Movement in Australia* (Nelson, 1989).

French, Susan F, 'Toward a Modern Law of Servitudes: Reweaving the Ancient Strands' (1981) 55 *Southern California Law Review* 1261.

French, Susan F, 'Design Proposal for the New Restatement of the Law of Property – Servitudes' (1987) 21 *UC Davis Law Review* 1213.

French, Susan F, 'Constitution of a Private Residential Government Should Include a Bill of Rights, The Symposium on the Uniform Real Property Acts' (1992) 27 *Wake Forest Law Review* 345.

French, Susan F, 'Tradition and Innovation in the New Restatement of Servitudes: A Report from Midpoint *Gallivan Conference*' (1994) 27 *Connecticut Law Review* 119.

French, Susan F, 'Highlights of the New Restatement (Third) of Property: Servitudes' (2000) 35 *Real Property, Probate and Trust Journal* 225.

Fritz, John-Patrick, 'Check Your Rights and Your Guns at the Door: Questioning the Validity of Restrictive Covenants against the Right to Bear Arms' (2005–2007) 35 *Southwestern University Law Review* 551.

Frug, Gerald E, 'The City As a Legal Concept' (1979) 93 *Harvard Law Review* 1057.

Gaus, Gerald and Shane D Courtland, 'Liberalism' in Edward N Zalta (ed.), *Stanford Encyclopedia of Philosophy* (2011).

Gillette, Clayton P, 'Courts, Covenants, and Communities' (1994) 61 *University of Chicago Law Review* 1375.

Glaezer, Edward, 'How Skyscrapers Can Save the City' *The Atlantic* (online), 9 February 2011 www.theatlantic.com/magazine/archive/2011/03/how-skyscrapers-can-save-the-city/308387/.

Glasze, Georg, Chris Webster and Klaus Frantz, *Private Cities: Global and Local Perspectives* (Routledge, 2006).

Gleave, Josie and Issy Cole-Hamilton, 'A World Without Play: A Literature Review' (Play England, 2012).

Gleeson, B, *Australian Heartlands: Making Space for Hope in the Suburbs* (Allen and Unwin, 2006).

Global Access Partners, 'Strata Laws On-line Consultation Final Report' (2012).

Goodman, Robin, and Kathy Douglas, 'Life in a Master Planned Estate – Community and Lifestyle or Conflict and Liability?' (2010) 28(4) *Urban Policy and Research* 451.

Grassmick, David E, 'Minding the Neighbor's Business: Just How Far Can Condominium Owners' Associations Go in Deciding Who Can Move into the Building' [2002] *University of Illinois Law Review* 185.

Gray, Kevin, 'Property in Thin Air' (1991) 50 *Cambridge Law Journal* 252.

Gray, Kevin and Susan Francis Gray, *Elements of Land Law* (Oxford University Press, 5th edn, 2009).

Growing Up in Australia, *Growing Up in Australia: The Longitudinal Study of Australian Children* www.growingupinaustralia.gov.au/.

Guernsey, Thomas F, 'The Mentally Retarded and Private Restrictive Covenants' (1983–1984) 25 *William and Mary Law Review* 430.

Gwyther, Gabrielle, 'Paradise Planned: Community Formation and the Master Planned Estate' (2006) 23(1) *Urban Policy and Research* 57.

Han, Esther and Natalie O'Brien, 'Living on the Edge in Danger High-Rises' *Sydney Morning Herald* (online), 30 September 2012 http://news.domain.com.au/domain/real-estate-news/living-on-the-edge-in-danger-highrises-20120929-26s6i.html.

Hardin, Garrett, 'The Tragedy of the Commons' (1968) 168 *Science* 1243.

Hardy, Louise, 'NSW Schools Physical Activity and Nutrition Survey: Executive Summary' (University of Sydney, 2010).

Harpum, Charles, *Megarry and Wade: The Law of Real Property* (Sweet and Maxwell, 6th edn, 2000).

Harrington Grove: Grove Life, *Community Documents* http://grovelife.com.au/community-documents.

Harris, Douglas C, 'The Condominium and the City: The Rise of Property in Vancouver' (2011) 36(3) *Law and Social Inquiry* 694.

Heller, Michael A, 'Tragedy of the Anticommons: Property in the Transition from Marx to Markets' (1998) 111 *Harvard Law Review* 621.

Heller, Michael A, 'The Boundaries of Private Property' (1999) 108 *Yale Law Journal* 1163.

Hohfeld, Wesley, 'Some Fundamental Legal Conceptions as Applied in Judicial Reasoning' (1913) 23 *Yale Law Journal* 16 (1917) 26 *Yale Law Journal* 710.

Holmes, O W, 'Path of the Law' (1896) 10 *Harvard Law Review* 457.

Howard, Ebenezer, *Garden Cities of To-morrow* (Swan Sonnenshein, 1902).

Hoyle, Brian, 'Global and Local Change on the Port-City Waterfront' (2000) 90(3) *Geographical Review* 395.

Hyatt, Wayne S, *Condominium and Homeowner Association Practice: Community Association Law* (American Law Institute-American Bar Association, Committee on Continuing Professional Education, 1981).

Hyatt, Wayne S, 'Common Interest Communities: Evolution and Reinvention' (1997) 31 *John Marshall Law Review* 303.

Hyatt, Wayne S and Susan F French, *Community Association Law: Cases and Materials on Common Interest Communities* (Carolina Academic Press, 2nd edn, 2008).

Hyatt, Wayne S and J Rhoads, 'Concepts of Liability in the Development and Administration of Condominium and Home Owner Associations' (1976) 12 *Wake Forest Law Review* 915.

Ilkin, Alex, *NSW Strata and Community Schemes Management and the Law* (Lawbook, 4th edn, 2007).

INHERITage, 'Final Heritage Assessment Report of Rosebery Estate' (City of Sydney, 2006) http://meetings.cityofsydney.nsw.gov.au/documents/meetings/2006/Committee/Planning/210806/21-08-06_pdtc_item10study-attachmenta.pdf.

ITV, *Downton Abbey*, 26 September 2010.

Jackson's Landing www.jacksonslanding.net.au/.

Jacobs, Jane, *The Death and Life of Great American Cities* (Penguin, 1965).

Joyce, Kelly, 'Man Accused of Shooting at HOA Board Members' (4 June 2013) myFox Orlando.com www.myfoxorlando.com/story/22491059/man-accused-of-shooting-at-hoa-board-members.

Judd, B, D Olsberg, J Quinn, L Groenhart and O Demirbilek, (2010). *Dwelling, Land and Neighbourhood Use by Older Home Owners, AHURI Final Report* (Australian Housing and Urban Research Institute, 2010).

Kaufman, Joanne, 'Because the Board Says So' *New York Times* (online), 24 May 2013 www.nytimes.com/2013/05/26/realestate/quirky-building-rules-because-the-board-says-so.html?pagewanted=1&_r=0&adxnnl=1&adxnnlx=1369784026-gTs8mMYCyH-Plbj6vMdeUA.

Kenna, Therese and Deborah Stevenson, 'Negotiating Community Title: Residents' Lived Experiences of Private Governance Arrangements in a Master Planned Estate' (2010) 28(4) *Urban Policy and Research* 435.

Kennedy, Duncan, 'Residential Associations as State Actors: Regulating the Impact of Gated Communities on Nonmembers' (1995) 105(3) *Yale Law Journal* 761.

Kim, Michael C, 'Involuntary Sale: Banishing an Owner from the Condominium Community' (1997–1998) 31 *John Marshall Law Review* 429.

King, Martin Luther, Jr, *I Have a Dream* (1963) Archives www.archives.gov/press/exhibits/dream-speech.pdf.

King & Wood Mallesons, 'Barangaroo: Inside a Deal' (Presentation, 8th Annual ACCAL Conference, 21 March 2013).

Kleinschmidt, Michael, 'Falling Short of the Target: Some Implications of Fiduciary Duties for Developer Practice in Queensland and New South Wales' (2011) 19 *Australian Property Law Journal* 263.

Kondos, Alex, 'The Hidden Faces of Power: A Sociological Analysis of Housing Legislation in Australia' in Roman Tomasic (ed.), *Legislation and Society in Australia* (Unwin Hyman, 1980).

Korngold, Gerald, 'Single Family Use Covenants: For Achieving a Balance Between Traditional Family Life and Individual Autonomy Symposium: The Changing Role of the Family in the Law' (1988) 22 *UC Davis Law Review* 513.

Korngold, Gerald, 'Reply: Resolving the Flaws of Residential Servitudes and Owners Associations: For Reformation not Termination' (1990) *Wisconsin Law Review* 513, 522.

Korngold, Gerald, 'Resolving the Flaws of Residential Servitudes and Owners Associations: For Reformation Not Termination Reply' (1990) Wisconsin Law Review 513.

Kozol, Jonathan, *The Shame of the Nation: The Restoration of Apartheid Schooling in America* (Crown Publishers, 2005).

Kushner, James A, 'Apartheid in America: An Historical and Legal Analysis of Contemporary Racial Residential Segregation in the United States' (1979) 22 *Howard Law Journal* 547.

Lasner, Matthew Gordon, *High Life: Condo Living the Suburban Century*, (Yale University Press, 2012).

LeGates, Richard T and Frederic Stout (eds), *The City Reader* (Routledge, 2003).

Leshinsky, Rebecca, Kathy Douglas, Peter Condliffe and Robin Goodman, 'Dispute Resolution under the Owners Corporations Act 2006 (Vic): Engaging with Conflict in Communal Living' (2012) 2(1) *Property Law Review* 39.

Lester, Stuart and Wendy Russell, 'Children's Right to Play: An Examination of the Importance of Play in the Lives of Children Worldwide' (Working Paper No. 57, Bernard van Leer Foundation, December 2010).

Liberty Grove Community Policies Handbook https://docs.google.com/viewer?a=v&pid=sites&srcid=bGliZXJ0eWdyb3ZlLm5ldC5hdXxob21lfGd4OjdjOThmYzhmYjgxZTljODg.

Liberty Grove Management Statement, DP 270137 (Land and Property Information, NSW).

Locke, John, 'Second Treatise of Government' in Thomas Hobbes and John Locke (eds), *On Sovereignty* (Collector's Library of Essential Thinkers, 2005).

Louv, Richard, *Childhood's Future: Listening to the American Family* (Houghton Mifflin, 1990).

Low, Setha M, *Behind the Gates: Life, Security, and the Pursuit of Happiness in Fortress America* (Routledge, 2003).

Macquarie Links Management Statement, DP270152, (Land and Property Information, NSW).

McDonald, Peter and Helen Brownlee, 'High-rise Parenting: Raising Children in Melbourne's High-rise Estates' in *Parenting in the 90s*, Family Matters No. 36, Australian Institute of Family Studies, 1993.

McKenzie, Evan, *Privatopia: Homeowner Associations and the Rise of Residential Private Government* (Yale University Press, 1994).

McKenzie, Evan, 'Emerging Regulatory Trends, Powers and Competing Interests in US Common Interest Housing Developments' in Ann Dupuis and Jennifer Dixon Sarah Blandy (eds), *Multi-owned Housing: Law, Power and Practice* (Ashgate, 2010) 53.

McKenzie, Evan, *Beyond Privatopia: Rethinking Residential Private Government* (Urban Institute Press, 2011).

Merrill, Thomas W and Henry E Smith, 'Optimal Standardization in the Law of Property: The Numerus Clausus Principle' (2000) 110 *Yale Law Journal* 1.

Merrill, Thomas W and Henry E Smith, 'What Happened to Property in Law and Economics?' (2001) 111 *Yale Law Journal* 357.

Merrill, Thomas W and Henry E Smith, 'The Morality of Property' (2007) 48 *William & Mary Law Review* 1849.

Merwe, C G van der, 'The Sectional Titles Act and the Wohnungseigentumsgesetz' (1974) 7 *Comparative and International Law Journal of Southern Africa* 165.

Michelman, Frank, 'Universal Resident Suffrage: A Liberal Defense' (1981) 130 *University of Pennsylvania Law Review* 1581.

Mill, John Stuart, *On Liberty* (Longman, Roberts & Green, 4th edn, 1869).

Monarch's Macquarie Links Estate, www.macquarielinks.net.

Monash University Accident Research Centre, *The Relationship Between Slips, Trips and Falls and the Design and Construction of Buildings* (Report No 281, 2008).

Munro, Kelsey, 'Parents Denied Safety Nets to Protect Kids in Apartments' *Sydney Morning Herald* (online), 24 January 2012 http://smh.domain.com.au/real-estate-news/parents-denied-safety-nets-to-protect-children-in-apartments-20120123-1qe3r.html.

Musemeche, C *et al.*, 'Pediatric Falls from Heights' (1991) 31(10) *Journal of Trauma* 1347.

Napolitano, Nicole, 'Fair Housing Act Amendments and Age Restrictive Covenants in Condominiums and Cooperatives, The Note' (1999) 73 *St John's Law Review* 273.

Natelson, Robert, 'Consent, Coercion, and "Reasonableness" in Private Law: The Special Case of the Property Owners Association' (1990) 51 *Ohio State Law Journal* 41.

Nelson, Robert M, 'Examining Cooperative Conversion: An Analysis of Recent New York Legislation' (1982) 11 *Fordham Urban Law Journal* 1089.

Nelson, Robert H, *Private Neighbourhoods and the Transformation of Local Government* (Urban Institute Press, 2005).

'New Flats Reflect Social Needs of the Age' *Daily News*, 12 March 1938, 9.

New South Wales Government, 'Draft Metropolitan Strategy for Sydney to 2031' (2013).

New South Wales Government, 'A New Planning System for New South Wales: White Paper' (2013).

New South Wales Government, 'A Plan for Growing Sydney' (2014).

New South Wales Law Reform Commission, *Disputes in Company Title Units*, Report No 115 (2007).

Nichols, Nick, *Frawleys Buys Q1 Management for $10m* (16 December 2011) goldcoast.com.au www.goldcoast.com.au/article/2011/12/16/374381_gold-coast-business.html.

North, Douglass C and Barry R Weingast, 'Constitutions and Commitment: The Evolution of Constitutional Governing Public Choice in Seventeenth-Century England' (1989) 49(4) *The Journal of Economic History* 803.

Notes, 'Condominium Regulation: Beyond Disclosure' (1975) 123(3) *University of Pennsylvania Law Review* 639.

Notes, 'Cooperative Housing Corporations and the Federal Securities Laws' (1971) 71 *Columbia Law Review* 118.

Nowra, Louis, *Kings Cross: A Biography* (NewSouth Publishing, 2013).

Nozick, Robert, *Anarchy, State, and Utopia* (Basic Books, 1974).

NSW Department of Lands & NSW Fair Trading, 'Review of NSW Community Schemes Legislation: Consultation Paper' (2006).

NSW Fair Trading, 'Making NSW No 1 Again: Shaping Future Communities: Strata and Community Title Law Reform Discussion Paper' (2012).

NSW Fair Trading, 'Children and Window Safety: Consultation Paper' (13 March 2013).

NSW Fair Trading, 'Strata & Community Title Law Reform Position Paper' (November 2013).

Nussbaum, Martha C, 'Capabilities as Fundamental Entitlements: Sen and Social Justice' (2003) 9(2–3) *Feminist Economics* 33.

Nussbaum, Martha C, *Frontiers of Justice: Disability, Nationality, Species Membership (The Tanner Lectures on Human Values)* (Harvard University Press, 2006).

Nussbaum, Martha C, 'Deliberation and Insight: Bloch v. Frischholz and the Chicago School of Judicial Behavior Essay' (2010) 77 *University of Chicago Law Review* 1139.

O'Connor, Pamela, 'Covenants as Regulation' (2011) 1 *Property Law Review* 145.

O'Connor, Pamela, Sharon Christensen and Bill Duncan, 'Legislating for Sustainability: A Framework for Managing Statutory Rights Obligations and Restrictions Affecting Private Land' (2009) 35 *Monash University Law Review* 233.

O'Hanlon, Seamus, *Together Apart: Boarding House, Hostel and Flat Life in Pre-war Melbourne* (Australian Scholarly Publishing, 2002).

Olsen, Donald J, *Town Planning in London: The Eighteenth and Nineteenth Centuries* (Yale University Press, 1964).

Orfield, Gary and Chungmei Lee, 'Why Segregation Matters: Poverty and Educational Inequality' (Civil Rights Project, Harvard University, 2005).

Page, John, 'Common Property and the Age of Aquarius' (2010) 19(2) *Griffith Law Review* 172.

Park, Ruth, *Harp in the South* (Houghton Mifflin, 1948).

Park, Ruth, *Poor Man's Orange* (Collins/Angus & Robertson, revised edn, 1992).

Parliament of New South Wales, *Parliamentary Debates*, Legislative Assembly, 13 November 1996, 5921 (Lo Po).

Peñalver, Eduardo M, 'Land Virtues' (2008) 94 *Cornell Law Review* 821.

Peñalver, Eduardo M, 'Property's Memories' (2011) 80 *Fordham Law Review* 107.

Perkins, Casey, 'Privatopia in Distress: The Impact of the Foreclosure Crisis on Homeowners' Associations' (2010) 10 *Nevada Law Journal* 561.

Pitt Town Water, *Central Park* http://pitttownwater.com.au/community/central-park/.

Profile.id, *Liberty Grove: Households with Children* http://profile.id.com.au/canada-bay/households-with-children?WebID=180.

Profile.id, *Liberty Grove: Service Aged Groups* http://profile.id.com.au/canada-bay/service-age-groups?WebID=180.

Putnam, Robert D, *Bowling Alone: The Collapse and Revival of American Community* (Simon & Schuster, 2000).

Queensland Government, 'South East Queensland Regional Plan 2009–2031' (2009).

Queensland Parliament, *Parliamentary Debates*, Legislative Assembly, 17 October 1985, 2190 (Hinze).

Radin, M, 'Property and Personhood' (1982) 34 *Stanford Law Review* 957.

Ramachandran, Arjun, 'Man Describes Horrific Moment when his Friends' Three-Year-Old Son Fell from Sydney Unit Window' *The Sydney Morning Herald* (online), 19 October 2009 www.smh.com.au/national/man-describes-horrific-moment-when-his-friends-threeyearold-son-fell-from-sydney-unit-window-20091019-h3mr.html.

Randolph, Bill, *Children in the Compact City: Fairfield as a Suburban Case Study* (City Futures, 2006).

Randolph, Bill, Darren Holloway and Kristian Ruming, 'Social Outcomes of Residential Development, Sydney Olympic Park Stage 1: Local Area Analysis' (City Futures Research, 2005).

Rath, A F, P J Grimes and J E Moore, *Strata Titles* (Lawbook, 1966).

Reichman, Uriel, 'Toward a Unified Concept of Servitudes' (1981) 55 *Southern California Law Review* 1177.

Reid, Sacha, Lisa Ruhanen and Nicole Johnston, 'Untangling the "Messy" Legislative Basis of Tourism Development Planning: Five Cases from Australia' (2012) 23(3) *Anatolia: An International Journal of Tourism and Hospitality Research* 413.

Reilly, J J and J Kelly, 'Long-Term Impact of Overweight and Obesity in Childhood and Adolescence on Morbidity and Premature Mortality in Adulthood: Systemic Review' (2011) 35 *International Journal of Obesity* 891.

Richmond, David T, New South Wales Department of Health, *Richmond Report Recommendations – Inquiry into Health Services for the Psychiatrically Ill and Developmentally Disabled* (March 1983).

Riley, Sophie and Grace Li 'Disclosure Requirements and Investor Protection: The Compatibility of Commonwealth, State and Territory Laws in Serviced Strata Schemes' (2009) 16(3) *Australian Property Law Journal* 262.

Roberts, Jan (ed.), *The Astor* (Ruskin Rowe, 2003).

Robinson, Glen, 'Communities' (1997) 83 *Virginia Law Review* 269.

Rose, Carol M, 'Servitudes, Security, and Assent: Some Comments on Professors French and Reichman Comments' (1981) 55 *Southern California Law Review* 1403.

Rose, Carol M, *Property and Persuasion: Essays on the History, Theory and Rhetoric of Ownership* (Westview Press, 1994).

Rose, Carol M, 'Property as the Keystone Right Propter Honoris Respectum' (1995) 71 *Notre Dame Law Review* 329.

Rose, Carol M, 'The Moral Subject of Property' (2007) 48 *William and Mary Law Review* 1897.

Rudden, Bernard, 'Economic Theory v. Property Law: The *Numerus Clausus* Problem' in John Eekelaar and John Bell (eds), *Oxford Essays on Jurisprudence* (Clarendon Press, 1987).

Shenassa, E, A Stubbendick A and M Brown, 'Social Disparities in Housing and Related Pediatric Injury: A Multilevel Study' (2004) 94(4) *American Journal of Public Health* 633.

Sherry, Cathy, 'Termination of Strata Schemes in New South Wales – Proposals for Reform' (2006) 13(3) *Australian Property Law Journal* 227.

Sherry, Cathy, 'The Legal Fundamentals of High-Rise Buildings and Master Planned Estates: Ownership, Governance and Living in Multi-owned Housing with a Case Study on Children's Play' (2008) 16(1) *Australian Property Law Journal* 1.

Sherry, Cathy, 'How Indefeasible Is Your Strata Title? Unresolved Torrens Problems in Strata and Community Title' (2009) 21(2) *Bond Law Review* 159.

Sherry, Cathy, 'The New South Wales Strata and Community Titles Acts: A Case Study of Legislatively Created High-rise and Master Planned Communities' (2009) 1(2) *International Journal of Law in the Built Environment* 130.

Sherry, Cathy, 'Long-Term Management Contracts as Developer Abuse' in S Blandy, J Dixon and A Dupuis (eds), *Multi-Owned Housing: Law, Power and Practice* (Ashgate Publishing, 2010).

Sherry, Cathy, Submission to the Office of Fair Trading, *Review of Strata and Community Title Scheme Laws*, November 2012, www.fairtrading.nsw.gov.au/biz_res/ftweb/pdfs/About_us/Have_your_say/Review_of_strata_community_scheme_laws/Individual_submissions/Sherry_cathy.pdf.

Sherry, Cathy, 'Building Management Statements and Strata Management Statements: An Unholy Mixing of Property and Contract' (2013) 87 *Australian Law Journal* 393.

Sherry, Cathy, 'Kids Can't Fly: The Legal Issues in Children's Falls from High-Rise Buildings' (2013) 2(1) *Property Law Review* 22.

Sherry, Cathy, 'Lessons in Personal Freedom and Functional Land Markets: What Strata and Community Title Can Learn From Traditional Doctrines of Property' (2013) 36(1) *University of New South Wales Law Journal* 280.

Sherry, Cathy, and Hazel Easthope 'Under-supply of Schooling in the Gentrified and Regenerated Inner City' (2016) 56 Cities 16–23.

Shipley, W E, 'Self-Dealing by Developers of Condominium Project as Affecting Contracts or Leases with Condominium Association' (1976) 73 *American Law Reports* 613.

Simpson, A W B, *A History of the Land Law* (Clarendon Press, 2nd edn, 1986).

Singer, Joseph William, *The Edges of the Field: Lessons on the Obligations of Ownership* (Beacon Press, 2000).

Singer, Joseph William, *Entitlement: The Paradoxes of Property* (Yale University Press, 2000).

Singer, Joseph William, 'Corporate Responsibility in a Free and Democratic Society' (2008) 58 *Case Western Reserve Law Review* 1031.

Singer, Joseph William, 'Democratic Estates: Property Law in a Free and Democratic Society Essay' (2008) 94 *Cornell Law Review* 1009.

Singer, Joseph William, 'Normative Methods for Lawyers' (2008) 56 *UCLA Law Review* 899.

Singer, Joseph William, 'How Property Norms Construct the Externalities of Ownership' in Gregory Alexander and Eduardo Peñalver (eds), *Property and Community* (Oxford University Press, 2010).

Slatter, Michele, 'From Anecdote to Evidence: The Next Challenge for Strata Reform?' (2009) 17 *Australian Property Law Journal* 179.

Smith, Alexandra, 'Sydney's Strata Surprises: Saga of Horse Next Door, the Trout Farm and "Niagara Falls"' *Sydney Morning Herald* (online), 21 March 2012 www.smh.com.au/business/property/sydneys-strata-surprises-saga-of-horse-next-door-the-trout-farm-and-niagara-falls-20120320-1vi2l.html.

Smith, Matthew J, 'Wages of Living in Sin: Discrimination in Housing Against Unmarried Couples, The Comment' (1991) 25 *UC Davis Law Review* 1055.

Smith, M, J Burrington and A Woolf, 'Injuries in Children Sustained in Free Falls: An Analysis of 66 Cases' (1975) 15(11) *Journal of Trauma* 987.

Soja, Edward W, *Seeking Spatial Justice* (University of Minnesota Press, 2010).

Spiegel, C and F Lindaman, 'Children Can't Fly: A Program to Prevent Childhood Morbidity and Mortality from Window Falls' (1977) 67(12) *American Journal of Public Health* 1143.

Steele, Cameron, *Cabarrus County, NC, Standoff Ends with Suspected Shooter and Two Others Dead* (7 April 2013) PoliceOne.com www.policeone.com/rural-law-enforcement/articles/6187650-3-dead-in-NC-standoff/.

Sterk, Stewart E, 'Freedom from Freedom of Contract: The Enduring Value of Servitude Restrictions' (1984) 70 *Iowa Law Review* 615.

Sterk, Stewart E, 'Foresight and the Law of Servitudes' (1987) 73 *Cornell Law Review* 956.

Stern, Stephanie M, 'The Dark Side of Town: The Social Capital Revolution in Residential Property' (Working Paper, Chicago-Kent College of Law, April 2013) http://ssrn.com/abstract=2243134 or http://dx.doi.org/10.2139/ssrn.2243134.

Stolzenberg, Nomi Maya, 'Return of the Repressed: Illiberal Groups in a Liberal State, The Liberalism and Illiberalism' (2001) 12 *Journal of Contemporary Legal Issues* 897.

Summerson, John, *Georgian London* (Penguin Books, 1978).

Sunstein, Cass R, 'Legal Interference with Private Preferences' (1986) 53(4) *University of Chicago Law Review* 1129.

Sunstein, Cass R, 'Social Norms and Social Roles' (1996) 96 *Columbia Law Review* 903.

Tarlock, Dan, 'Touch and Concern is Dead, Long Live the Doctrine' (1998) 77 *Nebraska Law Review* 804.

Taylor, C, 'What's Wrong With Negative Liberty?' in A Ryan (ed.), *The Idea of Freedom* (Oxford University Press, 1979).

Tekle, Asmara M, 'Safe: Restrictive Covenants and the Next Wave of Sex Offender Legislation' (2009) 62 *Southern Methodist Law Review* 1817.

The Bedford Estates, www.bedfordestates.com/.

The Children's Hospital at Westmead, *Working Party for the Prevention of Children Falling from Residential Buildings Outcomes Report* (February 2011).

'The Flat', *The Age* 15 February 1919, 6.

Tiebout, Charles M, 'A Pure Theory of Local Expenditures' (1956) 64(5) *Journal of Political Economy* 416.

Toohey, Lisa and Dan Toohey, 'Achieving Quality Outcomes in Community Title Disputes: A Therapeutic Jurisprudence Approach' (2010) *Monash University Law Review* 298.

'Trial Date Set for Man Accused of Fatally Shooting 2 Men at HOA Meeting' (3 June 2013) WLKY.com www.wlky.com/news/local-news/louisville-news/trial-date-set-for-man-accused-of-fatally-shooting-2-men-at-hoa-meeting/-/9718340/20401906/-/rb4fhe/-/index.html.

Troy, L, H Easthope, B Randolph and S Pinnegar, *Renewing the Compact City: Interim Report* (City Futures Research Centre, 2015).

Underkuffler, Laura S 'Property: A Special Right' (1996) 71 *Notre Dame Law Review* 1033.

Underkuffler, Laura S, *The Idea of Property: Its Meaning and Power* (Oxford University Press, 2003).

Underkuffler, Laura S, 'A Theoretical Approach: The Lens of Progressive Property' in Susan Bright and Sarah Blandy (eds), *Researching Property Law*, Palgrave, 2016.

Urban Land Institute and Community Association Institute, *Creating a Community Association: The Developer's Role in Condominium and Home Owner Associations* (Urban Land Institute, 1977).

van Vliet, W, 'Families in Apartment Buildings: "Sad Storey for Children?"' (1983) 15(2) *Environment and Behaviour* 211.

Vesk, Kristina (Chief Executive Officer, Cat Protection Society NSW Inc): Email to Cathy Sherry, 5 November 2013.

Victorian Government (Department of Infrastructure), 'Melbourne 2030: Planning for Sustainable Growth' (2002).

Victorian Government, 'Plan Melbourne' (2014).

Victorian Law Reform Commission, *Easements and Covenants: Final Report*, Report No 22 (2010).

Waghmode, Vishwas, 'Refused a Flat in Mumbai Because You Eat Meat? BMC Will Rescue You' *First Post*, 28 November 2014, www.firstpost.com/india/refused-a-flat-in-mumbai-because-you-eat-meat-bmc-will-rescue-you-1825497.html.

Waldron, Jeremy, 'Homelessness and the Issue of Freedom' (1991) 39 *UCLA Law Review* 295.

Wallace, Anne, Michael Weir and Les McCrimmon, *Real Property Law in Queensland*, (Lawbook Co., 2015). *Review* 295.

Weir, Michael 'Public and Private Planning – Implications for Legal Processes' (1994) 11 *Environmental and Planning Law Journal* 137.

Weir, Michael, 'Easements for Storage and Parking: A Time for a Rethink' in Lyria Bennett Moses, Brendan Edgeworth and Cathy Sherry (eds), *Property and Security: Selected Essays* (Lawbook Co, 2010).

Weir, Michael, 'The Westfield Case: A Change for the Better?' (2010) 21(2) *Bond Law Review* 182.

Weir, Michael, 'Wither Covenants in Queensland in the Name of Sustainability' in Penny Carruthers, Sharon Mascher and Natalie Skead (eds), *Property and Sustainability: Selected Essays* (Lawbook Co, 2011) 145.

Weir, Michael, 'Westfield 5 Years On' (2012) *21 Australian Property Law Journal* 166.

Whitzman, C and D Mizrachi, 'Creating Child-friendly High-rise Environments: Beyond Wastelands and Glasshouses' (2012) 30(3) *Urban Policy and Research* 233–49.

Wilhelm, Katherine, 'Rethinking Property Rights in Urban China' (2004) 9 *UCLA Journal of International Law and Foreign Affairs* 227.

Williams, Philippa and Barbara Pocock, 'Building "Community" for Different Stages of Life: Physical and Social Infrastructure in Master Planned Communities' (2010) 13(1) *Community, Work and Family* 71.

Winokur, James L, 'The Mixed Blessing of Promissory Servitudes: Toward Optimizing Economic Utility, Individual Liberty, and Personal Identity' (1989) *Wisconsin Law Review* 1.

Wizard of Oz, The (Directed by Victor Fleming and George Cukor) Metro-Goldwyn-Mayer, 1939.

Woolcock, G, B Gleeson and B Randolph, 'Urban Research and Child-friendly Cities: A New Australian Outline' (2010) 8(2) *Children's Geographies* 177.

Wulff, M, E Healy and M Reynolds, 'Why Don't Small Households Live in Small Dwellings?' (2004) 12(1) *People and Place* 56–71.

Yates, Judith, 'The Rhetoric and Reality of Housing Choice: The Role of Urban Consolidation' (2001) 19(4) *Urban Policy and Research* 491.

Young, J, 'Areas of Dispute in Condominium Law' (1976) 12 *Wake Forest Law Review* 983.

B Cases

Albrecht v. *Ainsworth* [2015] QCA 220.

Anthony Hordern & Sons Ltd v. *Amalgamated Clothing and Allied Trades Union of Australia* (1932) 47 CLR 1.

Application of Fox (1981) 2 BPR 9310.

Avranik Pty Ltd v. *Lloyd* [2012] VSC 306.

Avranik Pty Ltd v. *Lloyd* [2013] VSCA 244.

Beattie Place South [2009] QBCCMCmr 27.

Body Corporate for River City Apartments CTS 31622 v. *McGarvey* [2012] QCATA 47.

Body Corporate for Village Green (Caloundra) [2015] QCAT 101.

Body Corporate Strata Plan No 4303 v. *Albion Insurance Co Ltd* [1982] VR 699.

Bondlake Pty Ltd v. *Owners – Strata Plan No 60285* (2005) 62 NSWLR 158.

Breskvar v. *Wall* [1971] HCA 70; (1971) 126 CLR 376.

Brookfield Multiplex Ltd v. *Owners Corporation Strata Plan 61288* (2014) 254 CLR 185.

Brotherton v. *Sydney City Council* [2004] NSWLEC 475.

Bursill Enterprises Pty Ltd v. *Berger Bros Trading Pty Ltd* (1971) 124 CLR 73.

C v. *A* [2005] QADT 14.

Caloundra Gardens Village Body Corporate Committee, Re [2012] QCAT 98.

Carey, Re: Ex parte Exclude Holdings Pty Ltd [2006] WASCA 219.

Casuarina Rec Club Pty Ltd v. Owners – Strata Plan 77971 (2011) 80 NSWLR 711.

Central Park Apartments Varsity Central [2011] QBCCMCmr 107.

Chauhan v. *Jaynrees Services Pty Ltd* [2008] NSWSC 969.

Clem Smith Nominees Pty Ltd v. *Farrelly* (1978) 20 SASR 227.

Clissold v. *Perry* (1904) 1 CLR 363.

Clos Farming Estates Pty Ltd v. *Easton* [2002] NSWCA 389.

Clos Farming Estates Pty Ltd (Rec and Man Appt) v. *Easton* [2001] NSWSC 525.

Codelfa Construction Pty Ltd v. *State Land Rail Authority (NSW)* [1982] HCA 24; (1982) 149 CLR 337).

Collector of Customs v. *Agfa-Gevaert Ltd* (1986) 186 CLR 389.

Community Assoc DP No 270180 v. *Arrow Asset Management Pty Ltd* [2007] NSWSC 527.

Corporate for Palm Springs Residences v. *J Patterson Holdings Pty Ltd* [2008] QDC 300.

Corporation of Owners Strata Plan No 31471 v. *Neave, Mrs M* [1999] NSWSSB 43.

Council of the City of Sydney v. *Oaks Hotels and Resorts (NSW) No 2 Pty Ltd* [2010] NSWLEC 181.
Dainford Ltd v. *Smith* (1985) 155 CLR 342.
Dandaloo Pty Ltd v. *Darwich* [2006] NSWSC 1305.
Drexler v. *Owners Corp SP 56117* (Strata and Community Schemes) [2012] NSWCTTT 338.
Eatock v. *Bolt* [2011] FCA 1103.
Engelman v. *Owners Corporation* (Strata & Community Schemes) [2003] NSWCTTT 778.
Forbes v. *New South Wales Trotting Club Ltd* [1979] HCA 27; (1979) 143 CLR 242.
Frazer v. *Walker* [1967] 1 AC 569.
Grant and Owners of Rosneath Farm – Strata Plan 35452 [2006] WASAT 162.
Gumland Property Holdings Pty Ltd v. *Duffy Bros Fruit Market (Campbelltown) Pty Ltd* (2008) 234 CLR 237.
Henderson v. *Body Corporate for Merrimac Heights* [2011] QSC 336.
Hillcrest Apartments [2013] QBCCMCmr 211.
Hinkley v. *Star City Pty Ltd* [2010] NSWSC 138.
Horizons Corporations Law Pty Ltd v. *Rizons Pty Ltd* [1999] NSWSC 691.
Houghton v. *Immer (No 155) Pty Ltd* (1997) 44 NSWLR 46.
Hu v. *Stansure Strata Pty Ltd* [2014] FCCA 905.
Hulena v. *Owners Corporation Strata Plan 13672* [2009] NSWADT 119.
Hulena v. *Owners Corporation Strata Plan 13672* [2010] NSWADTAP 27.
Humphries v. *Proprietors 'Surfers Palms North' Group Titles Plan 1955* (1994) 179 CLR 597.
Italian Forum Ltd v. *Owners – Strata Plan 60919* [2012] NSWSC 895.
Jennifer Elizabeth James v. *Owners Strata Plan No SP 11478 (No 4)* [2012] NSWSC 590.
Karacominakis v. *Big Country Developments* (2000) 10 BPR 18, 235.
Lawandi v. *Owners Corporation 21842D (Owners Corporations)* [2015] VCAT 1810.
Levi v. *Spicer* [2003] NSWSC 183.
Lin v. *Owners – Strata Plan No 50276* (2004) 11 BPR 21,463.
Lion Nathan Australia Pty Ltd v. *Coopers Brewery Ltd* (2005) 56 ACSR 263.
Lion Nathan Australia Pty Ltd v. *Coopers Brewery Ltd* [2006] FCAFC 144.
Lynch, Glenn and Jenny Owners of Lot 75a v. *Owners Strata Plan No 36458* [1999] NSWSSB 55 (17 August 1999).
Mabo v. *Queensland (No 2)* (1992) 175 CLR 1.
Mackie v. *Henderson* [2011] WASC 197.
Mary Erling v. *Owners Strata Plan No 8891* [2010] NSWSC 824.
Meriton Apartments Pty Ltd v. *Industrial Court of NSW* (2011) 284 ALR 130.
Meriton Apartments Pty Ltd v. *Owners Strata Plan No. 72381* [2015] NSWSC 202.
Mineralogy Pty Ltd v. *Body Corporate for the 'The Lakes Coolum'* [2002] QCA 550.
Montagna v. *Owners Corporation* (Strata & Community Schemes) [2003] NSWCTTT 783.
Morgan v. *Owners Strata Plan 13937* [2006] NSWSC 1019.
Mornington Quays [2012] QBCCMCmr 576.
Morrish v. *Republic Tower Body Corporate Strata Plan No 341293D* [2004] VSC 56.
North Wind Pty Ltd v. *Proprietors – Strata Plan 3143* [1981] 2 NSWLR 809.
Oaks Hotels and Resorts (NSW) No 2 Pty Ltd v. *Council of the City of Sydney* [2011] NSWLEC 1054.

OC SP 74565 v. *Mantra Ettalong Pty Ltd* (Strata and Community Schemes) [2012] NSWCTTT 116.

Ondrich v. *Kookaburra Park Eco-Village* [2009] FMCA 260.

Owners Corporation PS331362S v. *Rhodes (Owners Corporations)* [2011] VCAT 642.

Owners Corporation PS501391P v. *Balcombe (Owners Corporations)* [2015] VCAT 956.

Owners Corporation SP 20856 v. *T S Management Pty Ltd* (Strata and Community Schemes) [2009] NSWCTTT 109.

Owners Corporation SP 49822 v. *May (Strata & Community Schemes)* [2006] NSWCTTT 739.

Owners Corporation SP 68751 v. *Community Assoc DP 270281* [2015] NSWCATCD 99.

Owners Corporation Strata Plan 70672 v. *Trustees of the Roman Catholic Church for the Archdiocese of Sydney* [2011] NSWSC 973.

Owners of 111 The Broadview Landsdale – Survey Strata Plan 38894 v. *Colavecchio* [2004] WASTR 15.

Owners of Pearl Beach Survey Strata Plan 49019 and Heyns [2011] WASAT 66.

Owners Strata Plan 3397 v. *Tate* (2007) 70 NSWLR 344.

Owners Strata Plan 35042 v. *Seiwa Australia Pty Ltd* [2007] NSWCA 272.

Owners Strata Plan 44999 v. *Premier Holdings Corp Pty Ltd and Gregg Ritchie and Partners Pty Ltd* [2012] NSWSC 171. *Paloma* [2012] QBCCMCmr 37.

Owners Strata Plan 50276 v. *Thoo* [2013] NSWCA 270.

Owners Strata Plan 50411 v. *Cameron North Sydney Investments Pty Ltd* [2003] NSWCA 5.

Owners Strata Plan 56117 v. *Drexler* [2013] NSWDC 67.

Owners Strata Plan 56443 v. *Regis Towers Real Estate Pty Ltd* (2003) 58 NSWLR, 78.

Owners Strata Plan 60285 v. *Bondlake Pty Ltd* [2004] NSWSC 1037.

Owners Strata Plan 60919 v. *Consumer Trader and Tenancy Tribunal* [2009] NSWSC 1158.

Owners Strata Plan 61643 v. *183 On Kent Management Pty Ltd* [2007] NSWSC 281.

Owners Strata Plan 64972 v. *Rinbac Pty Ltd* [2009] NSWSC 745.

Owners Strata Plan 78102 v. *Owners – Strata Plan 78101* [2010] NSWSC 973.

Paris, Arthur v. *Owners Strata Plan 16973* [1998] NSWSSB 12.

Pirie v. *Registrar General* (1962) 109 CLR 619; [1965] ALR 860.

Pivotal Point Residential [2008] QBCCMCmr 55.

Proprietors Cathedral Village Building Units Plan No 106957 v. *Cathedral Place Community Body Corporate* [2012] QSC 301.

Proprietors Cathedral Village Building Units Plan No 106957 v. *Cathedral Place Community Body Corporate* [2013] QCA 264.

Proprietors Strata Plan No 6522 v. *Furney* [1976] 1 NSWLR 412.

Proprietors Strata Plan No 9968 v. *Proprietors Strata Plan No 11 173* [1979] 2 NSWLR 605.

Quest Rose Hill Pty Ltd v. *Owners Corporation of Strata Plan 64025* [2012] NSWSC 1548.

Radaich v. *Smith* (1959) 101 CLR 209.

Renshaw, Ms – Owner of Lot 3 Strata Plan SP12963 v. *Owners Corporation of Strata Plan No SP12963* [1999] NSWSSB 40.

Richardson v. *OC SP 56695* (Strata and Community Schemes) [2008] NSWCTTT 928.

Ridis v. *Strata Plan 10308* (2005) 63 NSWLR 449.

Rinbac Pty Ltd v. *Owners Corporation SP 64972* (Strata and Community Schemes) [2009] NSWCTTT 302.
Ryledar Pty Ltd v. *Euphoric Pty Ltd* (2007) 69 NSWLR 603.
Salerno v. *Proprietors of Strata Plan No 42724* (1997) 8 BPR 15,457.
Santai v. *Owners – Strata Plan No 77971* [2010] NSWSC 628.
Scholars Cover [2011] QBCCMCmr 310.
Seiwa Pty Ltd v. *Owners Strata Plan 35042* [2006] NSWSC 1157.
Sutherland v. *Tallong Park Assoc Inc* [2006] NSWADT 163.
Sydney Diagnostic Services Pty Ltd v. *Hamlena Pty Ltd* (1991) 5 BPR 11,432.
Taunton Enterprises Pty Ltd Appeal Against Order No C579394 Strata Title Act [2005] TASRMPAT 76.
Taylor, Re [1995] 2 Qd R 564.
Thornton & Farnham v. *Owners SP 30653 (Strata & Community Schemes)* [2010] NSWCTTT 511.
Tooth & Co Ltd v. *Barker* (1960) 77 WN (NSW) 231.
Travis v. *Proprietors of Strata Plan 374* (1969) 90 WN (Pt 1) (NSW) 711.
Tutton v. *Body Corporate for Pivotal Point Residential CTS 33550* [2008] QCCTB-CCM 12.
Waldorf Apartment Hotel, The Entrance Pty Ltd v. *Owners Corp SP 71623* [2010] NSWCA 226.
West v. *Owners Strata Plan No 54288* [2009] NSWIRComm 30.
Westfield Management Ltd v. *Perpetual Trustee Co Ltd* (2007) 233 CLR 528.
White v. *Betalli* (2007) 71 NSWLR 381.
Williams v. *City of Melbourne* [1933] HCA 56; (1933) 49 CLR 142.
Wilson v. Meudon Pty Ltd [2005] NSWCA 448.
Yedway Pty Ltd v. *Owners Corporation of Strata Plan 62871* [2009] NSWSC 8.
Young v. *Owners S/P 3529* (2001) 54 NSWLR 60.

C Cases – other jurisdictions

United Kingdom

Austerberry v. *Corporation of Oldham* (1885) 29 Ch D 750 (CA).
Copeland v. *Greenhalf* [1952] Ch 488.
Egyptian Salt and Soda Co Ltd v. *Port Said Salt Assoc Ltd* [1931] AC 677.
Ellenborough Park, Re [1956] Ch 131; [1955] 3 All ER 667.
Elliston v. *Reacher* [1908] 2 Ch 374.
Haywood v. *Brunswick Permanent Benefit Building Society* (1881) 8 QBD 403.
Hill v. *Tupper* (1834) 2 My & K 517; 39 ER 1042.
Kennedy v. *De Trafford* [1897] AC 180.
Keppell v. *Bailey* (1836) 2 H & C 121; 159 ER 51.
London & South Western Railway Company v. *Gomm* (1882) 20 Ch D 562.
London Assoc of Shipowners and Brokers v. *London and India Docks Joint Committee* [1892] 3 Ch 242.
London County Council v. *Allen* [1914–15] All ER Rep 1008; [1914] 3 KB 642.
Moncrieff v. *Jamieson* [2007] UKHL 42.
Rhone v. *Stephens* [1994] 2 All ER 65.
Spencer's Case (1583) 5 Co Rep 16a; 77 ER 72.

Tulk v. *Moxhay* (1848) 2 Ph 774; 41 ER 1143.

United States

Bennett v. *Behring Corp*, 466 F Supp 689 (SD Fla, 1979).
Brown v. *Board of Education of Topeka*, 347 US 483 (1954).
Buchanan v. *Warley*, 245 US 60 (1917).
City of New York v. *Delafield 246 Corp*, 662 NYS 2d 286, 294 (1997).
Committee for a Better Twin Rivers v. *Twin Rivers Homeowners' Assoc*, 890 A 2d 947 (NJ Super Ct App Div, 2006), revd, 929 A 2d 1060 (NJ, 2007).
Eagle Enterprises, Inc v. *Gross*, 39 NY 2d 505 (1976).
Franklin v. *White Egret Condominium, Inc*, 358 So 2d 1084 (1977).
Hidden Harbour Estates, Inc v. *Basso*, 393 So 2d 637 (Fla Dist Ct App, 1981).
Hidden Harbour Estates, Inc v. *Norman* 309 So 2d 180 (Fla Dist Ct App, 1975).
Marsh v. *Alabama*, 326 US 501 (1945).
Nahrstedt v. *Lakeside Village Condominium Assoc Inc* 8 Cal 4th 361, (1994).
Neponsit Property Owners' Assoc Inc v. *Emigrant Industrial Savings Bank* 278 NY 248 (NY App, 1938).
People ex rel Moloney v. *Pullman's Palace-Car Co*, 51 NE 664 (1898).
People ex rel Moloney v. *Pullman's Palace-Car Co*, 175 Ill 125 (1898).
Shelley v. *Kraemer*, 334 US 1 (1948).
Simovits v. *Chanticleer Condominium Assoc* 933 F Supp 1394 (ND Ill, 1996).
State v. *Glassel*, 211 Ariz 33 (2005).
Streams Sports Club Ltd v. *Richmond*, 99 Ill 2D 182 (1983).
Village of Euclid, Ohio v. *Ambler Realty Co*, 272 US 365 (1926).
Wisconsin v. *Yoder*, 406 US 205 (1972).
Wolsieffer v. *Lakes of the Four Seasons Prop Owners Assn* (Ind Lake Co Cir, No 45C01–0303-CT-00045, 25 February 2011).

D Legislation and regulations

Age Discrimination Act 2004 (Cth).
Anti-Discrimination Act 1977 (NSW).
Anti-Discrimination Act 1991 (Qld).
Ashfield Development Control Plan 2007 (NSW).
Body Corporate and Community Management Act 1997 (Qld).
Body Corporate and Community Management (Accommodation Module) Regulation 2008 (Qld).
Building Act 1975 (Qld).
Building Units and Group Titles Act 1980 (Qld).
Building Units Titles Act 1965 (Qld).
Casino Control Act 1992 (NSW).
Civil and Administrative Tribunal Act 2013 (NSW).
Cluster Titles Act 1974 (Vic).
Community Land Development Act 1989 (NSW).
Community Land Management Act 1989 (NSW).
Community Title Act 2001 (ACT).
Community Titles Act 1996 (SA).

Companion Animals Act 1998 (NSW).
Conveyancing Act 1919 (NSW).
Conveyancing (Strata Titles) Bill 1959 (NSW).
Conveyancing (Strata Titles) Act 1961 (NSW).
Corporations Act 2001 (Cth).
Crimes Act 1900 (NSW).
Disability Discrimination Act 1992 (Cth).
Environmental Planning and Assessment Act 1979 (NSW).
Group Titles Bill 1972 (Qld).
Group Titles Act 1973 (Qld).
Integrated Resort Development Act 1987 (Qld).
Interpretation Act 1987 (NSW).
Land Title Act 1994 (Qld).
Landlord and Tenant (Amendment) Act (1948) (NSW).
Legislation Act 2001 (ACT).
Mixed Use Development Act 1993 (Qld).
National Construction Code, BCA 2013, Vol 1.
Owners Corporations Act 2006 (Vic).
Property Law Act 1958 (Vic).
Property, Stock and Business Agents Act 2008 (NSW).
Protection of the Environment Operations (Noise Control) Regulation 2008 (NSW).
Racial Discrimination Act 1975 (Cth).
Real Property Act 1900 (NSW).
Registration of Plans (HSP (Nominees) Pty Ltd) Enabling Act 1980 (Qld).
Registration of Plans (Stage 2) (HSP (Nominees) Pty Ltd) Enabling Act 1984 (Qld).
Residential Tenancies Act 2010 (NSW).
Sanctuary Cove Resort Act 1985 (Qld).
South Bank Corporation Act 1989 (Qld).
State Environmental Planning Policy (Housing for Seniors or People with a Disability) 2004 (NSW).
Strata Schemes Development Act 2015 (NSW).
Strata Schemes (Freehold Development) Act 1973 (NSW).
Strata Schemes (Leasehold Development) Act 1986 (NSW).
Strata Schemes Management Act 1996 (NSW).
Strata Schemes Management Amendment (Child Window Safety Devices) Act 2013 (NSW).
Strata Schemes Management Bill 2015 (NSW).
Strata Schemes Management Regulation 2010 (NSW).
Strata Titles Act 1973 (NSW).
Strata Titles Act 1985 (WA).
Strata Titles Act 1988 (SA).
Subdivision Act 1988 (Vic).
Transfer of Land (Stratum Estates) Act 1960 (Vic).
Unit Title Schemes Act (NT).
Unit Titles Act 2001 (ACT).
Unit Titles (Management) Act 2011 (ACT).
University of New South Wales Act 1989 (NSW).

E Legislation and regulations – other jurisdictions

Commonhold and Leasehold Reform Act 2002 (UK).
Davis-Stirling Common Interest Development Act, Pt 5, California Civil Code (US).
De Donis Conditionalibus 1285, 13 Edw 1 (UK).
Fair Housing Act, 42 USCA §3601 (1968) (US).
Law of Property Act 1925 (UK).
Local Law of the City of New York, No 57, 2011 (US).
National Housing Act (1962) (US).
New York City Health Code (US).
Statute of Quia Emptores 1290 (UK).
Strata Title Law (DIFC Law No 5 of 2007) (Dubai International Financial Centre).
Tenures Abolition Act 1660 (UK).
Uniform Common Interest Ownership Act (2008) (US).
Uniform Planned Community Act 7A ULA (Supp 43, 1981) (US).
Wohnungseigentumsgesetz [German Condominium Act] (Germany) 1951.

Index

Page numbers in **bold** denote figures.

For Product Safety Concerns and Information please contact our EU
representative GPSR@taylorandfrancis.com
Taylor & Francis Verlag GmbH, Kaufingerstraße 24, 80331 München, Germany

www.ingramcontent.com/pod-product-compliance
Ingram Content Group UK Ltd.
Pitfield, Milton Keynes, MK11 3LW, UK
UKHW022022180926
475701UK00005B/984

* 9 7 8 0 3 6 7 3 5 5 6 8 5 *